molecular mechanisms of genetic processes

N. P. Dubinin and D. M. Gol'dfarb, Editors

molecular mechanisms of genetic processes

Translated from Russian by A. Mercado
Translation edited by P. Harry

A HALSTED PRESS BOOK

JOHN WILEY & SONS
New York · Toronto

ISRAEL PROGRAM FOR SCIENTIFIC TRANSLATIONS
Jerusalem · London

Sole distributors for the Western Hemisphere
HALSTED PRESS, a division of
JOHN WILEY & SONS, INC., NEW YORK

Library of Congress Cataloging in Publication Data
Main entry under title:

Molecular mechanisms of genetic processes.

Translation of Molekuli͡arnye mekhanizmy geneticheskikh protśessov.
"A Halsted Press book."
Includes bibliographies
1. Molecular genetics—Addresses, essays, lectures.
I. Dubinin, Nikolaĭ Petrovich, ed. II. Gol'dfarb, David Moiseevich, ed. [DNLM: 1. Cytogenetics. 2. Genetics, Biochemical. QH431 M718]
QH438.M6413 1975 575.2'1 74-30205
0 470-22330-8

Distributors for the U.K., Europe, Africa and the Middle East
JOHN WILEY & SONS, LTD., CHICHESTER

Distributors for Japan, Southeast Asia and India
TOPPAN COMPANY, LTD., TOKYO AND SINGAPORE

Distributed in the rest of the world by
KETER PUBLISHING HOUSE JERUSALEM LTD.
ISBN 0 7065 1426 2
IPST cat. no. 22108

This book is a translation from Russian of
MOLEKULYARNYE MEKHANIZMY GENETICHESKIKH PROTSESSOV
Izdatel'stvo "Nauka"
Moscow, 1972

Printed and bound by Keterpress Enterprises, Jerusalem
Printed in Israel

CONTENTS

MOLECULAR MECHANISMS OF MUTAGENESIS

TRANSCRIPTION AND TRANSLATION OF GENETIC INFORMATION

MOLECULAR MECHANISMS OF GENETIC RECOMBINATION

REPAIR OF DNA LESIONS

THE MOLECULAR MECHANISMS OF MUTAGENESIS

UDC 575.24

UNSOLVED PROBLEMS IN THE MODERN MOLECULAR THEORY OF MUTATIONS

N.P. DUBININ
Institute of General Genetics, Academy of Sciences of the USSR, Moscow

The molecular theory of mutations encounters difficulties in the solution of the following fundamental problems. 1. The nature of the origin of complete mutations remains obscure. It is necessary to determine the mechanism by which a lesion in one strand is transmitted to the other strand in the case of complete mutations. 2. The nature of the molecular mechanisms which, after injury of one strand of the DNA, subsequently cause a complete breakage leading to the appearance of various structural changes in the chromosomes, is still poorly understood. 3. The molecular nature of the potential changes in genes and chromosomes occurring when primary lesions in the DNA molecule develop into durable states which can either be repaired back to the original normal condition or fixed into mutations both in the initial generation and in a number of subsequent generations is not yet clear. The proposed repair hypothesis of the molecular mechanisms leading to the appearance of mutations after formation of a primary lesion in one strand of the DNA suggests a possible unity of the molecular mechanisms accomplishing the fixation of both gene mutations and structural changes in the chromosomes.

INTRODUCTION

Molecular genetics has solved a number of the fundamental problems of mutagenesis following the discovery of the structure of the DNA molecule by Watson and Crick (1953a). The physical nature of the gene and the main types of alteration in the genetic code constituting the chemical mechanism of mutations have been established (Watson and Crick, 1953b; Freese, 1959; Crick et al., 1961). It was shown that gene mutations are associated with a variety of single-site changes in the DNA molecule in the form of individual base substitution, deletion, insertion, extended mutations, translocation, and inversion. Base substitution may be of two types: transition and transversion. The significance of the changes in the genetic code resulting from mutations was also determined.

Crick et al. (1961) established that each codon consists of a triplet of nucleotides and that one codon represents a single amino acid. Experiments carried out by a number of scientists revealed the structure of all 64 codons and according to the characteristics of these a functional classification of the mutations into the following categories was established: 1) missense mutations, involving the introduction of a wrong amino acid; 2) nonsense mutations, leading to the formation of nonsense codons (as a rule, the codons UAG, UAA, and UGA); 3) base switches, not resulting in a change in the sense of the codon; 4) frame shift mutations, arising through the insertion or deletion of nucleotides and which lead to a shift in the reading of the genetic code.

Apart from gene mutations, the chromosomes of eukaryotes and the DNA molecules of prokaryotes undergo mutations in the form of structural changes. The macromolecular principles determining the mechanism of these structural mutations (translocations, deletions, insertions, circular chromosomes and inversions) up to now remain obscure. The molecular approach has permitted elucidation of the major types of primary injury of DNA molecules as a result of radiation and chemical factors exerting a mutagenic effect and also the action of repair systems in the form of photoreactivation, various systems operating in the dark, and a multitude of recombination processes on these injuries.

PRIMARY LESIONS AND THE OPERATION OF REPAIR SYSTEMS

Primary lesions

The primary lesions studied in most detail are these caused by UV (ultraviolet) irradiation of the DNA molecule. The basic type of lesion here is the creation of pyrimidine dimers (Beukers and Berends, 1960; Wacker et al., 1962) arising as a result of the formation of covalent bonds between the carbon atoms of adjacent pyrimidine bases. Thymine dimers arise most frequently. However, cytosine dimers are also formed (Wacker, 1963; Smith, 1963; Setlow et al., 1965 a, b). It was shown that all strains of E. coli regardless of their sensitivity to UV, are almost equally susceptible to UV-induced primary injury of DNA molecules, the rate being about 5 dimers per genome per erg/mm^2 (Swenson and Setlow, 1966).

Ionizing radiation causes photochemical reactions with adenine, single or double breaks in the sugar-phosphate backbone of DNA molecules, cross-linking of strands within or between DNA molecules or the interlinking of these with protein molecules, and denaturation of DNA. The basic type of primary lesion in this case apparently consists of single-strand breaks in the DNA (Peacocke and Preston, 1960); these may be either simple breaks in the sugar-phosphate chain or breaks turned into gaps by the action of exonucleases.

Alkylating compounds chiefly alkylate the nitrogen of guanine in the 7th position (Brookes and Lawley, 1964). Bifunctional alkylating agents such as nitrogen and sulfur mustards induce sutures of the strands in the DNA molecule (Lawley and Brookes, 1965); methylmethane sulfonate causes single breaks in DNA (Strauss, 1968).

Altogether the various chemical mutagens cause deamination of bases, alkylation of bases or sugars in the DNA chain, depurination or depyrimidination of DNA, insertion or elimination of nucleotides, breaks in the polynucleotide strand, etc.

Thymine starvation prevents the incorporation of thymine during the synthesis of DNA, thus creating small gaps opposite the adenine bases. The DNA of bacteria grown in the absence of thymine breaks down into short fragments (Mennigmann and Szybalski, 1962).

A characteristic feature of the primary changes induced by any type of mutagen is that they appear in only one of the polynucleotide strands of the DNA molecules affected. The primarily injured DNA molecule represents a heteroduplex: one of its strands contains the primary lesion whereas the other complementary strand retains the normal structure of its characteristic genetic code. This circumstance is essential for an understanding of the mechanism of mutagenesis.

Mechanisms for the repair of primary lesions

Primary lesions arising in a DNA molecule do not necessarily turn into mutational changes. The cell is equipped with powerful protective mechanisms in the form of enzymatic repair systems which free the DNA molecules from the induced changes disrupting the normal transcription of the genetic information. Three such enzymatic repair systems are known at present: photoreactivation; dark repair by excision and patching, and a system associated with recombination mechanisms. These are only some of the potentialities of the cell for self-repair. Other as yet unknown repair systems undoubtedly exist.

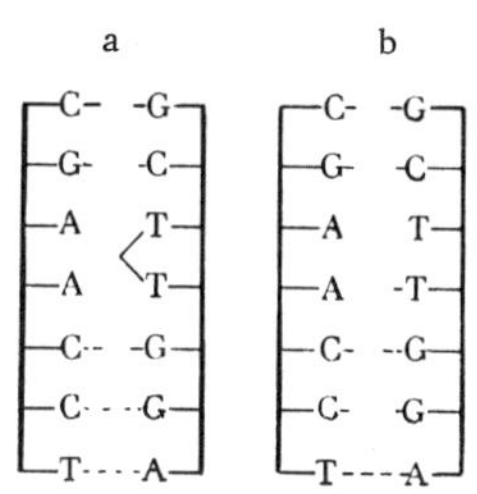

FIGURE 1. The nature of the photoreactivation effect (Wulff and Rupert, 1962):

a) dimerization of thymine in the DNA molecule; b) monomerization in the presence of light with the participation of the photoreactivating enzyme.

The phenomenon of photoreactivation was discovered by Kelner (1949). The essence of it is that visible light with a wavelength of 4,000Å activates a photoreactivating enzyme (Wulff and Rupert, 1962) which splits the thymine dimers (Figure 1). The photoreactivation process apparently has a purely protective significance: it sharply reduces the number of mutations by splitting the dimers. This reaction is strictly specific to pyrimidine dimers (Setlow, 1966). Splitting of the dimers is not observed in strains with mutations in the Phr locus (Phr^+ and Phr^-) having lost the ability to synthesize the photoreactivating enzyme (Harm and Hillebrandt, 1962). Such strains show a high mutability on exposure to UV irradiation. The photoreactivating enzyme restores exactly the original structure of the DNA, so that this process itself is not responsible for the appearance of mutations. Photoreactivation affects more than 90% of the pyrimidine dimers.

The systems of dark and recombinational repair not only serve to eliminate primary lesions but also include mechanisms enabling the fixation of mutations. Dark repair involves a whole sequence of events. It begins with the removal of thymine dimers from the affected DNA molecules by means of endonucleases (Setlow and Carrier, 1964; Boyce and Howard-Flanders, 1964a). The resulting gaps are filled according to the normal template of the opposite strand (Pettijohn and Hanawalt, 1964). This process of material synthesis, when the individual gaps in the DNA are repaired, bears the name of nonreplicational or repair synthesis since it occurs not during the replication of the DNA molecule but during the resting stage. This synthesis is achieved by the action of the enzyme DNA polymerase, isolated in 1958 by Kornberg and his group (Lehman et al., 1958). Later it was shown that this enzyme restores the complete double structure of the DNA molecule after partial injury of this by exonuclease III (Richardson et al., 1964).

Howard-Flanders and Boyce (1966) outlined the following four major stages of dark repair: 1) excision by endonuclease of the trinucleotide containing a pyrimidine dimer or other lesions; 2) widening of the gap with the loss of up to 500 nucleotides by the action of the enzyme exonuclease, which "bites off" nucleotides from the free end of the cut; 3) repair replication by means of DNA polymerase with the complementary area opposite the gap serving as template; 4) sealing of the sugar-phosphate chain of the DNA molecule by the action of ligases (Gellert, 1967; Olivera and Lehman, 1967). This sequence of events is shown in Figure 2.

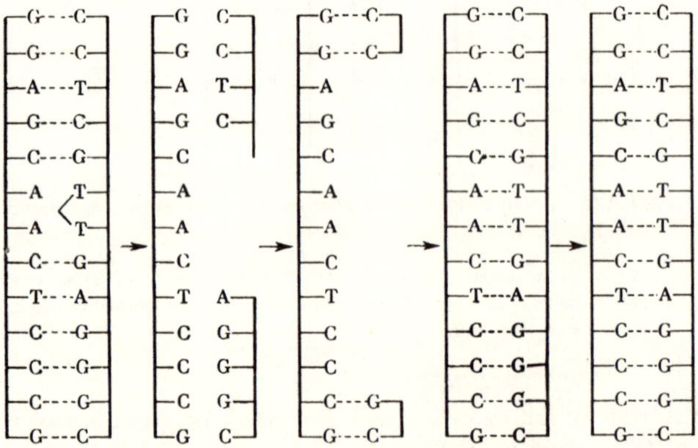

FIGURE 2. Scheme of dark repair (Howard-Flanders and Boyce, 1966).

The theory of the mechanism of crossing over was discussed for a long time at the cytogenetic level. Now this problem is being tackled at the level of the events associated with the molecular structure of DNA and the operation of enzyme systems. Using autoradiographic techniques, Taylor (1965) produced the first proofs that crossing over during meiosis involves breakage and exchange. Holliday (1964) proposed a hypothesis of crossing over based on the idea of hybridization of the DNA molecules. He proposed that breaks in two sister chromatids occur in polynucleotide chains with the same polarity and then the second strands break, the result being a complete exchange between the two molecules of DNA. This model does not require a nonreplicative synthesis during the recombinations. Hastings and Whitehouse (1964) proposed that the original breakage points in the DNA chain have a fixed position and are recognized by enzymes. Although this conception similarly treats crossing over as a hybridization of the DNA molecules, it differs from the former in postulating that the exchange requires repair synthesis. It was shown autoradiographically by the use of tritium-labeled thymidine that nonreplicative synthesis takes place in the pachytene stage of meiosis in Triturus viridescens (Wimber and Prensky, 1963). Experiments with radioactive phosphorus (^{32}P) have demonstrated that about 0.3% of all the DNA of Lilium longiflorum and Trillium erectum is synthesized during the zygotene and pachytene stages of meiosis in the anthers (Hotta et al., 1966).

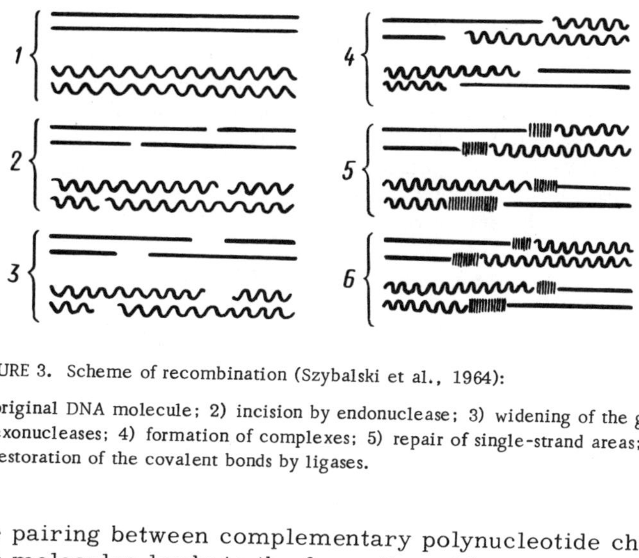

FIGURE 3. Scheme of recombination (Szybalski et al., 1964):

1) original DNA molecule; 2) incision by endonuclease; 3) widening of the gaps by exonucleases; 4) formation of complexes; 5) repair of single-strand areas; 6) restoration of the covalent bonds by ligases.

That base pairing between complementary polynucleotide chains from two different molecules leads to the formation of heteroduplexes of DNA is evident from numerous observations on recombination in bacteria and viruses. Such hypotheses were put forward by Szybalski (1964) and Meselson (1964). Experiments on Diplococcus pneumoniae (Fox and Allen, 1964) and Bacillus subtilis (Bodmer and Ganesan, 1964) using heavy isotopes showed that genetic transformation is associated with the physical introduction of a fragment of donor DNA into the DNA of the recipient. Further experiments involving the use of isotopes on D. pneumoniae showed that the resulting DNA molecule is a true hybrid both physically and in the genetic sense, and that the insertion of a segment of donor DNA is based on a mechanism of breakage and reunion. Kozinski and co-workers (1967) assume that the recombination of T4 begins with the appearance of gaps caused by the action of the enzyme on one of the DNA strands.

According to the hypothesis of Holliday and Whitehouse, the hybridization of DNA molecules during repair processes is based on the integrity of the entire conjugation complex. In the recombination sequence put forward by Szybalski (1964) and Howard-Flanders and Boyce (1966) it is assumed that the exchange begins with the appearance of fragments which are later combined via single-strand areas. On the whole, this sequence of events has much in common with the dark repair (Figure 3). In both cases the process begins with a break in the sugar-phosphate backbone of the molecule, after which the exonucleases widen the single-strand areas. This is followed by a complementary pairing of the molecules in the synapsis zone with the formation of a heteroduplex; the single-strand gaps are rebuilt and the fragments are finally sutured by ligases.

The recombination events associated with the formation of a heteroduplex from two DNA molecules can bring about a repair of primary lesions. Actual elimination of the gap necessitates its substitution with a segment which had as its template during synthesis the normal part of the strand in the heteroduplex.

The crucial element in reconstitution mechanisms both in the case of dark repair and in recombination processes is nonreplicative synthesis. This mechanism, however, contains potentialities not only for reconstruction but also for mutagenesis. Mutations may appear as a result of possible errors in the operation of DNA-synthesizing mechanisms causing the newly synthesized segment of the strand to deviate from its template because of erratic functioning of the enzymes (Hall and Lehman, 1968). Incorrect base pairing can obviously occur on the site of synthesis owing to an error in the operation of polymerases after the excision of the primary lesion by endonucleases and exonucleases (Figure 4). Similar considerations apply to nonreplicative synthesis during recombination processes according to both the scheme outlined by Szybalski and Howard-Flanders and that proposed by Whitehouse.

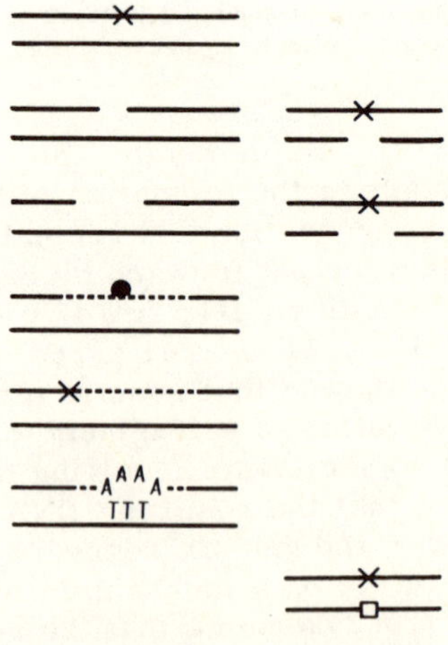

FIGURE 4. Scheme showing how mutations appear as a result of errors in the function of the DNA polymerase.

The effect of recombination processes on the frequency of mutations was established by Magni and von Borstel (Magni and von Borstel, 1962; Magni, 1963, 1964, 1969). These authors found that in yeast spontaneous mutations occur more frequently during meiosis than in cells where there is no recombination. Demerec (1962, 1963) discovered the phenomenon of selfing, namely that transduction by phages from auxotrophs to the same auxotrophic strain involving recombination with the transducing fragment causes the appearance of prototrophs and at the same time increases the mutability of various other genes.

In some cases the frequency of errors during nonreplicative synthesis can be significantly increased if the template determining synthesis is itself impaired. Such a situation could arise if the endonuclease did not invariably excise the primary lesion but sometimes excised instead the opposite part of the normal strand. This suggestion was put forward by a number of researchers (Holliday, 1962; Setlow, 1964; Nasim and Auerbach, 1967; Dubinin, 1968) and is of crucial importance for the whole theory of the fixation of mutations, as will be shown below. It was based on independent data and was used in different ways to explain the processes leading to the fixation of different types of mutation. The basis for the hypothesis was provided by the work of Hanawalt and Haynes (1965), who found that the enzyme recognizes the excision site not by the nature of the primary lesion proper but by the manner in which it distorts the secondary structure of the DNA molecule in this section. Since the rupture of the hydrogen bonds in the distortion of the secondary structure involves not only the damaged section but also the normal section opposite it, each of these may be recognized by the endonuclease.

If the excised section is that situated opposite the primary lesion, the nonreplicative synthesis acquires an impaired template. It was shown that the presence of a dimer alters the template properties of thymines, which under such circumstances can pair with guanine instead of adenine (Bollum and Setlow, 1963; Josse et al., 1961). An analogous pairing error occurs also in other primary lesions undergoing excision by endonucleases, for example, alkylated sites and lesions caused by nitrous acid. Obviously by excising the section opposite the lesion endonucleases manifest a highly mutagenic function.

All these cases of mutagenic injuries of DNA molecules concern a single strand. The fixation of such lesions naturally takes place during the synthesis of DNA. In other words, such mutations necessarily yield mosaics. By functioning as a template, the damaged strand creates a complete mutation consisting of the complementarily changed mutant DNA molecule in one half of the progeny, whereas the normal strand serves as template for the normal DNA molecules in the second half. However, complete mutations occur also in resting double-stranded DNA molecules.

A modern molecular theory of mutations must answer all the questions relating to the principles governing the appearance of mutations not only during the replication of DNA molecules but also during their fixation in resting double-stranded DNA molecules. Especially perplexing is the appearance of complete mutations in the absence of DNA replication.

THE PRINCIPAL MODES OF INTERACTION OF PRIMARY LESIONS WITH THE RESTORATION PROCESS AND WITH MUTAGENESIS

UV-induced primary lesions in the form of pyrimidine dimers are repaired by means of photoreactivation and dark repair enzymes operating by excision and patching. Mutant strains that have lost the ability to synthesize endonucleases (uvr^-, hcr^-) are distinguished by a greatly increased sensitivity (Hill, 1965; Witkin, 1966a, b; Kondo and Kato, 1966, 1968; Witkin, 1967). The same is true of mutants that have lost the capacity for recombination (rec^-) — they likewise display high mutability (Clark and Margulies, 1965; Howard-Flanders and Boyce, 1966; Willetts et al., 1969). If the pyrimidine dimers survive until the replication of the DNA molecules, as they do in the hcr^- strains, they can impair the template synthesis in the growing DNA molecule so that the new polynucleotide strand will have gaps in all sections opposite the dimers (Rupp and Howard-Flanders, 1968; Howard-Flanders et al., 1968). All the gaps are sealed after an hour of incubation (Figure 5). Since the gaps themselves appear during replication and they are filled after the replication processes are completed, Rupp and Howard-Flanders refer to this process as "post-replication repair." It does not affect the thymine dimers themselves, which persist in the DNA of the hcr^- strains for several cell generations (Bridges and Munson, 1968). The very process of post-replication repair of the gaps is based on the transmission of recombinations in the DNA molecules.

Ionizing radiations cause mostly single-strand breaks in DNA (Peacocke and Preston, 1960). It is known that in many cases these are repaired (McGrath and Williams, 1966). If the break is not turned into a gap as a result of the action of exonucleases it can be repaired by ligases. Once a gap is formed, it can only be sealed by the combined action of the enzymes involved in replicative synthesis and ligases.

However, excision has not been detected in single-strand breaks caused by ionizing radiations. It was shown (Bridges and Munson, 1966; Howard-Flanders and Boyce, 1966; Kondo, 1968) that *E. coli* strains which lack the ability to excise (hcr^-) do not show a higher frequency of mutations on exposure to X rays than strains capable of excision repair (hcr^+). At the same time X rays have a markedly greater mutagenic effect on rec^- strains which have lost the capacity for recombinations (Howard-Flanders and Boyce, 1966; Van de Putte et al., 1966). The same applies to lines showing mutations of the exr locus (Mattern et al., 1966). Thus, the enzymes responsible for recombination processes at the level of hybridization of DNA molecules apparently perform a major role in the repair of primary radiation injuries.

The primary lesions inflicted by alkylating compounds — alkylated bases and cross-links — are excised by endonucleases. The hcr^- strains, lacking the ability to excise injuries, are more sensitive to mustard gas (Bridges and Munson, 1966). Biochemical evidence for the excision of sutures was presented by Lawley and Brookes (1965). Solitary breaks caused by methylmethane sulfonate are not excised by endonucleases (Bridges and Munson, 1966). The same is true of lesions caused by ethylmethane sulfonate (Kondo, 1968). There are indications that lesions caused by nitrosoguanidine are similarly unexcisable by endonculeases (Witkin, 1967; Kondo, 1968).

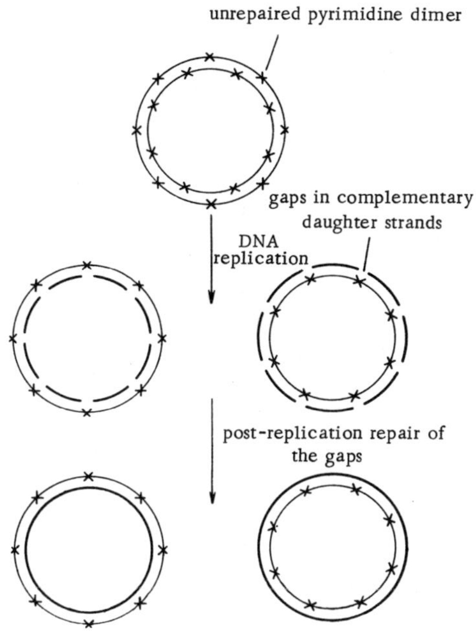

FIGURE 5. "Post-replication repair" of gaps (Rupp and Howard-Flanders, 1968).

Lesions caused by mitomycin C are excised by endonculeases, judging from the fact that hcr+ strains show fewer mutations than hcr⁻ strains (Boyce and Howard-Flanders, 1964 b; Kondo, 1968).

It must be noted, however, that we still know very little about the role of the different systems in repair processes and the appearance of mutations in the various types of primary lesion. The relationship between the known types of primary lesions and the known types of repair systems has been little studied. Intensive research in these fields is under way, and major advances towards the resolution of this important problem may be expected in the very near future.

SOME CENTRAL PROBLEMS IN THE MODERN THEORY OF MUTATIONS

Despite the fundamental achievements of the theory of mutations, expressed in its foundation on modern molecular principles, many of the most important problems associated with this complex phenomenon remain obscure. The following problems appear to be central to the theory of mutations at the present time.

1. Resonance mutagenesis and the appearance of complete gene mutations

As a rule, a multitude of mutational events can be traced back to the appearance of a primary lesion at the corresponding locus of one polynucleotide strand of the DNA molecule. However, the majority of mutations are complete — that is, they involve a complementary change of both polynucleotide strands of the resting DNA molecule in the absence of replication. The question arises: how is the single primary lesion transferred to the complementary bases of the second strand? In 1965, N. P. Dubinin referred to this phenomenon as resonance mutagenesis. If a complete mutation is to occur, the primary alteration of one strand must be transmitted to the other strand, which means that the primary injury serves merely as an initial potential change for the future mutation (Figure 6). Until very recently, the nature of the resonance mutagenesis remained a puzzle. The first steps in the analysis of this problem are just now being made. A basic assumption is that the area of the normal strand opposite the lesion can be excised, leading to a secondary induced lesion. Deviations from the normal code can occur as the secondary gap is being filled during the processes of nonreplicative synthesis.

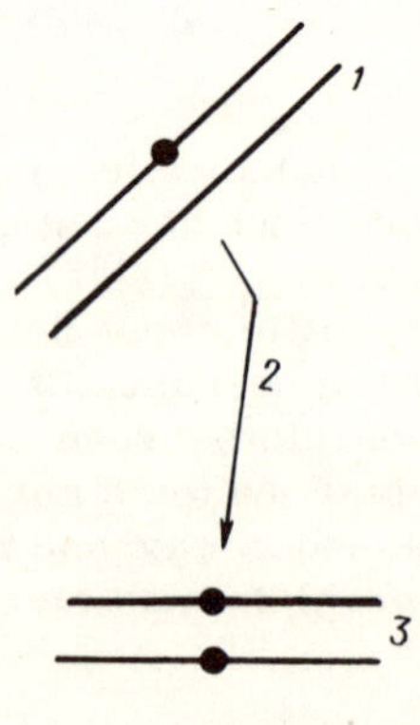

FIGURE 6. Resonance mutagenesis and the appearance of a complete mutation (Dubinin, 1965):

1) primary lesion in one DNA strand; 2) the phenomenon of resonance mutagenesis; 3) appearance of changes in both DNA strands — complete mutation.

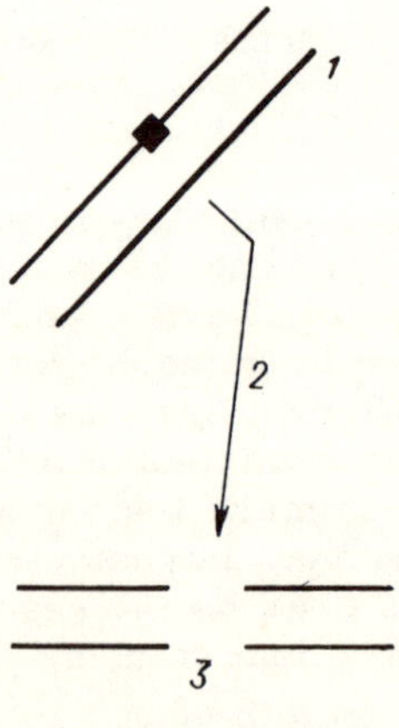

FIGURE 7. Resonance mutagenesis involving cutting of the double-helix DNA molecule:

1) primary lesion in one DNA strand; 2) the phenomenon of resonance mutagenesis; 3) cutting of both DNA strands.

2. Resonance mutagenesis involving the cutting of a double-helix DNA molecule

The primary injuries leading to alterations in the structure of the chromosomes during the resting stage of the DNA molecule must be more or less analogous in principle to those which cause gene mutations. At any rate, the origin of the process in both cases is often a lesion in one polynucleotide strand, which then necessitates the transmission of the lesion to the second strand. This process, however, does not result in a complete gene mutation but leads to complete breakage of the double-helix DNA molecule (Figure 7). This creates conditions conducive to exchange rearrangements or the formation of free fragments. What is the mechanism of this sectioning of DNA molecules? What further mutagenic changes can result in structural mutations of chromosomes? What is the relationship between the recombination processes and the operation of the enzymes of dark repair in these phenomena? These are only some of the important questions which arise in any discussion of this problem.

3. Realization of the primary lesions during DNA synthesis

The phenomenon of mosaicism concerns not only gene mutations but also the rebuilding of chromosomes. In both cases, after the original cell has been treated with the mutagen during the presynthesis or synthesis stages, it proves to have one mutant chromatid and one normal chromatid at the postsynthesis stage. This gives rise to a mosaic since one chromatid yields the mutant chromosome of one daughter cell and the other chromatid the normal chromosome of the other cell. In other words, mosaicism may arise when the lesion situated in one of the strands of the original DNA molecule is neither repaired nor transmitted by resonance to the second strand. The mosaicism persists and subsequently during replication of the molecule one strand yields a template with a locally altered genetic code whereas the other strand gives rise to a normal template (Figure 8). Mosaicism can also appear when the mutagen acts upon the DNA molecule during the process of synthesis, i. e. when the molecule is already split into two templates. A lesion of the code appearing in one of the templates leads to the appearance of one normal and one mutant daughter cell. Finally, mosaicism can arise when the mutagen acts upon the nucleus during the postsynthesis stage when the two chromatids are fully formed. Under such circumstances a complete mutation arises in one of the chromatids and as a result one of the daughter cells is a mutant and the other cell normal (Figure 9). In principle, however, this case falls into the category of complete mutations. The molecular basis of mosaicism resulting from exposure to mutagens during the presynthesis and synthesis stages remains largely obscure. From a methodological viewpoint, the study of chromosomal reorganizations in the analysis of this problem has the great advantage in the case of higher forms that both mosaics and complete mutations can be recognized at an early stage, during the metaphase of the treated cell, by the presence of chromatidal and chromosomal rearrangements respectively.

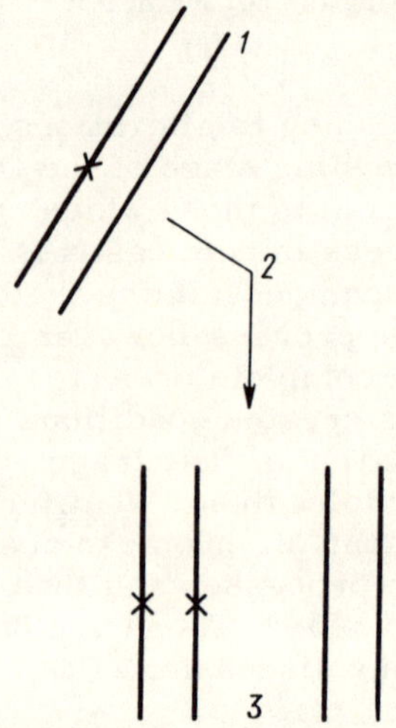

FIGURE 8. Appearance of mosaics as a result of replication:

1) primary lesion in one DNA strand; 2) replication; 3) mosaicism.

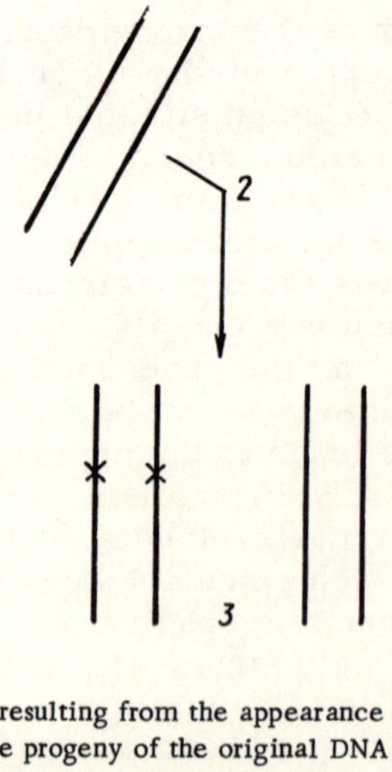

FIGURE 9. Mosaicism resulting from the appearance of a complete mutation among molecules of the progeny of the original DNA molecule:

1) original DNA molecule; 2) DNA synthesis; 3) appearance of a complete mutation in one of the DNA progeny molecules.

4. Repair systems and processes of mutagenesis

A mere increase of the mutation frequency in strains which have largely lost their capacity for excision does not constitute proof of a direct relationship between mutagenesis and the functioning of the enzymes of the dark repair system. These cases represent simply a change in the initial pool of primary lesions. An actual indication of the role of repair processes in the fixation of mutations can be obtained when strains sensitive to the mutagen show not only a change in the total number of mutations but also a change in the ratio between the yield of mosaic mutations and that of complete mutations. The sensitive strains (uvr^-, hcr^-) are unable to excise the primary lesions but retain some residual capacity for repair. Obviously a marked decline in the capacity for excision sharply increases the total yield of mosaic mutations and causes a relative decrease in the number of complete mutations. Precisely these results were obtained by Nasim (1968) in experiments with yeast. This is a convincing indication that the dark repair system is directly linked not only with the processes of protection against primary injuries but also with the processes of fixation of mutations. However, on the whole the question of the role of repair processes in the mutations of genes and chromosomes has been far from adequately studied.

5. Recombination systems and processes of mutagenesis

Mutant strains in which the processes of recombination are blocked (rec^-) may similarly display a greatly increased frequency of mutations. Here the recombination processes are linked with the repair of lesions arising in the molecular structure of the DNA. There is some evidence, on the other hand, that the occurrence of crossing over during meiosis increases the frequency of mutations. Witkin (1968a, b, c) assumes that erratic construction during post-replication synthesis involving recombination processes may be a source of mutations.

Recombination phenomena possibly play an important role both in repair and in the appearance of mutations induced by ionizing radiation.

6. Secondary induced changes in DNA molecules

Both the excision of a section of the normal strand opposite the lesion and the delay in the processes of replicative synthesis in the areas of the complementary strand opposite the lesions result in the formation of gaps which persist for some time as secondary induced lesions (Figure 10). The concept of post-replication repair of gaps was introduced by Rupp and Howard-Flanders (1968) and Witkin (1969b). There is no doubt, however, that all repair and mutation processes taking place both before replication and after DNA synthesis are based on uniform events occurring at the level of the double-stranded DNA molecule. What has to be determined is not the time at which the repair processes take place but the nature of the

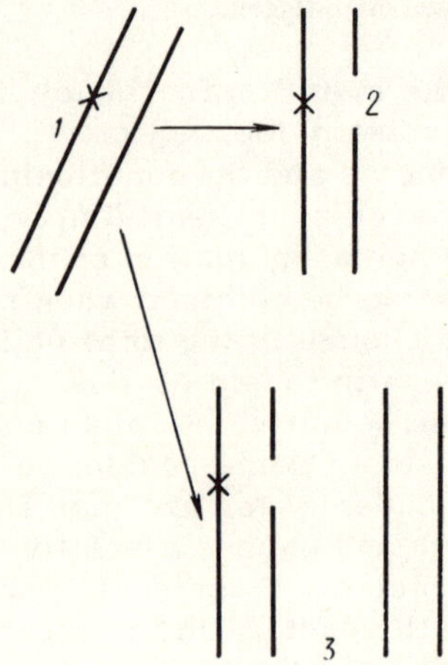

FIGURE 10. Appearance of gaps opposite primary lesions:

1) primary lesion in one strand of the DNA molecule; 2) gap opened by excision of the area opposite the lesion prior to replication of the DNA molecule; 3) gap formed in the newly synthesized strand during replication.

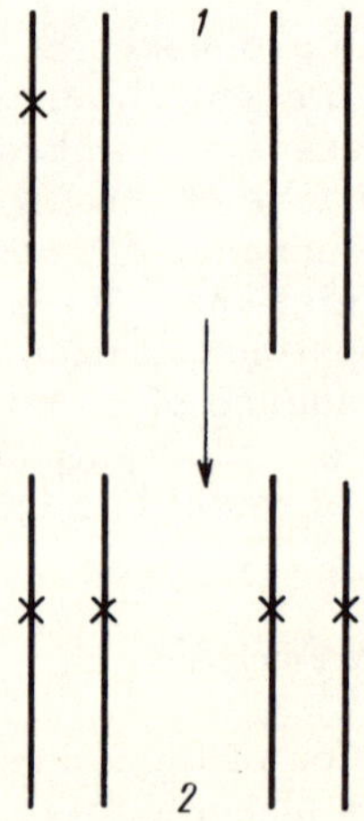

FIGURE 11. Resonance mutation in the four strands of two homologous chromosomes:

1) two pairs of homologous chromosomes, one of them containing a primary single-strand lesion in the DNA;
2) homozygous recessive mutation involving a change of all four DNA strands in both homologous chromosomes.

secondary change and the type of repair processes it involves. These secondary changes may be replicational, i. e. induced during DNA synthesis, or associated with repair, i. e. induced at a time when the DNA molecule has a double-stranded structure. The existence of induced replicational secondary changes was demonstrated by Rupp and Howard-Flanders (1968). Evidence for the presence of secondary induced repair-linked changes in higher forms was presented by Painter and his colleagues (Rasmussen and Painter, 1966; Painter, 1967; Painter and Cleaver, 1967; Painter and Schaefer, 1969), who found an additional repair synthesis of DNA in the presynthesis (G_1) and postsynthesis (G_2) stages of human (HeLa) cells following irradiation with UV and X rays, exactly as observed in bacteria after UV irradiation. The role of induced secondary changes in the DNA molecule in the processes of mutagenesis deserves particular attention. The existence of secondary induced changes (gaps) appears to be one of the central factors involved in the fixation of mutations. In this case, dark repair mechanisms and recombination processes might provide a clue to the origin of mutations.

7. Transmission of the homologous mutation to several double-helix DNA molecules

The simplest example of this kind of mutation is the appearance of isochromatid breaks where two double-stranded molecules are broken at the same level. The question of the existence of molecular bonds in the homologous loci and their equivalent breakage then arises. Proximity between the sister chromatids appears to be essential for the appearance of such breaks.

Even more difficult to explain theoretically are cases involving the appearance of somatic mosaicism in recessive alleles in the tissues of first-generation plants after exposure of the seeds to mutagens. Thus, the treatment of barley seeds with nitrosoethyl urea gives rise to sectors homozygous for different albinism genes in nearly 100% of the M_1 plants. Here the injury inflicted on one nucleotide strand persists as a potential alteration and is transmitted through a number of replications of the DNA. The mechanism of resonance mutation then begins to function and transfers the lesion not only to the second strand of this DNA molecule but also to both strands of another independent homologous chromosome (Figure 11). What are the actual mechanisms of these striking phenomena in the processes associated with mutation? The answer can be found through analysis of the processes of homologous resonance mutation in a multi-stranded structure since it appears that temporary complexes of four chromatids or two independent homologous chromosomes are formed at the moment of mutation. Holliday (1970) has proposed a scheme of mutation in a multi-stranded chromosome proceeding from the assumption that only one of the strands of the DNA molecule transmits its information to both newly synthesized strands — a possibility indicated by several authors (Jehle, 1965; Barricelli and Del Zoppo, 1968; Kubitschek, 1964; Jehle and Parke, 1968; Stahl, 1969). The development of hypotheses of resonance mutation based on repair or replication mechanisms at the sites of molecular associations of double-stranded DNA molecules is fundamental in this respect (Dubinin, 1968, 1969a, b; Dubinin and Soyfer, 1969).

8. One of the most enigmatic aspects of mutagenesis – the existence of a phase of potential changes

All states of the locus between the appearance of the primary lesion there and the fixation of mutations fall into this category. The phase of potential change is unstable in the sense that it may develop into a fixed mutational alteration or be repaired, in which case the locus reverts to its normal state. Potential changes may be of short duration or prolonged. In the latter case mutagenesis is protracted and the potential changes persist for days, months and even years in cells not undergoing DNA replication before they turn into mutations. This was established in plant seeds treated with a mutagen and then stored (Dubinin and Dubinina, 1968) and in rat liver cells (Dubinin et al., 1968). The existence of a special category of potential changes is evident from the phenomenon of replicative instability, discovered by Auerbach (Auerbach and Robson, 1946) in experiments on the effect of mustard gas on the spermatozoa of *Drosophila*. This category can now be interpreted as one of the main types in the system of potential changes (Dubinin, 1969a, b; Dubinin and Akif'ev, 1970). In the case of replicative instability the potential changes pass through a series of DNA replications. The potential change is apparently inherent in the primary structure of the DNA molecule since it is reproduced on replication of the DNA. Data relating to the treatment of loach spermatozoa with radiation and thio-TEPA (Romashov et al., 1962; Dubinin and Tarasov, 1969) indicate that these potential changes appear originally in a quantity characteristic of all the cells of the embryo. These changes may also arise as complete mutations, i. e. by the mechanisms of resonance mutagenesis. The realization of this type of potential change is labile and varies among different plant and animal tissues. It appears that the mechanism of fixation of the mutation for this type of potential change, taking the form of repair or recombination processes, depends on the conditions of development. The molecular basis of protracted mutagenesis and replicative instability remains obscure.

NEW APPROACHES TO THE PROBLEM OF THE ORIGIN OF COMPLETE GENE MUTATIONS AND THE BREAKAGE OF DOUBLE-HELIX DNA MOLECULES

During the period 1958—1969 scientists at the Institute of General Genetics of the USSR Academy of Sciences worked out a hypothesis for the repair mechanism of fixation of complete mutations in the double-stranded DNA molecule following the appearance of a primary lesion in one of the strands (Dubinin, 1969a, b; Dubinin, 1969; Dubinin and Soifer, 1969; Dubinin and Soyfer, 1969; Soifer, 1969). This hypothesis envisages the possibility that mechanisms of the dark repair system participate in mutagenesis. The relationship between repair mechanisms involving the enzymes of dark repair and mutagenesis had been discussed previously by a number of authors (Holliday, 1962; Setlow, 1964; Witkin, 1966a; Kimball, 1966; E. B. Freese and E. Freese, 1966; Nasim and Auerbach, 1967; Howard-Flanders, 1968; Sobels, 1969; Witkin, 1969a). However, a fully worked out hypothesis was presented only in the publications of the Institute of General Genetics of the USSR Academy of Sciences. Our concepts are based on the idea that if a primary lesion is present in one of the DNA strands and impairs the secondary structure of the molecule, the endonuclease may excise the normal section of the complementary strand opposite the lesion, either routinely or in error. Such an excision may trigger off a chain of events culminating in fixation of the complete mutation. Obviously if the excision affects the primary lesion proper this leads to repair, while if the primary lesion persists until chromosomal replication mosaicism ensues.

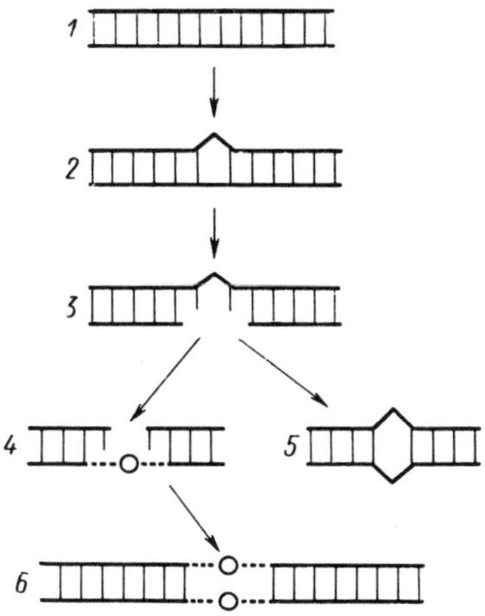

FIGURE 12. Appearance of a complete mutation as a result of excision of the area opposite the lesion:

1) original DNA molecule; 2) primary lesion in one strand; 3) excision in the strand opposite the primary lesion; 4) irregular synthesis in the gap and excision of the damaged area; 5) erroneous synthesis in the gap without a secondary excision; 6) fixation of the mutation.

Excision of the area opposite the lesion creates a gap which then has the damaged section as its template for nonreplicative synthesis. As a result, during closure of the gap by polymerase there may be errors in the process of nonreplicative synthesis. After closure of the first gap, a second action of endonuclease may excise the area actually containing the lesion and closure of this second gap will transmit to this section of the DNA the error in synthesis committed during the first repair (Figure 12).

This concept of the fixation of mutations in terms of repair processes involving transmission of the lesion from one DNA strand to two complementary strands lends itself to an extensive experimental verification. At the genetic and cytogenetic level, good results may be expected in experiments on the regulation of complete and mosaic mutations following the action of a mutagen on the resting double-helix DNA molecule. In this case by using different mutagens or test conditions modifying the effect of the particular mutagen, we can actually regulate the situation either to preserve the primary lesion until the moment of DNA synthesis (mosaicism), or to bring about some degree of transmission of the primary lesion to two complementary strands (complete mutations).

Soyfer (1969) has shown that the treatment of λ phage particles with a 1 M solution of hydroxylamine does not cause differences in the survival rate of cells possessing a normal system of dark repair and cells deficient in this respect. In other words, hydroxylamine belongs to the category of mutagens which cause irreparable primary lesions under the given conditions. In this particular case the result appears to be a direct deamination of cytosine (its conversion into uracil), which does not disrupt the secondary structure of DNA. UV irradiation, on the other hand, gives rise to numerous thymine dimers and other reparable lesions, and mutability under such circumstances may be expressed in the form of complete mutations. In a study of yeast colonies arising from single UV-treated cells Haefner (1966) proved conclusively that complete gene mutations appear under these conditions.

Numerous data indicate that mutations can arise as complete alterations, i.e. as a result of complementary changes in both DNA strands. This principle was clearly shown in *E. coli* (Bridges and Munson, 1968) as well as in higher organisms. Such alterations have been established in *Drosophila* (Muller et al., 1961; Altenburg and Browning, 1961; Inagaki and Nakao, 1966), *Ustilago* (Holliday, 1962), *Schizosaccharomyces* (Nasim and Clarke, 1965; Nasim and Auerbach, 1967), *Saccharomyces* (Ito et al., 1962; Resnick, 1969), and infusorians (Kimball, 1964).

Depending on the test conditions, a given mutagen may yield either mostly complete mutations or mostly mosaic mutations. The treatment of dry seeds of *Crepis capillaris* yields a mixture of complete mutations (chromosomal alterations) and mosaic mutations (chromatidal alterations). If the cells in growing rootlets are treated, all the alterations are of the chromatid type (Dubinina and Dubinin, 1968).

It is quite conceivable also that genetically determined changes in the function of dark repair enzymes must alter in a definite sense the ratio between complete and mosaic mutations in comparison with the spectrum in normal strains. Experiments with yeast (Nasim, 1968) have shown that the frequency of complete mutations is higher in normal cells than in cells of a sensitive strain in which operation of the excision system is partly blocked.

It must be recognized that the repair hypothesis of the origin of complete mutations has not yet been sufficiently tested experimentally. Direct proofs that endonuclease excises the normal strand opposite the lesion were thought to be lacking while the indirect proofs were considered insufficient. There was no evidence for a second excision, eliminating the lesion itself under these conditions.

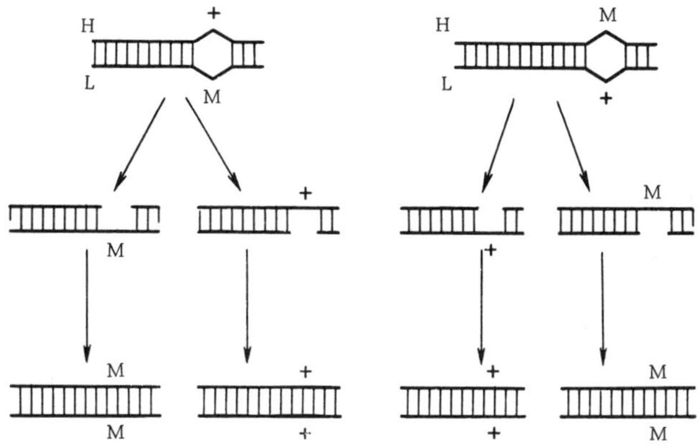

FIGURE 13. Diagram showing the formation of a homoduplex from a heteroduplex by excision and repair (Hogness et al., 1966):

H) heavy strand, transcribed; L) light strand, not transcribed; +) normal code; M) mutant code inactivating the phage.

Suddenly it was realized, however, that the principle of excision of the normal area opposite the lesion, suggested by a number of authors, had already been proved by Hogness and his colleagues in 1966 (Hogness et al., 1966).

These results had until then remained unnoticed in discussions on the relationship between repair processes and mutagenesis. They were derived from an analysis of corrections in heteroduplexes and were not regarded by their authors in terms of mutagenesis. However, the data obtained by Hogness and his colleagues are directly related to the question of the origin of complete mutations by the mechanism of resonance mutation. Hogness and his group obtained a double-stranded heteroduplex from two single-stranded DNA molecules of which one, taken from one phage, was heavy while the other, taken from another phage, was light. Only the heavy strand in this heteroduplex was transcribed (Figure 13). Two heteroduplexes were obtained, differing in the site of the mutation (M). In one case the mutated code was situated in the light DNA strand, while in the other case it was in the heavy DNA strand. The mutated phage itself, being a homoduplex with respect to M, turned out to be inactive under the test conditions. Hogness and his colleagues analyzed all three of the theoretically possible ways in which the heteroduplexes could have been formed. First it was assumed that excision did not occur. In this case, for the heteroduplex H+ /LM all 100% of the phage particles must be active since the strand H+ is the one transcribed. Conversely, for the HM/L+ heteroduplex, all 100% of the phage particles must be inactive. The second possibility is that only the lesion is excised. Here all 100% of the phage particles will be active in both cases. Third, both the lesion and the opposite part of the normal strand may be excised. In this case, assuming an equal probability for these excisions, both for the H+ /LM heteroduplex (left-hand side of Figure 13) and the HM/L+ heteroduplex (right-hand side of Figure 13), 50% of the phage particles must be inactive and 50% active. The equal probability of excision of the lesion and of the opposite normal part in the heteroduplex has been confirmed experimentally. The specific role of the enzymes of dark repair in this case was studied separately. Recipient bacterial cells were treated with UV rays and pyrimidine dimers induced in their chromosomes. The results obtained showed that the operation of the repair enzymes with respect to phage heteroduplexes was markedly altered under these conditions.

Thus, the results obtained by Hogness and his group (1966) provide direct proof of the central principle of the hypothesis proposed for the mechanism by which complete mutations arise, namely, the principle of excision of the normal section opposite the lesion. These authors propose a method for determining the relationship between the different types of primary lesions and the operation of dark repair systems, and this should be extensively used.

Wolff and Scott (1969) reject categorically the participation of dark repair systems in the emergence of mutations. Their main arguments can be summed up as follows. First, the study of mutational events should not be confined to processes occurring in the DNA molecule, since the chromosomes of higher forms also contain protein and often show a multi-stranded structure under the light or electron microscope. This remark is indeed true, but nevertheless the greatest advances in the analysis of chromosome replication and the phenomena associated with mutation recently have been obtained precisely by authors who regard the DNA molecule as the genetic basis of the chromosome. Even with respect to chromosomal alterations there can be no doubt as to the importance of the polarity of the DNA molecule, which determines the type of reunion of the fragments (Brewen and Peacock, 1969). The second objection deserves greater attention. The authors cited examined the formation of chromosome rearrangements in cells of the normal strain of the Chinese hamster and in normal human leukocytes, both having a fully functional system of dark repair, in comparison with strains in which this system is apparently lacking (rootlets of *Vicia faba*, the B_{14} strain of the Chinese hamster, and leukocytes of people suffering from xeroderma). It was shown that X rays cause breakage of chromosomes and reunion of the fragments in all the cells. Treatment with caffeine during stepwise irradiation of *V. faba* cells similarly did not affect the process of formation of rearrangements. All these data suggest that the dark repair systems are not associated with chromosomal mutations. The approach of Wolff and Scott to this problem is undoubtedly correct. However, the volume of material presented in their work is small. It does not allow real evaluation of the differences in the quantitative yield of mutations and, more important, it does not permit determination of the changes in the ratio of complete and mosaic mutations, a crucial shortcoming in this case. This discussion must be continued as new experimental data are obtained. Yamamoto and Yamaguchi (1969) found, for example, that caffeine influences the processes of formation of chromosomal rearrangements induced by gamma rays in barley cells. They concluded that the view of Wolff and Scott as regards the independence of chromosomal breakages from the systems of dark repair is not justified.

Witkin (1969a, b) similarly rules out the participation of dark repair systems in mutagenesis. In her view, the gap left after excision of the lesion by endonuclease is repaired so accurately that no mutation is possible. According to her calculations, the dimers which are not excised but yield post-replication gaps cause mutations 500 times more frequently than the excised dimers. This conclusion, however, is not irreproachable. The data of Witkin apply only to small doses of ultraviolet (20 ergs). A different picture emerges if one compares the mutability of the hcr^+ and hcr^- strains at equal survival levels. At a dose of about 800 ergs/mm^2, for example, the frequency of mutations in the hcr^+ strain is 12—13 times higher than in the hcr^- strain. All this indicates that the relationship between the operation of repair systems and mutations is more complex than any assumption of an extraordinarily accurate excision and rebuilding of the gap in the form of complete repair.

The second major problem in the modern theory of mutations is the nature of the formation of chromosomal rearrangements. The solution of this problem is likewise associated with the questions of resonance mutagenesis since rearrangements may also appear in cases in which the primary lesion arises in one strand of the DNA molecule. We can extend the repair hypothesis to the process involved in the sectioning of a double-helix DNA molecule by assuming a slight modification of the repair mechanism leading to complete mutations (Figure 14). The excision of the section of the normal strand opposite the lesion may give rise to a long-term gap. Excision of the section containing the primary lesion occurs during the existence of this gap. The result is a superimposition of gaps, i. e. a sectioning of the double-helix DNA molecule. The existence of such a mechanism enables the formation of chromosomal rearrangements both by contact (with the formation of intermolecular bonds) and by fragmentation.

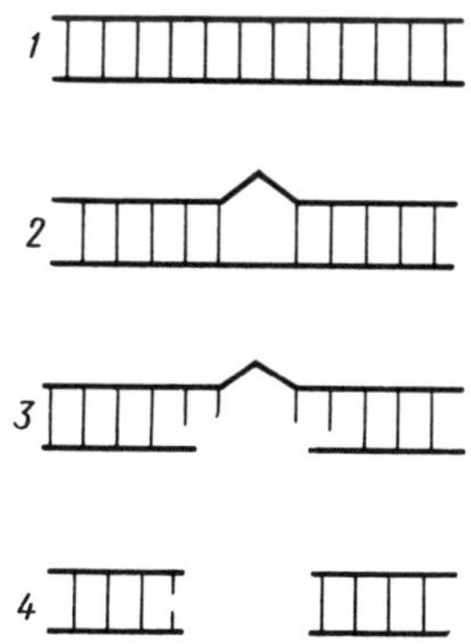

FIGURE 14. Diagram of repair sectioning of a double-helix DNA molecule:

1) original DNA molecule; 2) appearance of a primary lesion in one strand; 3) excision of the area of the normal strand opposite the primary lesion; 4) excision of the primary lesion, leading to a sectioning of the DNA molecule.

It is very difficult to prove that sectioning of double-helix DNA molecules occurs in the chromosomes of higher organisms. However, scientists at the Institute of General Genetics of the USSR Academy of Sciences have obtained some data relevant to this question in experiments on a classic cytological object, the plant C. capillaris concerning the phenomenon of "threading" of circular chromosomes.

For many decades, cytogeneticists have examined chromosomal preparations from different organisms after irradiation of the cells or their treatment with chemical and other mutagens, and it was generally believed that all the basic types of structural mutations of chromosomes had already been analyzed. However, L. S. Nemtseva and the present author were able to shed some light on the nature of a particular kind of chromosomal rearrangement which had been largely neglected previously, namely the phenomenon of "threading." In this a circular chromosome is physically strung on normal rodlike chromosomes. How does such a strange phenomenon arise? The data obtained rule out the possibility that the rings are slipped onto the chromosomes mechanically and also provide a clue as to the molecular basis of this peculiar behavior of the circular chromosomes. The simple fact of "threading" was noted by Darlington and La Cour (1945), Newcombe (1942), and others.

Our data show the existence of various regular types of "threading." The mechanism by which the ring is slipped onto the chromosome may be understood if we assume the initial formation of a rigid molecular loop and the operation of a repair mechanism cutting all four polynucleotide strands at the point where the loop is twisted (Figure 15). The study of different forms of "threading" of circular chromosomes has led to the conclusion that after the initial threading the threading pattern is duplicated during DNA synthesis and that the two resulting threaded sister ring chromosomes often participate in a mitotic crossing over. The chromatids surrounded by the circular chromosomes are able to slip out during the crossing over. This conclusion is based on the deviations from the basic type of threading observed in our experiments (Figure 16). Theoretical analysis shows that ten different types of threading of circular chromosomes are possible (Figure 17). All ten predicted types of threading were found to occur in reality. These facts can be explained by assuming that the double-helix DNA molecules are split into fragments during the process of crossing over. Such an explanation seriously contradicts the hypotheses on crossing over proposed by Holliday (1964) and Hastings and Whitehouse (1964).

We found that the chromatids actually split during crossing over into fragments capable of being either reunited or exchanged. The very fact that double-helix DNA molecules split into fragments gives substantial support to the repair hypothesis of the origin of chromosomal rearrangements. The data obtained from studies of threading of circular forms agree with the views of Szybalski, Howard-Flanders, and Boyce. Moreover, the repair hypothesis indicates how the initial sectioning of double-helix DNA molecules can lead to the formation of recombination fragments.

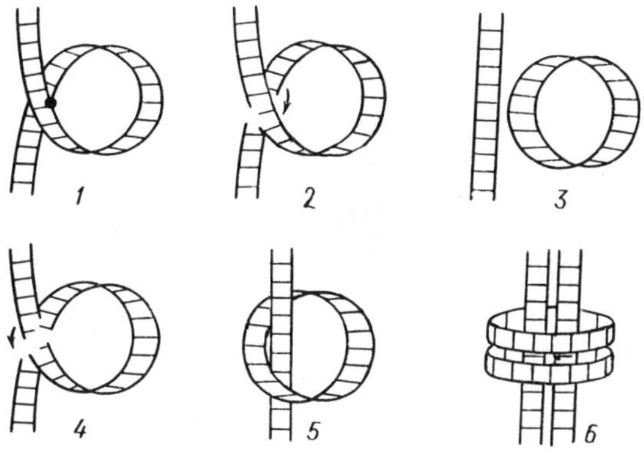

FIGURE 15. Scheme of the formation of a circular chromosome and its "threading" on chromatids:

1) loop reinforced by molecular cross-linking at the point of contact; 2) sectioning of the two double-helix molecules; 3) free ring and fragment resulting from external fusion of the ends of the middle fragment (arrow in illustration 2); 4) closure of the ring around the fragment; 5) ring threaded on a rodlike fragment; 6) double threading pattern after DNA synthesis.

Thus, the repair hypothesis assumes the possibility of a common pattern in the fixation of the main types of mutation, i. e. both gene mutations and structural mutations of chromosomes. This hypothesis is, of course, not the only possible explanation of the mechanism of fixation of mutations.

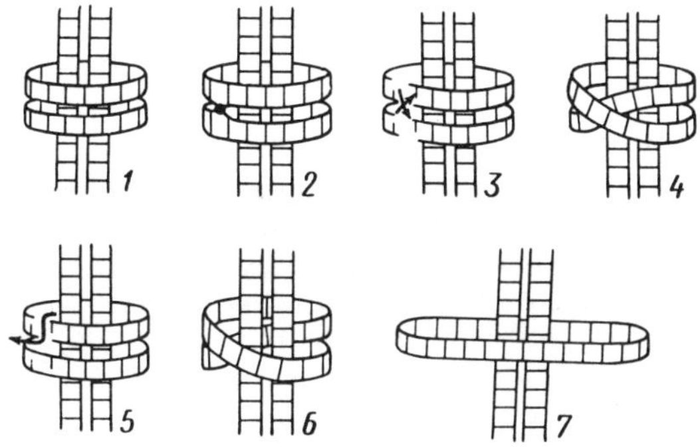

FIGURE 16. Mitotic crossing over in threaded circular chromosomes:

1) two sister chromatid rings, threaded on two sister chromatids; 2) molecular sectioning at the site of conjugation of homologous loci of sister DNA molecules in the rings; 3) sectioning of the rings in such a way that the cut ends can be joined via single-strand regions; 4) formation of a single ring in the form of a figure of eight on two sister strands; 5) escape of the two inner strands through a gap in the upper ring; 6) single threaded ring prior to uncoiling; 7) single threaded ring after uncoiling.

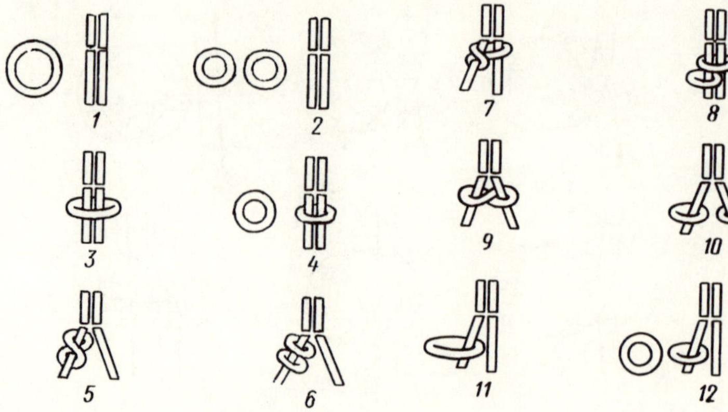

FIGURE 17. Possible types of "threading" of circular chromosomes, predicted theoretically on the assumption that the double-helix DNA molecules are cut at the point of crossover exchange and the inner chromatids escape from the rings via a gap at the site of the cut:

1) both chromatids slip out of both rings during crossing over, leading to the formation of a single ring; 2) two rings slip off as the fragments lock into the original position; 3) both inner chromatids slip through the hole of one ring during crossing over, so that a single ring is threaded on both chromatids; 4) locking of the initial fragments retains both rings, one threaded on both chromatids and the other free; 5) one chromatid slips into the holes of both rings during crossing over, giving a single ring threaded on one chromatid in the form of a figure of eight; 6) locking of the original fragments leads to the threading of both rings on a single chromatid; 7) one chromatid slips into the hole of one ring and crossing over yields a single ring which is threaded on both chromatids and then again threaded on one chromatid forming a figure of eight; 8) as the fragments close together in the initial position, one ring surrounds both chromatids and the other a single chromatid; 9) the chromatids slip into the holes of different rings, and crossing over gives a single ring threaded in turn on each chromatid in the form of a figure of eight; 10) one ring is threaded on each chromatid as the fragments reunite in the original position; 11) both chromatids slip into the hole of one ring, and one chromatid also passes through the hole of the other ring; crossing over creates a single ring threaded on one chromatid; 12) the fragments are reunited in the original position so that both rings are retained, one of them threaded on one chromatid and the other free.

Recombination processes at the molecular level clearly play a major role in mutagenesis. A number of variants of the repair hypothesis are possible. For example, with the breakage of one polynucleotide strand, characteristic of a number of chemical mutagens, excision of the opposite normal section immediately cuts the double-helix DNA molecule and the fragments are subsequently completed by nonreplicative synthesis if the positions of the breaks in the two chains do not coincide.

It must be recognized that research on the intracellular systems of enzymatic repair is just beginning and much remains unknown, especially the role of these systems in mutagenesis.

Despite the difficulties confronting the hypothesis of repair fixation of gene mutations and complete sectioning of double-helix DNA molecules, it is firmly supported by three groups of facts, as shown above. The first is the proof that for a single event there is an equal probability of excision of the primary lesion and excision of the opposite part of the normal strand, as demonstrated by Hogness and his colleagues. The second is the discovery that Buvr⁻ strains differ from Cuvr⁺ strains not only in the overall number of mutations but also in the ratio between complete and mosaic mutations (Nasim, 1968). The third category of proofs is based on the evidence that double-helix DNA molecules are actually cut into fragments, obtained in studies of the "threading" of circular chromosomes (Dubinin and Nemtseva, 1969, 1970). All these findings open up promising paths for the experimental analysis of the problem of the fixation of mutations in terms of the repair hypothesis.

PROBLEMS ASSOCIATED WITH THE MOLECULAR BASIS OF PROTRACTED MUTAGENESIS AND REPLICATIVE INSTABILITY

The molecular principles determining the prolonged survival of the potential changes both transmitted through the reproduction of DNA and realized during the resting stage of the molecule or the first replication of the DNA so far remain completely obscure. Nevertheless, the study of these phenomena by cytogenetic and genetic methods has revealed some interesting facts.

The study of chromosomal rearrangements has shown that the potential changes vary in duration: they may persist until the next stage of the cell cycle (Dubinina and Dubinin, 1968; Dubinin et al., 1968) or continue through the replication of the DNA molecules, in which case they are realized one or several generations later. The latter was established in tissue cultures of human fibroblasts (Dubinin and Saprykina, 1964) and in transplanted mouse cells (Akif'ev et al., 1965). The existence of potential changes of extraordinarily long duration was shown in experiments involving the storage of seeds treated with alkylating compounds in such a way that the action of the mutagen could be interrupted at any time from the onset of the G_1 and S (DNA synthesis) stages (Dubinin and Dubinina, 1968). In similar experiments Garina and Romanova (1970) established a parallelism between the production of chromosomal rearrangements and the appearance of gene mutations (Figure 18). It was also found that the duration of the presynthesis stage affects the fixation of chromosome-type mutations (Dubinina, 1970). As regards the mechanisms of potential changes, the only available data concern the transfer of thymine dimers through DNA synthesis (Bridges and Munson, 1968).

A radically new and still unexplained phenomenon is the wavelike, rhythmic character of the changes occurring during the transition of the potential changes into mutations in dry seeds (Dubinin and Dubinina, 1968) at different times after their treatment with an alkylating compound (Figure 19). It is noteworthy that the ratio between chromosomal and chromatidal rearrangements (i. e. between complete and mosaic mutations) is not the same during the periods of maximal and minimal emergence of mutations. During the peak periods there is a shift toward chromosome-type mutations, whereas the proportion of chromatid-type mutations is greater during the periods of minimal mutation yields.

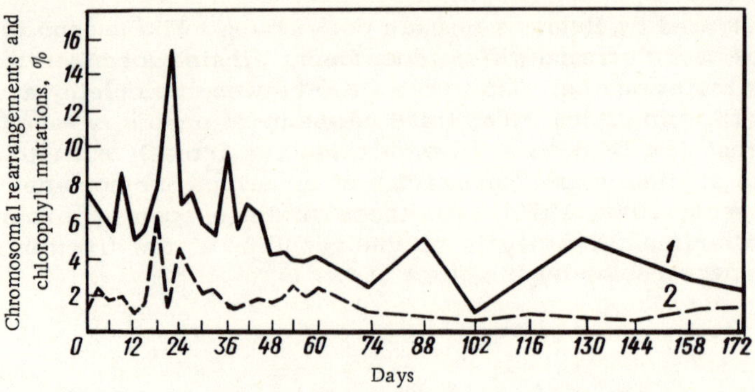

FIGURE 18. Parallelism between the wavelike kinetics of chromosomal rearrangements and gene mutations in stored barley seeds treated with alkylating compounds:

1) chromosomal rearrangements; 2) chlorophyll mutations.

Although it is not yet known whether the repair system of the cells participates in the establishment of these rhythms of mutagenesis, the correlation between the rhythms and the changes in the ratio between complete and mosaic mutations is highly significant.

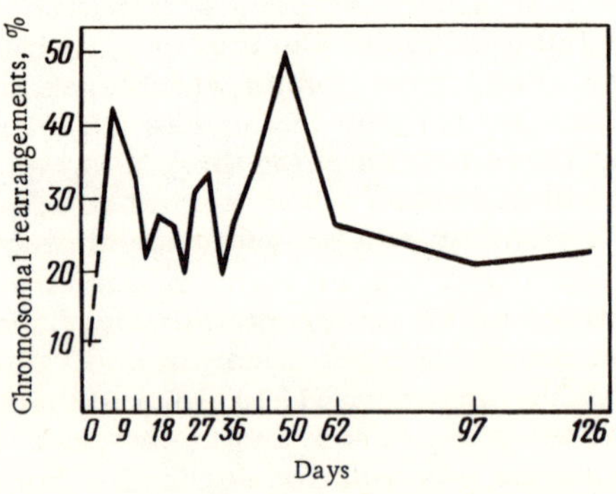

FIGURE 19. Wavelike kinetics of the yield of structural mutations of chromosomes at different times during the storage of seeds of *Crepis capillaris* treated with alkylating compounds.

The explanation of replicative instability in molecular terms is especially difficult. It was assumed in many cases that the strand of the treated DNA molecule becomes capable of inducing the phenomenon of replicative instability during the replication of the molecules into daughter strands which in turn induce this instability throughout a number of cell generations (Dubinin, Akif'ev and Vyshtynetskaya, 1971). Experiments with cultured cells of the Chinese hamster involving a twice-repeated synchronized selective removal of the cells in metaphase have shown that the primary lesions induced by mustard gas pass through two replications of the DNA and are realized during the second mitosis. The original templates were labeled with ^{3}H-thymine. This permitted a differential followup of the fate of the original templates which possibly bear the initial primary lesions, and the fate of the DNA molecules composed of freshly synthesized strands (Figure 20). It was found that mutations arise both in the labeled and in the unlabeled chromatids in the second generation. Clearly the induction of replicative instability in newly synthesized DNA strands, demonstrated in these experiments, lies at the basis of the instability of genes, the instability of chromosomes in sperm or embryos treated with mutagens, and the instability of human fibroblasts in tissue culture.

The transmission of these concealed potential changes to daughter polynucleotide strands suggests that they represent a particular type of disturbance of the primary structure of DNA — namely, disturbances of the nucleotide sequence. It is in this sense that such changes fall into the general category of gene mutations. These changes, however, are distinguished by the fact that they are manifested as various types of mutation a long time after the action of the mutagen and after they have passed through a number of DNA replications. In the case of structural alterations, the manifestation of these potential changes of the gene mutation type causes sectioning of the double-helix DNA molecule. The actual molecular basis of these phenomena, however, is so far totally unknown.

CONCLUSION

We have seen that there are many problems in the modern molecular theory of mutations which remain unsolved. Among these the transformation of the primary lesions arising in one strand of the DNA molecule into actual mutations is particularly important. This is a complex process and is of course not confined solely to the action of the dark repair system. The fixation of mutations in the double-stranded DNA molecule may be associated not only with the action of the various systems of dark repair but also with recombination-type nonreplicative processes. This applies to reparable primary lesions. Irreparable primary lesions persist until DNA synthesis and are fixed into mutations by the replication mechanism.

The explanation of how mutations are fixed in the double-stranded DNA molecule in terms of dark repair or recombination systems, and the established existence of different types of potential change have placed the modern theory of mutations on a different level from that of the classic concepts of Watson, Crick and Freese. According to the early view, the mechanism of mutagenesis consists in the appearance of primary lesions in the DNA molecule and their fixation as such or by an exchange during the replication of DNA molecules. These early concepts could not explain the appearance of complete gene mutations or the molecular nature of the structural mutations of chromosomes. By including repair systems, involving either dark repair enzymes or recombination processes, in the mechanism of mutagenesis, the recent approaches provide a new conception of how mutations are fixed. These modern views bring mutagenesis into the framework of the enzymatic events occurring in the cell. They reveal another aspect of the processes securing preservation of the genetic information and also indicate the role of these phenomena in mutagenesis and recombinations.

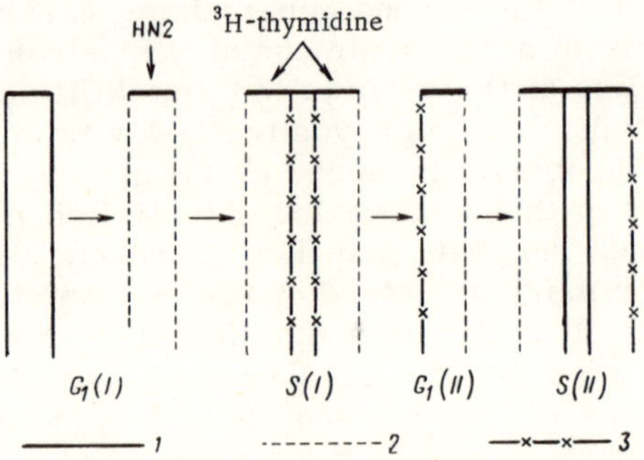

FIGURE 20. Scheme of an experiment showing the induction of replicative instability in newly synthesized strands of the DNA molecule in higher organisms:

1) normal strand of the initial DNA molecule; 2) strands of the DNA molecule after treatment with mustard gas; 3) newly synthesized strands labeled with ^{3}H-thymidine; G_1(I) — presynthesis stage of the first mitosis (treatment with mustard gas); S(I) — synthesis stage of the first mitosis; G_1(II) — presynthesis stage of the second mitosis; S(II) — synthesis stage of the second mitosis with the appearance of chromatid deletions (designated by a line with x's) both in the labeled chromatids (both strands new) and in the nonlabeled ones (one strand initially treated with mustard gas, the other strand new).

All this signifies that the development of the theory of mutations has now entered a new stage. With the realization that the conversion of the primary lesions in DNA molecules into actual mutations involves the repair enzymes operating in dark repair or recombination mechanisms, geneticists have gained a clue as to the molecular basis of the fixation of gene mutations and structural changes in the chromosomes. It is especially encouraging that this approach to the problem is developing into a mainstream of experimental research. The repair hypothesis provides promising new pathways for the experimental study of mutations. These new insights in the theory of mutations must form the basis of any attempts to resolve the problem of the control of mutation.

It should be stressed that research into the molecular basis of persistent potential changes in genes and chromosomes and replicative instability is just beginning. The very principles determining our understanding of the molecular mechanisms of the origin of mutations may therefore have to be modified in the future.

BIBLIOGRAPHY

Akif'ev, A.P., V.B. Makarov, V.A. Polunovskii, and V.V. Yurchenko. — Genetika, No.3 (1965), 19.
Altenburg, E. and L.S. Browning. — Genetics **46**, No.2 (1961), 203.
Auerbach, C. and J.M. Robson. — Nature **157**, No.3984 (1946), 302.
Barricelli, N.A. and G. Del Zoppo. — Molec. and Gen. Genet. **101**, No.1 (1968), 51.
Beukers, R. and W. Berends. — Biochim. biophys. Acta **41**, No.3 (1960), 550.
Bodmer, W.F. and A.T. Ganesan. — Genetics **50**, No.4 (1964), 717.
Bollum, F.J. and R.B. Setlov. — Biochim. biophys. Acta **68**, No.4 (1963), 599.
Boyce, R.P. and P. Howard-Flanders. — Proc. Natn. Acad. Sci. U.S.A. **51**, No.2 (1964a), 293.
Boyce, R.P. and P. Howard-Flanders. — Z. VererbLehre 95, No.4 (1964b), 345.
Brewen, J.G. and W.J. Peacock. — Mutation Res. **7**, No.3 (1969), 433.
Bridges, B.A. and R.J. Munson. — Biochem. Biophys. Res. Commun. **22**, No.3 (1966), 268.
Bridges, B.A. and R.J. Munson. — Proc. R. Soc., B, **171**, No. 1023 (1968), 213.
Brookes, P. and P.D. Lawley. — J. Cell. Comp. Physiol., Vol.64, Suppl. 1,(1964), 3
Clark, A.J. and A.D. Margulies. — Proc. Natn. Acad. Sci. U.S.A. **53**, No.2 (1965), 451.
Crick, F.H.C., L. Barnett, S. Brenner, and R.J. Watts-Tobin. — Nature **192**, No. 4809 (1961), 1227.
Darlington, C.D. and L.E. La Cour. — J. Genet. **46**, Nos. 2,3 (1945), 180.
Demerec, M. — Proc. Natn. Acad. Sci. U.S.A. **48**, No.10 (1962), 1969.
Demerec, M. — Genetics **48**, No. 11 (1963), 1519.
Dubinin, N.P. — Proc. 12th Intern. Congr. Genetics, Vol.1 (1968), 2.
Dubinin, N.P. — Proc. 12th Intern. Congr. Genetics, Vol.3 (1969), 79.
Dubinin, N.P. — Genetika **5**, No.8 (1969a), 5.
Dubinin, N.P. — Vest. Akad. Nauk SSSR, Vol.6 (1969b), 97.
Dubinin, N.P. and A.P. Akif'ev. — Usp. Sovrem. Biol. 69, No.2 (1970), 272.
Dubinin, N.P., A.P. Akif'ev, V.A. Polunovskii, I.E. Kondrat'eva, and E.D. Aingorn. — Genetika **4**, No.6 (1968), 22.
Dubinin, N.P., A.P. Akif'ev, and T.A. Vishtynetskaya. — Genetika, Vol.7 (1971), 29.
Dubinin, N.P. and L.G. Dubinina. —Doklady AN SSSR **179**, No.5 (1968), 1221.
Dubinin, N.P. and L.S. Nemtseva. — Doklady AN SSSR **189**, No.3 (1969), 643.
Dubinin, N.P. and L.A. Nemtseva. — Proc. Natn. Acad. Sci. U.S.A. **66**, No.1 (1970), 211.
Dubinin, N.P. and E.G. Saprykina. — Doklady AN SSSR **158**, No.4 (1964), 956.
Dubinin, N.P. and V.N. Soifer. — Izv. AN SSSR, biol. ser., No.5 (1969), 637.
Dubinin, N.P. and V.N. Soyfer. — Mutation Res. **8**, No.2 (1969), 353.
Dubinin, N.P. and V.A. Tarasov. — Usp. Sovrem. Genet., No.2, p.3. Moscow, "Nauka," 1969.
Dubinina, L.G. — Genetika **6**, No.7 (1970), 61.
Dubinina, L.G. and N.P. Dubinin. — Genetika **4**, No.2 (1968), 6.
Fox, M.S. and M.K. Allen. — Proc. Natn. Acad. Sci. U.S.A. **52**, No.2 (1964), 412.
Freese, E. — Proc. Natn. Acad. Sci. U.S.A. **45**, No.4 (1959), 622.
Freese, E.B. and E. Freese. — Genetics **54**, No.5 (1966), 587.
Garina, K.P. and N.I. Romanova. — Molec. and Gen. Genet. 106, No.2 (1970), 93.
Gefter, M.L., A. Becker, and J. Hurwitz. — Proc. Natn. Acad. Sci. U.S.A. **58**, No.1 (1967), 240.
Gellert, M. — Proc. Natn. Acad. Sci. U.S.A. **57**, No.1 (1967), 148.
Haefner, K. — Photochem. Photobiol. **5**, No.7 (1966), 587.
Hall, Z.W. and I.R. Lehman. — J. Molec. Biol. **36**, No.3 (1968), 321.
Hanawalt, P.C. and R. H. Haynes. — Biochem. Biophys. Res. Commun. **19**, No.4 (1965), 462.
Harm, W. and B. Hillebrandt. — Photochem. Photobiol. 1, No.2 (1962), 271.
Hastings, P.J. and H.L.K. Whitehouse. — Nature **201**, No. 4923 (1964), 1052.
Hill, R.F. — Photochem. Photobiol. **4**, No.3 (1965), 563.
Hogness, D.S., W. Doerfler, J.B. Egan, and L.W. Black. — Cold Spring Harbor Sympos. Quant. Biol., Vol.31 (1966), 129.
Holliday, R. — Genet. Res. **3**, No.3 (1962), 472.
Holliday, R. — Genet. Res. **5**, No.2 (1964), 282.

Holliday, R. — In: Organization and Control in Prokaryotic and Eukaryotic Cells. Sympos. 20. Soc. Gen. Microbiol. London, p.359. Cambridge Univ. Press., 1970.
Hotta, Y., M. Ito, and H. Stern. — Proc. Natn. Acad. Sci. U.S.A. 56, No.4 (1966), 1184.
Howard-Flanders, P. — A. Rev. Biochem., Vol. 37 (1968), 175.
Howard-Flanders, P. and R.P. Boyce. — Radiat. Res., Suppl. 6 (1966), 156.
Howard-Flanders, P., B.M. Wilkins, and W.D. Rupp. — In: Molecular Genetics, p.161. Berlin, Heidelberg, N.Y., Springer Verlag, 1968.
Inagaki, E. and Y. Nakao. — Mutation Res. 3, No.3 (1966), 268.
Ito, T., T. Yamasaki, and Y. Matsudaira. — Jap. J. Genet. 37, No.4 (1962), 276.
Jehle, H. — Proc. Natn. Acad. Sci. U.S.A. 53, No.6 (1965), 1451.
Jehle, H. and W.C. Parke. — In: Structural Chemistry and Molecular Biology, p.399. San Francisco — London, W. H. Freeman, 1968.
Josse, J., A.D. Kaiser, and A. Kornberg. — J. Biol. Chem. 236, No.3 (1961), 864.
Kelner, A. — J. Bact. 58, No.4 (1959), 511.
Kimball, R.F. — Mutation Res. 1, No.2 (1964), 129.
Kimball, R.F. — Adv. Radiat. Biol., Vol. 2 (1966), 135.
Kondo, S. — Proc. 13th Intern. Congr. Genetics. Vol. 2 (1968), 126.
Kondo, S. and T. Kato. — Photochem. Photobiol. 5, Nos. 11—12 (1966), 827.
Kondo, S. and T. Kato. — Adv. Biol. Med. Phys., Vol.12 (1968), 283.
Kozinski, A.W., P.B. Kozinski, and R. James. — J. Virol. 1, No.4 (1967), 758.
Kubitschek, H.E. — Proc. Natn. Acad. Sci. U.S.A. 52, No.6 (1964), 1374.
Lawley, P.D. and P. Brookes. — Nature 206, No.4983 (1965), 480.
Lehman, I.R., M.J. Bessman, E.S. Simms, and A. Kornberg. — J. Biol. Chem. 233, No.1 (1958), 163.
McGrath, R.A. and R.W. Williams. — Nature 212, No.5061 (1966), 534.
Magni, G.E. — Proc. Natn. Acad. Sci. U.S.A. 50, No.5 (1963), 975.
Magni, G.E. — J. Cell. Comp. Physiol., Vol.64, Suppl. 1 (1964), 165.
Magni, G.E. — Proc. 12th Intern. Congr. Genetics, Vol.3 (1969), 27.
Magni, G.E. and R.C. von Borstel. — Genetics 47, No.8 (1962), 1097.
Mattern, I.E., H. Zwenk, and A. Rörsch. — Mutation Res. 3, No.5 (1966), 374.
Mennigmann, H.D. and W. Szybalski. — Biochem. Biophys. Res. Commun. 9, No.5 (1962), 398.
Meselson, M. — J. Molec. Biol. 9, No.3 (1964), 734.
Muller, H.J., E. Carlson, and A. Schalet. — Genetics 46, No.2 (1961), 213.
Nasim, A. — Genetics 59, No.3 (1968), 327.
Nasim, A. and C. Auerbach. — Mutation Res. 4, No.1 (1967), 1.
Nasim, A. and C. H. Clarke. — Mutation Res. 2, No.5 (1965), 395.
Newcombe, H.B. — J. Genet. 43, Nos.1, 2 (1942), 145.
Olivera, B.M. and I.R. Lehman. — Proc. Natn. Acad. Sci. U.S.A. 57, No.5 (1967), 1426.
Painter, R.B. — Radiat. Res. 31, No. 3 (1967), 608.
Painter, R.B. and J. E. Cleaver. — Nature 216, No. 5113 (1967), 369.
Painter, R. B. and A. Schaefer. — Nature 221, No. 5187 (1969), 1215.
Peacocke, A.R. and B. N. Preston. — Proc. R. Soc., B, 153, No. 950 (1960), 90.
Pettijohn, D. and P. Hanawalt. — J. Molec. Biol. 9, No.2 (1964), 395.
Rasmussen, R.E. and R. B. Painter. — J. Cell. Biol. 29, No.1 (1966), 11.
Resnick, M.A. — Mutation Res. 7, No.3 (1969), 315.
Richardson, C.C., R.B. Inman, and A. Kornberg. — J. Molec. Biol. 9, No. 1 (1964), 46.
Romashov, D.D., V.N. Belyaeva, K.A. Golovinskaya, and A.A. Prokof'eva-Bel'govskaya. — In: "Radiatsionnaya genetika," p.247. Moscow, AN SSSR, 1962.
Rupp, W.D. and P. Howard-Flanders. — J. Molec. Biol. 31, No.2 (1968), 291.
Setlow, J.K. — Radiat. Res., Suppl.6 (1966), 141.
Setlow, R.B. — J. Cell. Comp. Physiol. Vol.64, Suppl. 1 (1964), 51.
Setlow, R.B. and W. L. Carrier. — Proc. Natn. Acad. Sci. U.S.A. 51, No.2 (1964), 226.
Setlow, R.B., W. L. Carrier, and F. J. Bollum. — Proc. Natn. Acad. Sci. U.S.A. 53, No.5 (1965a), 1111.
Setlow, R.B., W. L. Carrier, and F. J. Bollum. — Abstr. Biophys. Soc., No.3, 1965b.

Smith, K.C. — Photochem. Photobiol. 2, No. 4 (1963), 503.
Sobels, F. H. — Proc. 12th Intern. Congr. Genetics, No.3 (1969), 203.
Soifer, V. N. Molecular Mechanism of Mutagenesis. — Moscow, "Nauka," 1969. (Russian)
Soyfer, V.N. — Microbiol. Genet. Bull., No.31 (1969), 18.
Stahl, F.W. The Mechanism of Inheritance. 2nd ed. — New Jersey, Prentice Hall, 1969.
Strauss, B.S. — Current Topics in Microbiol. and Immunol., Vol. 44, No.1, 1968.
Swenson, P. A. and R. B. Setlow. — J. Molec. Biol. 15, No. 1 (1966), 201.
Szybalski, W. — Abh. dt. Akad. Wiss. Berl., No. 4 (1964), 1.
Taylor, J. H. — J. Cell. 25, No. 2, Pt. 2 (1965), 57.
Van de Putte, P., H. Zwenk, and A. Rörsch. — Mutation Res. 3, No.5 (1966), 381.
Wacker, A. — Progress in Nucl. Acid Res., Vol. 1 (1963), 369.
Wacker, A., H. Dellweg, and D. Jacherts. — J. Molec. Biol. 4, No.5 (1962), 410.
Watson, J.D. and F.H.C.Crick. — Nature 171, No.4356 (1953a), 737.
Watson, J. D. and F. H. C. Crick. — Nature 171, No. 4361 (1953b), 964.
Willetts, N.S., A. J. Clark, and B. Low. — J. Bact. 97, No.1 (1969), 244.
Wimber, D. E. and W. Prensky. — Genetics 48, No. 12 (1963), 1731.
Witkin, E. M. — Science 152, No. 3727 (1963), 1345.
Witkin, E. M. — Radiat. Res., Suppl. 6 (1966b), 30.
Witkin, E. M. — Brookhaven Sympos. Biol., No. 20 (1967), 17.
Witkin, E. M. — Proc. 13th Intern. Congr. Genetics, Vol. 2 (1969a), 30.
Witkin, E. M. — Mutation Res. 8, No. 1 (1969b), 9.
Witkin, E. M. — Annual Rev. Genet., Vol. 3 (1969c), 525.
Wolff, S. and D. Scott. — Expl. Cell Res. 55, No. 1 (1969), 9.
Wulff, D. L. and C. S. Rupert. — Biochem. Biophys. Res. Commun. 7, No.3 (1962), 237.
Yamamoto, K. and H. Yamaguchi. — Mutation Res. 8, No. 2 (1969), 428.

UDC 575.24:576.858.2

THE INFLUENCE OF CHEMICAL AGENTS ON THE LETHAL AND MUTAGENIC EFFECT OF UV IRRADIATION ON THE EXTRACELLULAR PHAGE

A.S. KRIVISKII, L.P. TIKHOMIROVA and V.V. ESIPOVA
Institute of Molecular Biology, Academy of Sciences of the USSR, Moscow

The lethal and mutagenic effects of UV irradiation after preliminary treatment of the S_d phage with hydroxylamine (HA) or acridine (AO), or after the substitution of thymine with 5-bromouracil (BU) were investigated. The combined action of HA and UV produced an additive mutagenic effect. The substitution of thymine with BU increased the sensitivity of the phage to the lethal action of UV whereas the mutagenic effect remained unchanged. Preliminary treatment with AO did not alter the mutagenic effect of irradiation within the dose range giving a linear increase in the number of mutations, but at higher doses the mutagenic effect was increased. The authors conclude that the departure of the mutation curve from a linear relationship can be attributed to an increased sensitivity of the permutation lesions of the DNA of the phage to subsequent irradiation.

The molecular mechanisms of the mutagenic effect of UV irradiation are still obscure. Today it is becoming increasingly clear that thymine dimers are the main photoproduct responsible for the lethal and mutagenic effects alike (Bridges and Munson, 1968; Witkin, 1968, 1969). According to other data the mutagenic effect of UV irradiation may also be due to the formation of cytosine photohydrates or dimers (Setlow and Carrier, 1966; Grossman and Rodgers, 1968). Extracellular bacteriophages represent a convenient model for the study of the primary mechanisms of the mutagenic effect of irradiation. Unfortunately, most phages are characterized by a very low mutability when irradiated in vitro. One exception is the virulent phage S_d of Escherichia coli CK, possessing the unique property of yielding a great number of plaque form mutants as a result of exposure not only to chemical agents but also to UV and X-irradiation in vitro. Studies of the lethal and mutagenic effects of UV on the phage S_d have shown that the mutation curve reaches a peak at a certain level of survival (Kriviskii et al., 1963; Zavil'gel'skii et al., 1965a, b). The mutation curves follow a similar course in a wide variety of organisms. Although a number of hypotheses have been proposed to explain the existence of a peak in the mutation curves, the nature of this peak remains unknown (Soifer, 1966).

It seemed to us that since the molecular mechanisms of the mutagenic effect of some chemical agents are better known than those underlying the effect of irradiation, the combined application of chemical agents and UV irradiation might shed light on the mechanisms involved in the mutagenic effect of UV irradiation on the S_d phage and explain the observed deviations of the mutation curve from the linearity postulated by the classical target theory.

First we examined the kinetics of the combined action of UV irradiation and hydroxylamine (HA) on the S_d phage. The results obtained from the exposure of the phage to these agents in different sequences revealed that the mutagenic effect of UV and HA is additive. The spectra of the induced mutants were the same in all three cases, i. e. with UV and HA separately and with a combination of these mutagens (for a more detailed account see Kriviskii and Tikhomirova, 1970). An additive action of mutagens may be observed if the replacement of one mutagen by the other does not alter the ratio between lethal and mutational changes, even though the agents may have totally different mutagenic and inactivating mechanisms. It appears, however, that the particular agents used have a similar mechanism of mutagenic action since the spectra of the induced mutations are similar and also recent evidence indicates that UV and HA attack cytosines and convert them to identical products. It is known that HA alters the cytosine bases and has practically no effect on thymines. It can be assumed, therefore, that the mutagenic effect of UV irradiation on the S_d phage can likewise be attributed to lesions of cytosines.

To obtain further data on the molecular mechanisms of the mutagenic effect of UV irradiation we irradiated phage in which the thymine of the DNA had been replaced with 5-bromouracil (BU).

The results of these experiments are shown in Figure 1. Comparison of the mutagenic effects of UV irradiation on normal and BU phage shows that the same dose of radiation induces a greater number of mutations in the BU phage than in the normal phage, but the sensitizing effect of BU is greater with respect to the lethal action of UV than in the case of its mutagenic action. Although the results obtained do not permit a firm conclusion concerning the molecular basis of the mutations, the unequal reinforcement of the inactivating and mutagenic effects of UV in the presence of BU indicates that inactivation and mutation are not caused by the same lesions.

Another possible approach to the investigation of the molecular basis of mutations induced by UV irradiation would be to study the possibilities of photoreactivation of the lethal and mutational lesions of the phage. If dimers are responsible not only for the inactivation of S_d phage but also for the appearance of mutations, photoreactivation should conceivably increase the survival of the phage and lower the number of mutants. However, in the case of the S_d phage no photoreactivation at all was detected under conditions in which the effectiveness of photoreactivation of lethal lesions in the T4 phage is quite high. Photoreactivation is also effective in the S_d phage-sensitive bacteria *E. coli* CK, with which the experiments of phage photoreactivation were carried out. The lack of photoreactivation in the S_d phage may be explained by assuming that the main cause of phage inactivation by UV irradiation is the formation of products which cannot be photoreactivated, although it cannot be ruled out that this phage has an inherent dark repair system which masks the effect of photoreactivation.

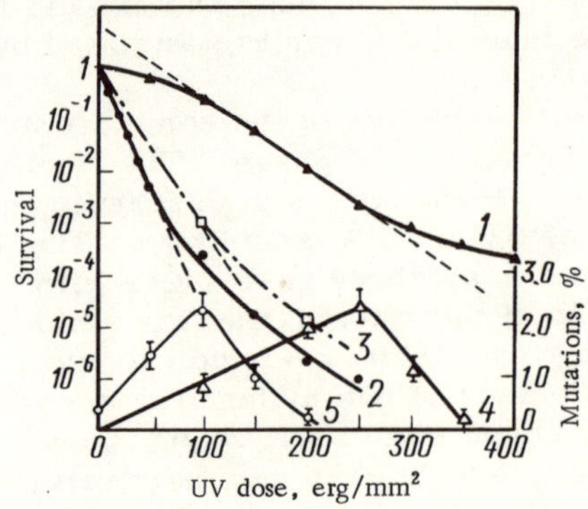

FIGURE 1. Kinetics of inactivation and mutagenesis in S_d phage exposed to UV irradiation:

1) inactivation of normal phage; 2) inactivation of BU phage; 3) inactivation of phage in which 70% of the thymine of the DNA has been replaced by BU; 4) mutagenesis of normal phage; 5) mutagenesis of BU phage.

It is known that acridine dyes protect DNA from lethal photoreactions induced by UV irradiation (Zavil'gel'skii et al., 1969; Setlow and Carrier, 1967). It seemed of interest to determine whether acridine confers a similar protection against mutational damage. The phage S_d was treated with acridine (AO) at a concentration not inducing mortality or mutations ($1 \cdot 10^{-5}$ M, 30 minutes). Subsequent UV irradiation revealed that this treatment gave moderate protection against the inactivating effect of the radiation, although the percentage of mutants did not decrease but actually rose (Figure 2). At double this concentration of AO the survival rate of the S_d phage decreased roughly one hundred fold during incubation in darkness, while the percentage of mutants increased. Subsequent UV irradiation revealed a stronger protective effect and a marked rise in mutagenesis (Figure 3). The results obtained lead to the conclusion that the inactivating and mutagenic effects of UV irradiation result from different lesions.

A study of the kinetics of mutagenesis in the extracellular phage S_d showed a decline in the relative mutation rate with all variations of treatment. In the studies forming the basis of the previously proposed explanation for the decline of the mutation curve of the S_d phage multiple reactivation was not ruled out (Zavil'gel'skii et al., 1965 a, b). In view of the morphological properties of most UV-induced mutants of the S_d phage, their disappearance as a result of multiple reactivation seemed quite possible. Under seeding conditions excluding multiple infection of the bacteria with irradiated phage, the observed number of UV-induced mutants was slightly higher but still significantly different from the level expected on the basis of a linear increase of the number of mutants. Thus, the decrease in the mutation rate following irradiation of extracellular phage constituting a nonmetabolizing system cannot be explained merely in terms of multiple reactivation.

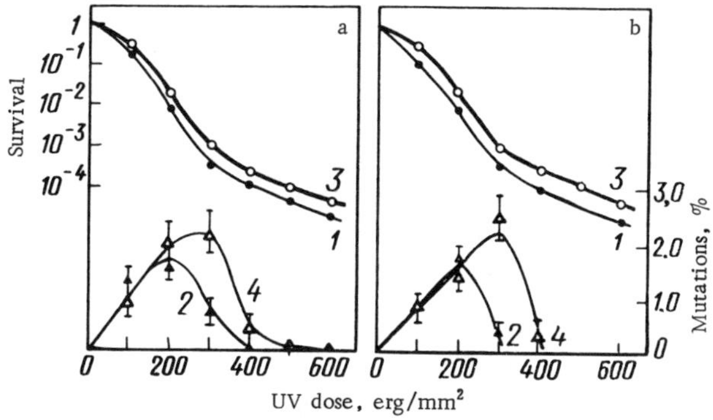

FIGURE 2. Inactivation and mutagenesis resulting from the UV irradiation of S_d phage previously treated with $1 \cdot 10^{-5}$ M AO:

a) first experiment; b) second experiment; 1 and 2) inactivation and mutagenesis of untreated phage; 3 and 4) inactivation and mutagenesis of phage previously treated with AO.

As long ago as 1938 one of us (Kriviskii, 1938) suggested that the decrease in the relative mutation rate of acetic acid bacteria might be due to an increased sensitivity to subsequent UV irradiation on the part of organisms in which nonlethal premutational lesions had already been induced. This hypothesis has been confirmed by a number of physicochemical studies. Data indicating an increased reactivity of units adjacent to the primarily damaged ones were obtained by the irradiation of synthetic polynucleotides (Johns, 1966). It was found that lesions impairing the stacking interaction of the base pairs of the native DNA widen the melting range of the double helix and thus increase the reactivity of the bases situated in this area (Berestetskaya et al., 1970). It can be assumed on these grounds that the departure of the mutation curves from linearity following UV irradiation may indeed be due to preferential formation of various photoproducts near those already present, i. e. to the higher sensitivity of organisms previous subjected to premutational and other nonlethal lesions to subsequent irradiation. The lesion alters the secondary structure of DNA in the area affected, thus strongly increasing the reactivity of neighboring nucleotides. A second lesion can conceivably inactivate the genome if inflicted in the vicinity of a premutational lesion.

Experiments with AO might confirm this assumption. By conferring protection against lethal photoreactions AO may also prevent the inactivation of organisms already subjected to a premutational lesion, thus causing the manifestation of a greater number of mutants. However, this topic requires further research.

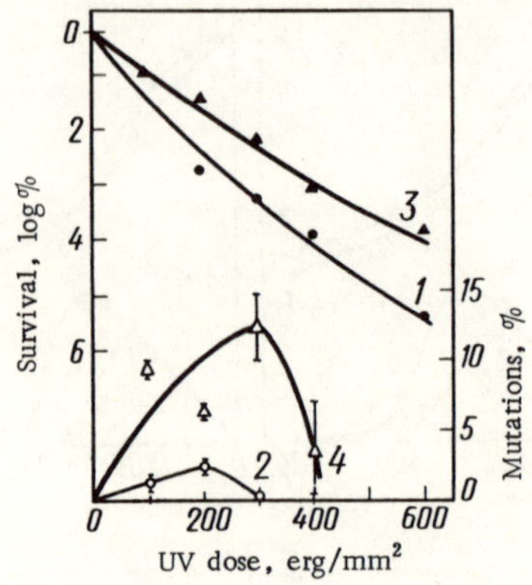

FIGURE 3. Inactivation and mutagenesis resulting from the UV irradiation of S_d phage previously treated with $2.5 \cdot 10^{-5}$ M AO :

1 and 2) inactivation and mutagenesis of untreated phage; 3 and 4) inactivation and mutagenesis of phage previously treated with AO.

BIBLIOGRAPHY

Berestetskaya, N.V., Yu. N. Kosaganov, Yu. S. Lazurkin, E.N. Trifonov, and M.D. Frank-Kamenetskii. — Molekulyarnaya Biologiya 4, No. 1 (1970), 137.

Bridges, B.A. and R.J. Munson. — Proc. R. Soc., Vol.171 (1968), 213.

Grossman, L. and S. Rodgers. — Biochem. Biophys. Res. Commun., Vol. 33 (1968), 957.

Johns, H.J. — Radiat. Res. Proc. 3rd Intern. Congr. Cortina d'Ampezo, p.733, 1966.

Kriviskii, A.S. — Vest. Rentg. Radiol., Vol. 20 (1938), 375.

Kriviskii, A.S. and L.P. Tikhomirova. — Genetika 6, No.5 (1970), 84.

Kriviskii, A.S., G.B. Zavil'gel'skii, and V.L. Ivanov. — Doklady AN SSSR 150, No.4 (1963), 399.

Setlow, R.B. and W.L. Carrier. — J. Molec. Biol., Vol. 17 (1966), 237.

Setlow, R.B. and W. L. Carrier. — Nature, Vol. 213 (1967), 906.

Soifer, V.N. — Bull. Moscow Society of Naturalists (MOIP), biol. sec. 71, No.5 (1966), 88.

Witkin, E.M. — Proc. 12th Intern. Congr. Genetics, Tokyo, Vol. 2 (1968), 30.

Witkin, E.M. — Mutation Res. 8, No.1 (1969), 9.

Zavil'gel'skii, G.B., A.S. Kriviskii, and V.I. Ivanov. — Izv. AN SSSR, biol. ser., Vol. 5 (1965a), 700.

Zavil'gel'skii, G.B., A.S. Kriviskii, and V.I. Ivanov. — Genetika, 1, No. 4 (1965b), 34.

Zavil'gel'skii, G.B., O.N. Rudchenko, and V.V. Danileichenko. — Biofizika 14, No.1 (1969), 34.

UDC 575.24:576.851.48

A STUDY OF ANTIMUTAGENS IN THE SYSTEM OF OCHRE SUPPRESSOR MUTATIONS IN ESCHERICHIA COLI

C.H. CLARKE
School of Biological Sciences, University of East Anglia, Norwich

The effect of compounds behaving as antimutagens in microbial systems on spontaneous and UV-induced try$^+$ reversions was investigated. In the E. coli B/r system some of these agents did not exert an antimutagenic effect on UV-induced reversions but rather, like caffeine, acted as inhibitors of the dark repair process.

INTRODUCTION

The scheme in Figure 1 indicates how mutations arise and are manifested (Clarke, 1967b). The left-hand side of the diagram represents the reaction (or reactions) between the mutagen and DNA (or RNA in the case of RNA-containing viruses). These reactions depend largely on such factors as the secondary structure of the nucleic acid (single-stranded or double-helix) (Schuster and Wilhelm, 1963; Schuster and Wittmann, 1963; Sehgal and Krause, 1968; Singer and Fraenkel-Conrat, 1969) and the repression or de-repression of the genes (Savic et al., 1969; Kölsch and Starlinger, 1965a). DNA damaged prior to its replication may be acted upon by repair systems which either eliminate the potentially mutagenic lesions or, frequently, generate errors in the DNA and thus act as mutagens (Witkin, 1967, 1968). The replication of DNA in itself may be mutagenic (Drake et al., 1969; Drake, 1970; Hall and Lehman, 1968; Yanofsky et al., 1966) or may induce mutations under the influence of certain agents (Orgel and Orgel, 1965; Hall and Lehman, 1968; Cerda-Olmedo et al., 1968; Baker and Tessman, 1968). On the other hand, in some cases DNA replication can be regarded as an antimutagenic process (Drake et al., 1969; Drake, 1970). The recombinational repair system also begins to operate after replication of the DNA. It is thought (Witkin, 1967, 1968, 1969a, b) that recombinational repair generates errors in the DNA and is thus a mutagenic factor. The repression or de-repression of genes affects their accessibility to the action of various repair systems and thus influences the efficiency of repair (Kölsch and Starlinger, 1965b; Savic et al., 1969; Witkin, 1969b). Apart from the direct reactions between the mutagen and the nucleic acid, mutagens can act upon DNA precursors. These modified precursors may then be used in the repair or replication of DNA, thus leading to mutations (Grossman and Brown, 1969). Repair inhibitors such as caffeine, $MnCl_2$

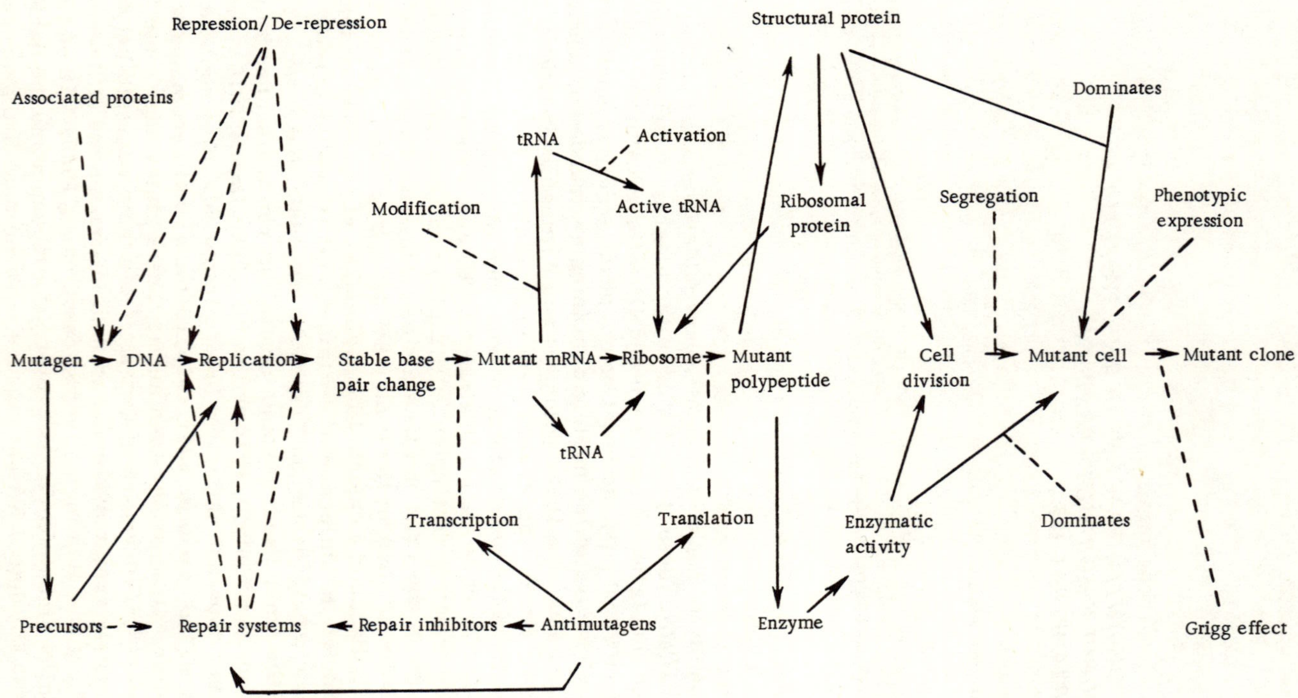

FIGURE 1. The mutational pathway.

and acridines (Witkin, 1969b; Clarke, 1967a) exert a marked influence not only on the survival of cells but also on the frequency of mutations (Witkin and Farquharson, 1969). The transcription of DNA containing a stable substitution of base pairs yields mutant DNA (informational DNA, transport DNA). Transcription (Grossman and Brown, 1969) and subsequent translation (Lewis, 1970) are the first stages toward the manifestation of errors and consequently of mutations, arising as a result of the operation of damaged repair enzymes or DNA polymerase. It must be borne in mind that the action of the mutagen within the cell is not confined to the DNA (Kilbey, 1969a, b). Also significant is the fact that errors can arise at any of the consecutive stages of the mutation sequence and not necessarily at the level of a chemical reaction between the mutagen and DNA.

If the mutant informational RNA is translated into a mutant polypeptide, this polypeptide may be a structural protein or its precursor, e. g. a ribosomal protein, or it may be an enzyme. In the case of a structural protein or an enzyme, one or more cell divisions must occur before the mutation is phenotypically expressed if the mutation affects a recessive character (for example, the loss of phage receptors, or streptomycin resistance). The necessity of segregation for the phenotypic expression of the mutation does not exist in the case of a dominant character (for example, in back mutations causing reversions to prototrophy). The next stage after the complete phenotypic expression of the mutation in one individual, i. e. the formation of a mutant clone, may be prevented by the Grigg effect or the phenomenon of the predominant utilization of growth factors by the majority of nonmutant cells (Corran, 1969).

Antimutagenic factors and the genetic background influence spontaneous and induced mutagenesis (see Figure 1). The mechanism of action of antimutagens is largely obscure, except in the case of certain repair inhibitors such as caffeine, which sometimes act as antimutagens (Grigg and Stuckey, 1966; Horneck-Witt and Kaplan, 1968; Witkin and Farquharson, 1969; Clarke, 1970). Figure 1 shows at least three possible stages of the mutation pathway at which antimutagens may intervene. Thus, antimutagens may reduce the frequency of errors in certain repair systems or increase the relative activity of other repair mechanisms operating faultlessly; in the latter case the antimutagen may either act directly on the repair system or suppress the inhibitor of this system. Some antimutagens, e. g. adenosine, guanosine and spermine, are normal components of the cell (Novick, 1956; Stevens, 1970).

The genetic background influences mutagenesis in a variety of ways. As noted above, mutations in some genes responsible for repair lead to an increased mutability. Damage to other genes controlling repair systems may decrease the mutability (Witkin, 1967, 1968, 1969a, b). It must be stressed that a difference in mutability between a highly mutable wild strain and a nonmutable mutant with a defective repair mechanism (and therefore sensitive to the mutagen) cannot be regarded as proof that the given repair system normally generates errors and is therefore responsible for the appearance of mutations (Witkin, 1968). Let us assume that a chemical mutagen reacts with DNA and induces two types of damage: A — a potentially lethal lesion, and B — a potentially mutagenic lesion.

Let us assume further that the initial ratio between these two types of lesions is 10A:1B, and that the cell can withstand up to 10 lesions of type A. The cell is treated with a dose of the mutagen inducing 200 A lesions and 20 B lesions and subsequently the differential action of the repair systems brings this ratio to 10 A : 10 B. Such a cell will survive, giving rise to mutations. In the case of a strain with a damaged repair system, however, the maximal nonlethal dose of mutagen to which the cell can be exposed will cause only 100 lesions of each type. Thus, the chances for the appearance of mutations in the surviving treated cells of this strain will be 10 times lower than in the cells of a strain with a normal repair system. Among the actual lesions of DNA which fit the above hypothesis are cross linkages (lethal lesions) and lesions of one of the DNA strands (mutagenic lesions) (Witkin, 1968; Clarke, 1970).

The most eloquent examples of the role of the genetic background in mutagenesis are lesions of mutator genes, at least some of which represent mutations of DNA polymerases (Drake, 1970; Yanofsky et al., 1966; Zamenhof, 1969). Finally, the influence of the genetic background on mutagenesis may be of the form described by Witkin and Theil (1960) in which the introduction of a streptomycin resistance marker led to a decrease in the frequency of UV-induced reversions with respect to the tryptophan marker. The most probable explanation of this effect is that the modification of the ribosomal protein responsible for resistance to streptomycin causes a decrease in the efficiency of the ochre suppressor mutations bringing about UV-induced reversions to try^+ (Kuwano et al., 1968). Many other similar observations permit a physiological or biochemical explanation of the effect on mutagenesis exerted by the genetic background (Clarke, 1962, 1963; Mohn and Kaplan, 1967). In other cases there is as yet no satisfactory explanation of the effect of certain genetic markers on the spontaneous or induced mutability of other markers (Glover, 1956; Chopra, 1967; Allison, 1969; Zetterberg, 1962).

We shall consider below the compounds showing an antimutagenic effect in microbial systems. This work is actually the first review dealing with these agents; it includes a theoretical scheme for determining their activity and data on their influence on the reversion system of *Escherichia coli* B/r.

ANTIMUTAGENS IN MICROBIAL SYSTEMS

Purine ribonucleotides

Nothing has been published on the antimutagenic properties of adenosine, guanosine and inosine since the appearance of the article by Novick (1956). When tryptophan-requiring cells of *E. coli* B/r were grown in a chemostat with tryptophan as growth limiting factor the following results were obtained. Purine ribonucleosides fully suppressed the mutagenic activity of caffeine, theophylline, paraxanthine, theobromine, azaguanine, 8-methoxycaffeine and adenine. On the other hand, there was only a partial suppression of spontaneous mutations and mutations induced by tetramethyl uric acid and benzimidazole. No antimutagenic effect was detected with respect to mutations caused by UV or gamma-ray irradiation. It would be interesting to find out whether the antimutagenic effect of purine ribonucleosides is confined to particular types of molecular change in the DNA.

Acridines

Data are available on the antimutagenic effect of acridines in microbial systems. Webb and Kubitschek (1963) found that acridine orange sharply decreases the level of spontaneous and caffeine-induced mutations associated with resistance to the T5 phage in E. coli B/r try$^-$, incubated in a chemostat with glucose as the growth limiting factor.

It was found that quinacrine (atabrine) suppresses the formation of spontaneous mutations affecting resistance to various antibiotics in the case of E. coli, Aerobacter aerogenes and Staphylococcus aureus (Sevag, 1964; Sevag and Ashton, 1964). Johnson and Bach (1966) showed that quinacrine possesses antimutagenic properties with regard to the streptomycin resistance system of E. coli affecting both the mutator gene described by Treffers which causes strictly AT — GC transversions (Yanofsky et al., 1966), and also mutations induced by 2-aminopurine. Quinacrine, being an antimutagen with respect to spontaneous mutations of E. coli K12, does not affect mutations caused by the ast 1 mutator gene (Zamenhof, 1969). It was also found that 5-aminoacridine is an antimutagen for Saccharomyces cerevisiae in which it reduces the number of spontaneous histidine reversions and mutations conferring resistance to canavanine during the process of mitotic growth (Magni et al., 1964; Puglisi, 1967). The antimutagenic properties of some methyl derivatives of acridines have also been demonstrated.

Polyamines

The antimutagenic effect of the polyamines spermine and spermidine was discovered by Sevag and Drabble (1962) in a study of spontaneous mutations conferring resistance to various antibiotics in S. aureus and A. aerogenes. These observations were confirmed by Johnson and Bach (1965, 1966), who found that spermine suppresses the appearance of streptomycin-resistant mutants and revertants to prototrophy with respect to tryptophan in S. aureus and E. coli. The antimutagenic effect of spermine covers mutations induced by caffeine, UV irradiation, 2-aminopurine and the mutator gene described by Treffers as well as spontaneous mutations. It was also found that although spermine is not antimutagenic with respect to the ast 1 mutator gene, it undoubtedly exerts an antimutagenic effect as regards the appearance of spontaneous mutations in a nonmutator strain of E. coli K12 (Zamenhof, 1969).

Caffeine

Grigg and Stuckey (1969) found that caffeine suppresses the formation of spontaneous histidine-independent reversions in auxotrophic cells of

E. coli 15 during the stationary phase of growth. Clarke (1967a) reported a slight antimutagenic effect of caffeine at a concentration of 500 mg/ml with respect to the UV-induction of tryoptophan-independent revertants in cells of the hcr^- strain (derived from E. coli B/r) during the stationary phase of growth. It was also shown that caffeine at a concentration of 1,000 mg/ml exerts an antimutagenic effect on the induction of reversions to prototrophy with respect to methionine by low doses of UV in dividing cells of Schizosaccharomyces pombe (Clarke, 1968); caffeine does not affect reversions caused by nitrous acid or nitrosomethyl urethane, however. Caffeine at concentrations greater than 0.1% is an antimutagen with respect to UV-induced mutations lowering the level of streptomycin resistance in the strain E. coli B phr^- (Horneck-Witt and Kaplan, 1968). Witkin and Farquharson (1969) found that caffeine at a concentration of 0.2 % exerts a strong antimutagenic effect on reversions to tryptophan prototrophy and on mutations to streptomycin resistance induced by UV irradiation in several hcr^- strains of E. coli. They assume that caffeine possibly interferes with the function of the exr gene and thus lowers the normal frequency of errors during repair. Finally, Clarke (1970) was able to demonstrate clearly the antimutagenic effect of caffeine at a concentration of 0.05% with respect to nitrous acid — induced reversions to tryptophan prototrophy in the hcr^- strain of E. coli B/r.

Other antimutagens

The literature contains a number of independent communications on the antimutagenic properties of various other compounds in microbial systems. Witkin (1961) found that pyronin at a concentration of 0.0005%, like chloramphenicol, reduces the stimulatory effect of mixtures of amino acids on the yield of reversions induced by UV irradiation. The tyrosine auxotroph WU-36 of E. coli B/r was used in this work, subjecting a broth culture in the late lag phase to irradiation. Arditti and Sermonti (1962) showed that $MnCl_2$ (5μ M) suppresses the appearance of mutations to azaguanine resistance in Penicillium chrysogenum previously treated with nitrogen mustard. $MnCl_2$ apparently did not exert any effect on mutations induced by UV or gamma rays, or by diethyl sulfate. Clarke (1962, 1963, 1965) found that 1-methionine is a broad-spectrum antimutagen for dividing cells of S. pombe with respect to reversions to adenine prototrophy arising either spontaneously or under the influence of various mutagens. Finally, Puglisi (1968) showed that actinomycin D and basic fuchsin are antimutagenic with respect to spontaneous mutations conferring resistance to canavanine during the mitotic growth of S. cerevisiae.

The following considerations must be taken into account in all studies of the antimutagenic effect of a compound concerning induced, as distinct from spontaneous, mutations. To begin with, a clear distinction must be made between the decline in the mutation yield resulting directly and inevitably from the change in the mutagen dose and the decline caused by a genuine, specific antimutagenic effect (Clarke, 1967a). Comparison of the mutability of two strains, or of the same strain in the presence and absence of the tested agent, with a constant dose of the mutagen may lead to incorrect conclusions if the difference in the survival rates of the two strains or the differences in the conditions of incubation of the cells are

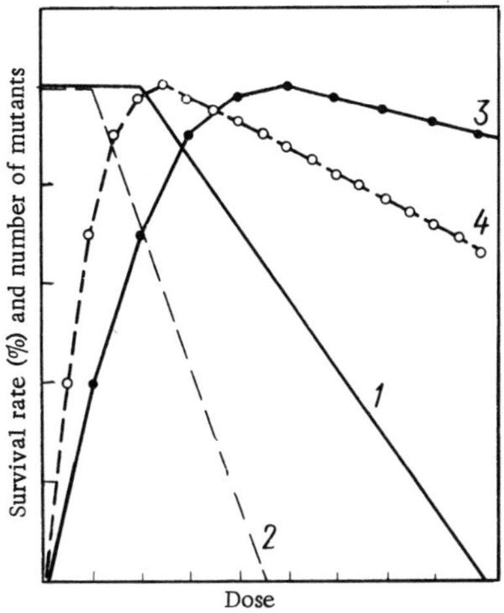

FIGURE 2. Effect of the mutagen dose on the survival rate of bacteria (%, logarithmic scale) and the absolute number of mutants (nonlogarithmic scale):

1) survival of A; 2) survival of B; 3) mutability of A; 4) mutability of B.

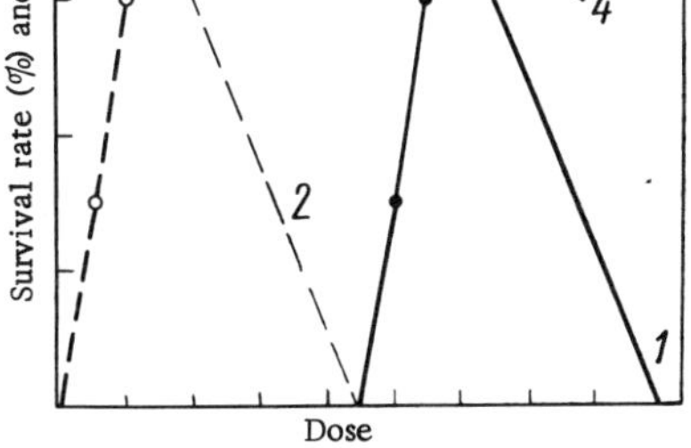

FIGURE 3. Effect of the mutagen dose on the survival rate (%, logarithmic scale) and the absolute number of mutants (nonlogarithmic scale). Key as in Figure 2. The mutability curve of A is based on the assumption that the absolute number of mutants does not increase until the start of cell mortality (threshold effect). If a threshold effect is lacking, the initial part of the mutability curve of A will obviously assume a different shape.

not considered. For this reason an increase in the absolute number of mutations in comparison with the spontaneous background can only be detected at high survival levels. On the other hand, it would be wrong to compare the mutability of two strains, one of which shows a normal resistance to UV while the other is highly sensitive to it. Here a given radiation dose will result in a survival rate of 50% in the wild strain (where the increase of the absolute number of mutations will be easily detected) and only 0.1% in the sensitive strain (where an appreciable increase in the absolute number of mutants will not be obtained anyway). Such considerations fully apply to those cases in which the potential antimutagen simultaneously acts as a sensitizer to the lethal effect of the mutagen (examples are caffeine and acridines, functioning as sensitizers to UV irradiation).

To overcome these difficulties in the interpretation of the data obtained it is necessary to plot the dose/effect curves for both strains tested or for incubation in the presence and absence of the tested agent. Examples of such curves are given in Figure 2, illustrating the case in which the mutation or the agent present during incubation cause a uniform decline in the survival rate at all doses, and in Figure 3, where the effect of the mutation or agent is expressed by the disappearance of the shoulder on the survival curve. Clearly the results obtained may be incorrectly attributed to an antimutagenic effect (apparent antimutagenesis) if the experiments are carried out using a single dose of the mutagen without reference to the different survival levels.

Apparent and true, specific antimutagenesis are best distinguished by comparing complete dose/effect curves (Figures 2, 3). Curves showing the relationship between the mutation frequency and the survival rate are very convenient for this purpose (Witkin, 1966; Clarke, 1967a, b, 1970). Changes in the mutation yield resulting from an increase or decrease in the dose are manifested as mere shifts of the same curve, whereas in the case of true, specific antimutagenesis two separate curves are obtained (Witkin, 1966; Clarke, 1970).

RESULTS

This work was carried out with agents previously shown to be antimutagenic in various microbial systems. We examined the properties of these agents in a single mutational system, namely, the UV-induced reversion to tryptophan prototrophy in cells of the tryptophan auxotroph E. coli WWP — $2hcr^+$ grown in broth and treated during the late lag phase (Hill, 1963; Clarke, 1967a, 1970). The high percentage of these reversions is attributed to the appearance of ochre suppressor mutations (Bridges et al., 1967; Clarke, 1970).

The potential antimutagenic agents were included in the media used for seeding the cells after irradiation; as a rule, a single concentration of the tested agent was used. The reversions were detected by inoculation on minimal medium containing a minute amount of tryptophan as primer and on the same medium with added casein hydrolysate (Clarke, 1967a,

1969, 1970). In the latter case this strain showed the stimulatory effect exerted by amino acid mixtures on the yield of try+ revertants — the "broth effect" (Witkin, 1966; Clarke, 1969). Thus, the use of two media for seeding — with and without casein hydrolysate — reveals the antimutagenic effect of any agent which acts like pyronin B and suppresses the broth effect (Witkin, 1961). The following agents were tested (μg/ml):

Putrescine — 2HCl	100	Chloro-IPC	50
Spermine — 4HCl	100	Basic fuchsin	20
Quinacrine — HCl	10	Pyronin B	50

Putrescine, spermine and chloro-IPC (De-Glovanni-Donnelly et al., 1967) do not affect the cell lethality, the yield of try+ revertants or the enhancing effect of casein hydrolysate following UV irradiation.

Quinacrine inhibits dark repair and lowers the survival rate after UV irradiation (Figure 4). Any observed effect of this agent on the mutation yield can be attributed to a mere dose increase, as may be seen from the results obtained with the medium containing casein hydrolysate (Figure 5).

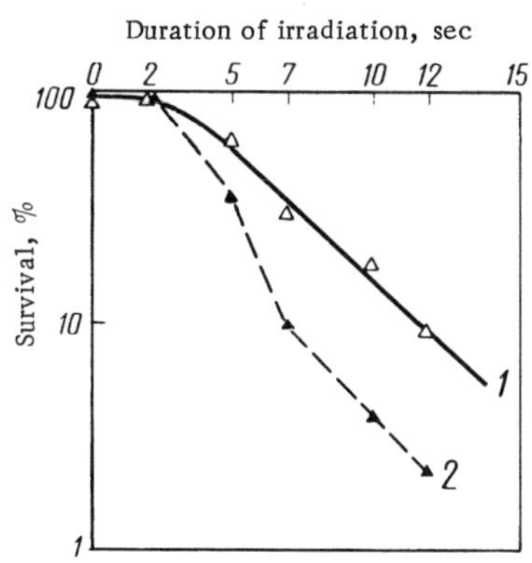

FIGURE 4. Effect of the UV irradiation dose ($\approx$ 38 erg/mm^2 · sec) on the survival rate of E.coli WWP-2hcr$^+$ try$^-$.

Survival rate in medium: 1) without quinacrine; 2) with quinacrine.

The graph in Figure 5 represents the effect of the survival rate on the frequency of reversions. Quinacrine obviously raises the level of spontaneous reversions in nonirradiated cultures by 50%. Thus, quinacrine possesses slight mutagenic properties — a fact already reported in the literature (Ames and Whitfield, 1966).

Basic fuchsin, like quinacrine, obviously inhibits dark repair (Figure 6); consequently, any effect on the yield of revertants can be attributed to a mere dose increase (Figure 7).

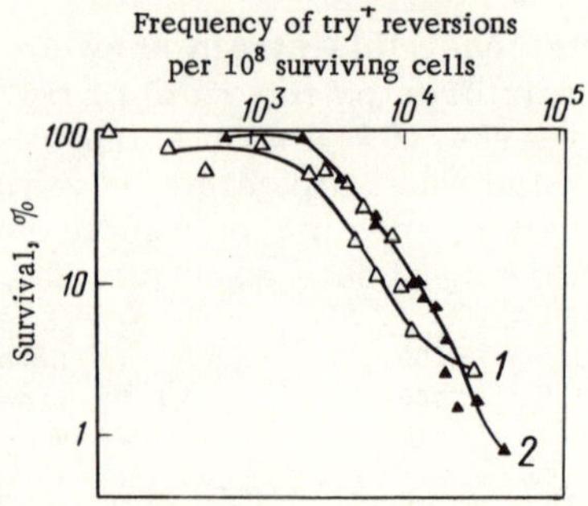

FIGURE 5. Effect of the survival rate on the frequency of try$^+$ reversions in E. coli WWP-2 hcr$^+$.

1) minimal medium containing a low concentration of tryptophan and casein hydrolysate; 2) same medium with quinacrine added.

Pyronin B similarly appears to be an inhibitor of dark repair, as are caffeine (Clarke, 1969), quinacrine, and basic fuchsin. Pyronin B did not exert any antimutagenic effect which could not be fully attributed to a dose increase. Of particular interest is the finding that pyronin B does not suppress the enhancing effect of casein hydrolysate on the yield of revertants.

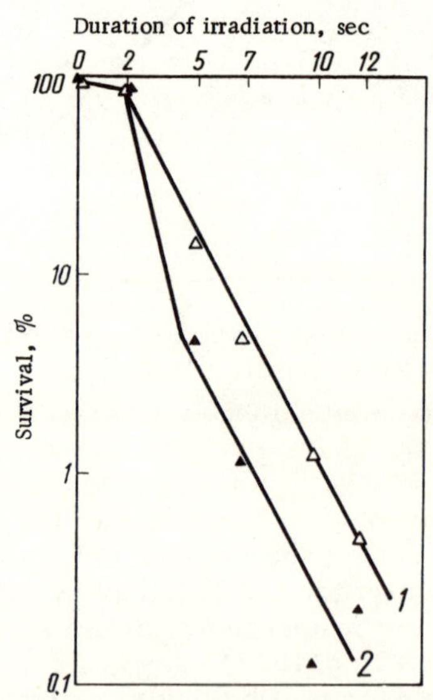

FIGURE 6. Effect of the UV radiation dose on the survival rate of cells of E. coli WWP-2 hcr$^+$.

1) survival in a medium with a low concentration of tryptophan and casein hydrolysate added; 2) same with basic fuchsin added.

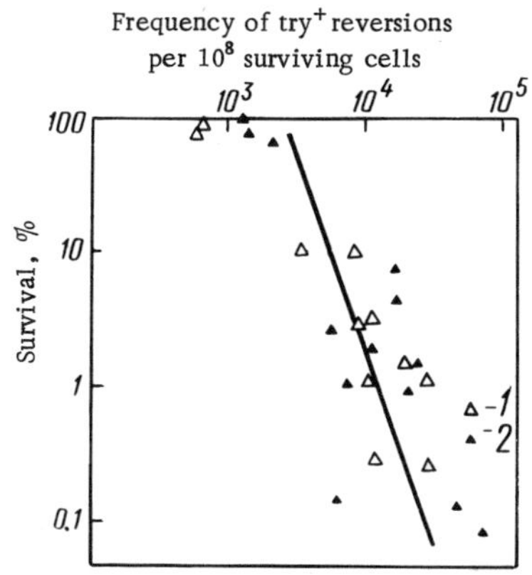

FIGURE 7. Relationship between the frequency of try$^+$ revertants and the survival rate.

1) minimal medium with a low concentration of tryptophan and casein hydrolysate added; 2) same medium with basic fuchsin added.

This finding contradicts the data of Witkin (1961), who worked with broth-grown cells of the tyrosine auxotroph E. coli B/r WU-36 in the late lag phase and used pyronin at a concentration of 0.0005%. In our system no suppression of the casein hydrolysate effect was detected even at pyronin B concentrations ten times as high (see above).

To sum up, these experiments did not reveal any true, specific antimutagenic effect of caffeine, putrescine, spermine, quinacrine, basic fuchsin, pyronin B and chloro-IPC on spontaneous or UV-induced try$^+$ reversions. Moreover, it was shown that quinacrine, basic fuchsin and pyronin B, like caffeine (Clarke, 1967a, 1969), probably inhibit dark repair. There are indications that quinacrine is itself slightly mutagenic. The results obtained demonstrate the specificity of the antimutagenic effect in the sense that the antimutagenicity of compounds such as caffeine, spermine or quinacrine can only be proved in certain mutational systems under particular physiological conditions.

We are grateful to Charlotte Auerbach, Ruth Hill, Evelyn Witkin and Brian Bridges for their friendly and helpful discussions. The proposed conception of the mutational pathway is largely based on several years of cooperation with Professor C. Auerbach in Edinburgh.

BIBLIOGRAPHY

Allison, M. J. — Mutation Res., Vol. 7 (1969), 297.
Ames, B.N. and H.J. Whitfield. — Cold Spring Harbor Sympos. Quant. Biol., Vol. 31 (1966), 221.
Arditti, R.R. and G. Sermonti. — Genetics, Vol. 47 (1962), 761.
Baker, R. and I. Tessman. — J. Molec. Biol., Vol. 35 (1968), 439.
Bridges, B.A., R.E. Dennis, and R. J. Munson. — Genetics, Vol. 57 (1967), 897.
Cerda-Olmedo, E., P.C. Hanawalt, and N. Guerola. — J. Molec. Biol., Vol. 33 (1968), 705.

Chopra, V.L. – Mutation Res., Vol.4 (1967), 382.
Clarke, C. H. – Z. VererbLehre, Vol. 93 (1962), 435.
Clarke, C. H. – J. Gen. Microbiol., Vol. 31 (1963), 353.
Clarke, C. H. – J. Gen. Microbiol., Vol. 39 (1965), 21.
Clarke, C. H. – Molec. and Gen. Genet., Vol. 99 (1967a), 97.
Clarke, C. H. – Molec. and Gen. Genet., Vol. 100 (1967b), 225.
Clarke, C. H. – Mutation Res., Vol. 5 (1968), 33.
Clarke, C. H. – Mutation Res., Vol. 8 (1969), 35.
Clarke, C. H. – Mutation Res., Vol. 9 (1970), 359.
Corran, J. – Mutation Res., Vol. 7 (1969), 287.
De Glovanni-Donnelly, R., S. M. Kolbye, and J. A. Dipaola. – Mutation Res., Vol. 4 (1967), 543.
Drake, J. W. – San Francisco, Holden-Day Inc., 1970.
Drake, J. W., E. F. Allen, S. A. Forsberg, R. M. Preparata, and E. O. Greening. – Nature, Vol. 221 (1969), 1128.
Freese, E. and H. B. Strack. – Proc. Natn. Acad. Sci. U.S.A., Vol. 48 (1962), 1796.
Glover, S. W. – Carnegie Inst. Wash. Publ., No. 612 (1956), 121.
Grigg, G. W. and J. Stuckey. – Genetics, Vol. 53 (1966), 823.
Grossman, L. and D. M. Brown. Mutation as Cellular Process. – Ciba Sympos., p. 109. London, J. and A. Churchill Ltd., 1969.
Hall, Z. W. and I. R. Lehman. – J. Molec. Biol., Vol.36 (1968), 321.
Hill, R.F. – J. Gen. Microbiol., Vol. 30 (1963), 281.
Horneck-Witt, G. and R. W. Kaplan. – Molec. and Gen. Genet., Vol. 101 (1968), 123.
Johnson, H.G. and M.K. Bach. – Nature, Vol.208 (1965), 408.
Johnson, H.G. and M.K. Bach. – Proc. Natn. Acad. Sci. U.S.A., Vol.55 (1966), 1453.
Kilbey, B.J. Mutation as Cellular Process, p.50. – Ciba Sympos. London, J. and A. Churchill Ltd., 1969a.
Kilbey, B.J. – Mutation Res., Vol. 8 (1969b), 73.
Kölsch, E. and P. Starlinger. – Z. VererbLehre, Vol. 96 (1965a), 297.
Kölsch, E. and P. Starlinger. – Z. VererbLehre, Vol. 96 (1965b), 304.
Kuwano, M., M. Ishizawa, and H. Endo. – J. Molec. Biol., Vol. 33 (1968), 513.
Lewis, C. M. – Heredity 25, No. 4 (1970), 686.
Magni, G.E., R.C. von Borstel, and S. Sora. – Mutation Res., Vol. 1 (1964), 227.
Mohn, G. and R.W. Kaplan. – Molec. and Gen. Genet., Vol. 99 (1967), 191.
Novick, A. – Brookhaven Sympos. Biol., Vol. 8 (1956), 201.
Orgel, A. and L. E. Orgel. – J. Molec. Biol., Vol. 14 (1965), 453.
Puglisi, P.P. – Mutation Res., Vol. 4 (1967), 289.
Puglisi, P.P. – Molec. and Gen. Genet., Vol. 103 (1968), 248.
Savic, D.J., D. Kanazir, and P.E. Hartman. – Genetics, Vol. 61, (Suppl. 52), 1969.
Schuster, H. and R.C. Wilhelm. – Biochim. biophys. Acta, Vol. 68 (1963), 554.
Schuster, H. and H.-G. Wittmann. – Virology, Vol.19 (1963), 421.
Sehgal, O.P. and G.F. Krause. – J. Virol., Vol.2 (1968), 966.
Sevag, M.G. – Archs. Biochem. Biophys., Vol.108 (1964), 85.
Sevag, M.G. and B. Ashton. – Nature, Vol.203 (1964), 1323.
Sevag, M.G. and W.T. Drabble. – Biochem. Biophys. Res. Commun., Vol.8 (1962), 446.
Singer, B. and H. Fraenkel-Conrat. – Virology, Vol. 39 (1969), 395.
Stevens, L. – Biol. Rev., Vol. 45 (1970), 1.
Webb, R.B. and H. E. Kubitschek. – Biochem. Biophys. Res. Commun., Vol. 13 (1963), 90.
Witkin, E.M. – J. Cell. Comp. Physiol., Suppl. 1, Vol. 58 (1961), 135.
Witkin, E.M. – Radiat. Res., Suppl., Vol. 6 (1966), 30.
Witkin, E.M. – Brookhaven Sympos. Biol., Vol. 20 (1967), 17.
Witkin, E.M. – Proc. 12th Intern. Congr. Genet., Tokyo, No.3, 1968.
Witkin, E.M. – Mutation Res. Vol. 8 (1969a), 9.
Witkin, E.M. – Annual Rev. Genet., Vol. 3 (1969b), 525.
Witkin, E.M. and E.L. Farquharson. Mutation as Cellular Process, Vol. 36. – Ciba Sympos., London, J. and A. Churchill Ltd., 1969.
Witkin, E.M. and E.C. Theil. – Proc. Natn. Acad. Sci. U.S.A., Vol. 46 (1960), 226.
Yanofsky, C., E.C. Cox, and V. Horn. – Proc. Natn. Acad. Sci. U.S.A., Vol. 55 (1966), 274.
Zamenhof, P.J. – Mutation Res., Vol. 7 (1969), 463.
Zetterberg, G. – Expl. Cell. Res., Vol. 27 (1962), 560.

UDC 575.24:576.851.48

THE FREQUENCY OF COMPLETE AND MOSAIC MUTATIONS IN CELLS OF ESCHERICHIA COLI UVR+ AND UVR- AT HIGH DOSES OF UV RADIATION

V.N. SOIFER and Yu.L.DOROKHOV,
Institute of General Genetics, Academy of Sciences of the USSR, Moscow

The kinetics of the appearance of E.coli K12 mutants incapable of assimilating glucose following exposure to UV radiation were studied. The bacteria used included a strain with a normal system of dark repair and a radiosensitive strain incapable of excising thymine dimers from irradiated DNA. In the former strain the frequency of induced mutations rose in proportion to the irradiation dose reaching a peak at a dose causing a 1,000-fold reduction in the survival rate. Further increase in the dose caused a decline in the mutation frequency. Such a peak was not obtained in the strain with a defective dark repair system. The existence of the peak is attributed to the influence of repair systems.

It is known that the curve of the relative frequency of mutations induced by UV irradiation reaches a plateau and then begins to decline. This type of mutational kinetics is referred to as mutation frequency peak.

The existence of this peak has been variously attributed to an increased sensitivity of the mutants (Russel and Russel, 1959); interference of mutations (Alikhanyan et al., 1960); cytoplasmic lesions (Roepke and Mercer, 1947); heterogeneity of the population (Emmons and Hollaender, 1939); a pronounced delay of cell division, enabling repair processes to occur (Emmons and Hollaender, 1939); the appearance of suppressor mutations (Soifer, 1965); and also changes affecting the secondary structure of DNA and thus preventing the migration of energy to the critical areas (Zavil'gel'skii et al., 1965).

With the discovery of the dark repair enzymes it became possible to explain some instances of peaks in the activity of repair systems. Assuming some finite number of lesions (thymine dimers, for example), if their elimination from the DNA structure were intensified in parallel with the dose this might lead to a peak in the mutation frequency (Soifer, 1966).

We tested this assumption by subjecting E. coli to UV irradiation, using cells with normal repair systems as well as cells defective as regards dark repair. The E. coli strains used were those of Howard-Flanders — K12 AB1157 and a derivative of this line uvr$^-$ A6 AB1886. The latter mutant was chosen because the defect in its repair system consists in an inability to excise dimers from the DNA structure (Howard-Flanders and Boyce, 1966). Mutants were recognized on the basis of the glucose uptake test. Triphenyltetrazolium chloride was added to the agar medium in order to reveal the mutant colonies.

The bacteria were grown to a density of $5 \cdot 10^7$ cells/ml on M9 medium (with added peptone, casein hydrolysate, yeast extract and tryptophan) and then irradiated with three BUV-15 lamps from a distance of 25 cm. Samples of the irradiated cultures were plated on agar containing peptone, yeast extract, meat extract, glucose, and triphenyltetrazolium chloride.

Figure 1 shows the survival rates of E. coli K12 AB1157 (uvr^+) and E. coli K12 AB1886 (uvr^-) following UV irradiation. The strain incapable of excising pyrimidine dimers is much more sensitive to UV light than the wild strain. Particularly striking is the difference in the slope of the inactivation curves for cells of the uvr^- A6 strain incubated after irradiation on normal medium and on medium containing tetrazolium.

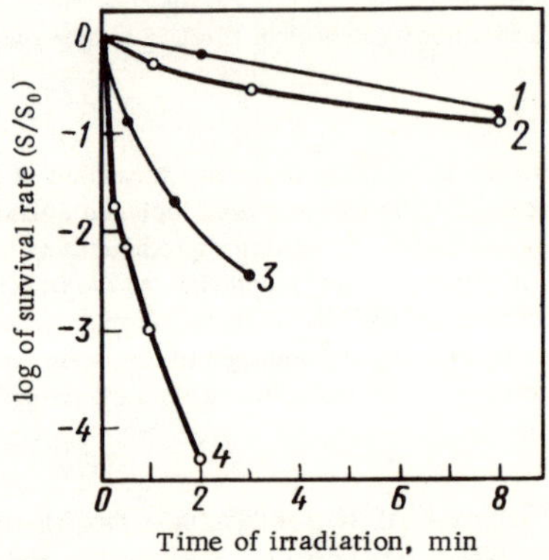

FIGURE 1. Inactivation curves of E.coli K12 AB 1157 (1,3) and E.coli K12 AB 1886 (2,4) after exposure to UV light on medium with (3,4) and without (1,2) triphenyltetrazolium chloride:

S_0) initial number of bacteria; S) number of bacteria after UV irradiation.

The mutant colonies formed included not only complete colonies but also sectored colonies (mosaic mutations) in which mutants accounted for $^3/_4$, $^1/_2$, $^1/_4$ or an even smaller proportion of the colony.

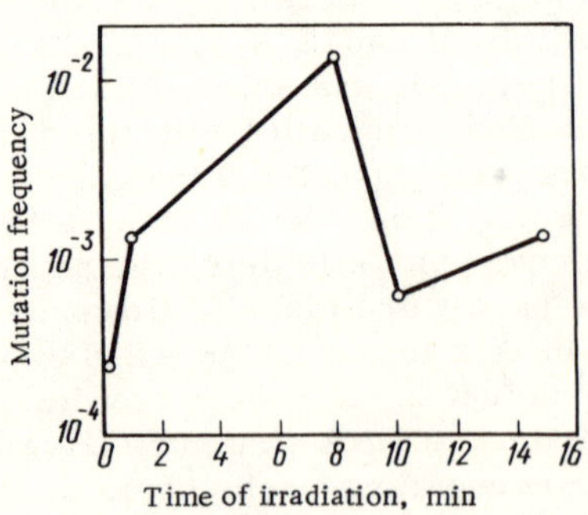

FIGURE 2. Induced mutation frequency in E.coli K12 uvr^+ after exposure to UV radiation.

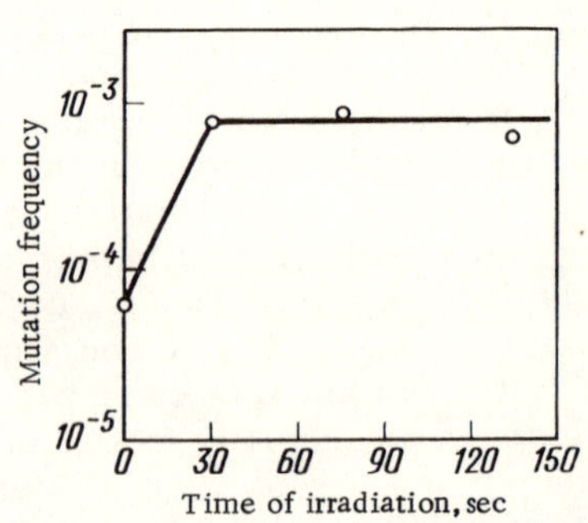

FIGURE 3. Induced mutation frequency in E.coli K12 uvr^-A6 after exposure to UV radiation.

As shown in the table, the frequency of spontaneous and induced mutations is greater in the uvr$^+$ strain than in the uvr$^-$ strain. The mutation frequency of the uvr$^+$ strain reached a peak of $1.2 \cdot 10^{-2}$ at an exposure time of 8 minutes. This frequency is 52 times higher than the spontaneous background. Further increase of the radiation dose lowered the mutation frequency. The frequency of UV-induced mutations in the strain with a defective dark repair system was 14 times higher than the spontaneous background.

Figure 2 shows the kinetics of mutation of the glucose marker in the strain E. coli K12 uvr$^+$. The mutation curve reaches a peak with an exposure of 8 minutes. Further exposure lowers the mutation frequency. In the case of the uvr$^-$ A6 strain, the mutation frequency curve reaches a plateau and does not decrease with further exposure to UV (Figure 3).

Frequency of glucose mutations induced by UV light in cells of E. coli K12 uvr$^+$and E. coli K12 uvr$^-$ A6

Strain	Duration of exposure of bacteria to radiation, min	Number		Mutation frequency
		colonies	mutants	
E. coli K 12 uvr$^+$ AB 1157	Control	8,523	2	$2.3 \cdot 10^{-4}$
	1	113,810	139	$1.2 \cdot 10^{-3}$
	8	19,952	234	$1.2 \cdot 10^{-2}$
	10	62,936	33	$5.2 \cdot 10^{-4}$
	15	31,692	39	$1.2 \cdot 10^{-3}$
E. coli K12 uvr$^-$ A6 AB 1886	Control	67,662	4	$5.9 \cdot 10^{-5}$
	0.50	27,210	22	$8.0 \cdot 10^{-4}$
	1.25	53,360	55	$9.4 \cdot 10^{-4}$
	2.25	26,043	16	$6.1 \cdot 10^{-4}$

The difference in the mutability of the uvr$^+$ and uvr$^-$ strains can be explained by assuming that in one of the strains the dimers are excised and the resulting gaps sealed by repair replication. Under such circumstances the survival rate would increase in parallel with the elimination of the premutational lesion.

It is known from the work of Howard-Flanders et al. that the excision of dimers in the uvr$^-$ A6 strain is impaired owing to a mutation in the uvr A gene. The kinetics of the dimerization process in the uvr$^+$ strain on exposure to high doses of UV radiation has not been studied.

Figure 4 shows our own data on the kinetics of dimerization in the uvr$^+$ strain. The dimerization curve rises linearly with the dose, reaching a peak at 8 minutes, and then declines.

The course of this curve resembles that of the mutation frequency curve.

The fall in the mutability curve and the curve of the frequency of thymine dimerization can be attributed to several factors.

The first and most natural explanation is that excision-type repair takes place immediately during the irradiation. The time available for repair would then obviously be proportional to the radiation dose (assuming a constant intensity of irradiation). The combined effect of the two processes: dimerization, increasing linearly with the dose, and repair, depending only on time might lead to the observed fall in the mutability curve.

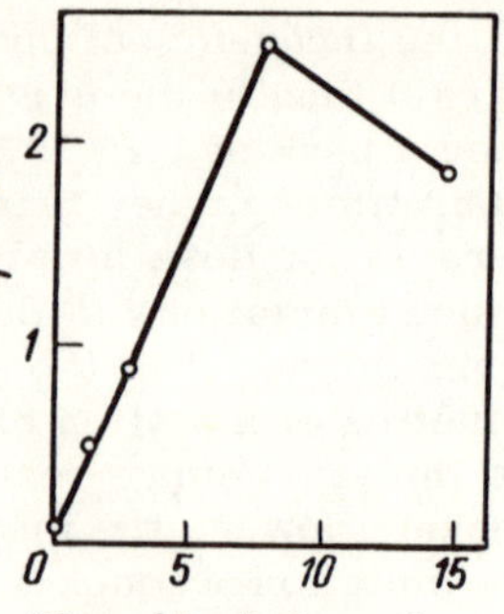

FIGURE 4. The kinetics of dimerization in E. coli K12 uvr⁺ with exposure to UV radiation.

However, this explanation does not apply to the irradiation of extracellular phage which also gives a mutation frequency peak (Zavil'gel'skii, Kriviskii and Ivanov, 1965).

Since the linear relationship between the concentration of thymine dimers in the DNA and the radiation dose breaks down when a certain concentration of dimers is reached, it follows that a progressive increase in the dose will not so easily cause further dimerization.

If we assume as before that the rate of the excision reaction is always constant and does not depend on the number of dimers, this may also be a reason for the peak.

Another possibility is that the rate of elimination of the dimers begins to rise with the radiation dose after a certain critical "density" is reached, e.g. through the excision of groups instead of single dimers.

Moreover, if lesions as well as dimers appear in the DNA, excision of the dimers — a process accelerated in the presence of high doses — may lead to a more progressive elimination of the lesions and thus create the peak effect.

BIBLIOGRAPHY

Alikhanyan, S.I., L.I. Erokhina, and S.I. Lyubinskaya. Features of Induced Mutagenic Process in Microorganisms. — Moscow, AN SSSR, 1960. (Russian)

Emmons, C.W. and A. Hollaender. — Am. J. Bot., Vol. 26 (1939), 467.

Howard-Flanders, P. and R.P. Boyce. — Radiat. Res., Suppl., Vol. 6 (1966), 156.

Roepke, R.R. and F.E. Mercer. — J. Bact., Vol. 54 (1947), 731.

Russel, W.L. and L.B. Russel. — In: Progress in Nuclear Energy, ser 6, Biol. Sci., Vol. 2 (1959), 179.

Soifer, V.N. — In: Eksperimental'nyi mutagenez zhivotnykh, rastenii i mikroorganizmov, p.137. Moscow, "Nauka," 1965.

Soifer, V.N. — Bull. Moscow Society of Naturalists (MOIP), biol. ser. 71, No.1 (1966), 88.

Zavil'gel'skii, G.B., A.S. Kriviskii, and V.I. Ivanov. — Izv. AN SSSR, biol. ser., No.5 (1965), 700.

UDC 575.172.3:576.851.48

PLEIOTROPIC DISTURBANCES OF BIOSYNTHETIC PROCESSES IN ESCHERICHIA COLI MUTANTS WITH LESIONS IN THE PEP-DEPENDENT PHOSPHOTRANSFERASE SYSTEM

V.N. GERSHANOVICH, N.V. YUROVITSKAYA, T.N. BOL'SHAKOVA, T.P. SAPRYKINA and G.I. BURD
Gamaleya Institute of Epidemiology and Microbiology
Academy of Medical Sciences of the USSR, Moscow

It was shown that ctr mutants of E. coli grown on a noncarbohydrate medium have a much slower rate of inducible synthesis of β-galactosidase, L-tryptophan synthase and D-serine deaminase in comparison with the wild type, regardless of which of the three components of the Roseman system is impaired. The biosynthetic activity is restored in the ctr revertants. It is assumed that the ctr mutation hampers the general regulatory mechanism associated with the structure of the bacterial membrane.

INTRODUCTION

Roseman and his colleagues found in 1964 that E. coli contains an enzyme system responsible for the phosphorylation of many carbohydrates using phosphoenolpyruvate (PEP) as phosphate donor (Kundig et al., 1964). This system consists of three components: enzyme I, enzyme II, and the thermostable protein HPr. It was found that the phosphorylation proceeds in two stages:

$$1)\ \text{PEP} + \text{HPr} \xrightleftharpoons[\text{Mg}^{++}]{\text{enzyme I}} \text{Phospho-HPr} + \text{pyruvate};$$

$$2)\ \text{Phospho-HPr} + \text{hexose} \xrightarrow[\text{Mg}^{++}]{\text{enzyme II}} \text{phosphohexose} + \text{HPr}.$$

Further studies suggested that the Roseman system participates in the transport of carbohydrates into the bacterial cell (Kundig et al., 1966). This hypothesis was confirmed in a number of works demonstrating that pleiotropic mutations involving loss of the ability of the bacteria to grow in mineral media with carbohydrates (ctr^- mutations) are accompanied by the loss of some components of the phosphotransferase system (Tanaka and Lin, 1967; Burd et al., 1967; Simoni et al., 1967). Such mutants lack the capacity for intracellular accumulation of carbohydrates. Finally, Kaback (1968) was able to demonstrate in preparations of isolated bacterial membranes that the carbohydrates are phosphorylated during their transfer from the external medium to the intramembranal pool.

It was accepted, therefore, that the role of the Roseman phosphotransferase system consists in active transport coupled with the phosphorylation of carbohydrates. However, evidence irreconcilable with this view gradually began to accumulate. It was shown, for example, that ctr⁻ mutants cease to grow on noncarbohydrate substrates such as acetate or succinate (Asensio et al., 1963; Wang and Morse, 1968). Some of these mutants were incapable of growth on glycerol (Simoni et al., 1967; Wang and Morse, 1968), which enters the cell by facilitated diffusion only (Sanno et al., 1968). There were no grounds for assuming that the Roseman system participates also in the transport of the above compounds since its function is closely coupled with the ability to phosphorylate the compounds transported. Compounds such as glycerol, acetate and succinate do not serve as phosphate acceptors in the phosphotransferase reaction.

These considerations led to the hypothesis that the decrease in the growth rate of the ctr⁻ mutants on noncarbohydrate substrates might be due to a disturbance of their biosynthetic processes. We present below evidence in support of this hypothesis.

RESULTS

Rate of enzyme induction in ctr⁻ mutants

First of all we examined the ctr⁻ mutants isolated and studied in our laboratory (Burd et al., 1969). We observed on many occasions that these mutants grow poorly in media containing noncarbohydrate substrates such as cassaminic acids showing a longer lag period and an early onset of the stationary phase. Such phenomena might result from a lowering of the

Differential rate of induction of β-galactosidase, L-tryptophan synthase and D-serine deaminase (in %) in E. coli mutants with a defective Roseman system*

Strain**	β-galactosidase †	L-tryptophanase synthase	D-serine deaminase
K12 (J-62)	100	100	100
P-34	38	14	21
GT-17	13	35	39
GT-17-46 R	—	108	100
ML-308-225	—	100	100
6E	—	70	76
6E-1R	—	121	129

* The conditions of the experiment are described in the work of Gershanovich et al. (1970).

** The P-34 and GT-17 mutants obtained from the J-62 strain have the genotypes (enzyme I + HPr)⁻ and (enzyme I, enzyme II)⁻ respectively. The mutant 6E, obtained from the strain ML-308-225, has the genotype (enzyme I)⁻. GT-17-46R and 6E-IR are revertants obtained from the mutants GT-17 and 6E respectively.

† The experiments were made with the lac⁺ recombinants HfrC × J-62, HfrC × P-34, and HfrC × GT-17.

induction rate of the enzymes required for utilization of the particular substrate. For this reason we examined this process using the following inducible enzymes as examples: β-galactosidase (Enzyme Catalog No. 3.2.1.23), L-tryptophan synthase (E. C. 4. 2. 1. 20), and D-serine deaminase (E. C. 4.2.1.13).

The data given in the table above show that the differential rate of the induction of these enzymes is actually lowered by some extent in the mutants and is restored in the revertants.

Synthesis of enzymes of the lac operon in the ctr⁻ mutant isolated from the constitutive (i⁻) strain ML-308

With a view to explaining the decline in the rate of inducible synthesis of the enzymes in the ctr⁻ mutants it seemed of interest to establish the following: a) whether this effect results from an interference of the ctr⁻ mutation with the interaction between inducer and repressor; b) whether the ctr⁻ mutation influences the transport of inducers into the cells, thus slowing down the rate of induction. Answers to these questions were obtained from a study of the ML-308/1 mutant (enzyme I)⁻, isolated from the strain ML-308 (i⁻) with a constitutive system for the synthesis of enzymes of the lac operon (Gershanovich et al., 1970). The following results were obtained:

Strain*	β-galactosidase, %	System of transport of β-galactosides, %
ML-308	100	100
ML-308/1	29	17
ML-308/1R	133	135

* The cells were incubated in medium M9 salt base with added casein hydrolysate. The other conditions are described in the work of Gershanovich et al. (1970).

As shown by these results, the activity of β-galactosidase in the ML-308/1 mutant is markedly lower than in the wild type. The same is true of the activity of the β-galactoside permease. Thus, the disturbances in the rate of the induction processes observed earlier in the GT-17, P-34 and 6E mutants can be attributed to an inhibitory effect on both the transport systems associated with the inductors and on the induced enzymes themselves as a result of the mutation. The reason for this is clearly the interference of the ctr⁻ mutations with protein biosynthesis, rather than a disturbance of the interaction between repressors and inductors. .

The next step was to answer the following questions: 1. Does the observed effect result from a loss of activity solely of the components of the Roseman system or from lesion of the genes closely linked with the ctr region? 2. Is the observed slowing down of protein synthesis due to a previous impairment of the synthesis of DNA or RNA?

Identification and properties of a mutant with a thermosensitive Roseman phosphotransferase system

To answer these questions we used a ctr⁻ mutant with a thermosensitive Roseman system. We assumed that the cells of such a mutant, incubated at a permissive temperature and then transferred to nonpermissive conditions, would show the direct effect of inactivation of the phosphotransferase system on biosynthetic processes. We selected such a mutant (ts 19) from nitrosoguanidine-treated cells of E. coli K12 (HfrC). At 37°, ts19 bacteria did not grow in M9 medium with added D-glucose, D-mannose, D-mannitol, maltose, D-fructose or D-galactose as the sole carbon sources. At 27°, however, the growth rate of the mutant in the presence of these compounds did not differ from that of the original HfrC strain. At 37° the ts 19 cells retained the ability to grow on glucose-6-phosphate, gluconate, and lactate. Further experiments showed that the differential rate of synthesis of L-tryptophan synthase in the mutant cells is lower than in the Hfr cells. Thus at 37° the ts 19 bacteria behaved like the other ctr⁻ mutants described in this work and in other publications, whereas at 27° they retained the phenotype of the original strain. As could be expected, the ctr⁻ phenotype at 37° resulted from thermoinactivation of the Roseman system. Figure 1 shows that the activity of the phosphotransferase system in extracts of heated ts 19 cells incubated at 27° is inversely related to the temperature of heating. A lowering of activity is also observed in proportion to the time of incubation of the cells at 37° (Figure 2). The determination of the activity of each of the three components of the Roseman system in extracts of ts 19 cells heated for one hour at 37° showed that such treatment inactivates enzyme I of this system. It is noteworthy that the higher temperature inactivated also the system responsible for the intracellular accumulation of glucose in ts 19 bacteria (see Figures 1 and 2). This series of experiments thus demonstrated that the thermosensitive protein in the ts 19 cells is enzyme I of the Roseman phosphotransferase system and that the thermoinactivation of the glucose transport system and the thermodenaturation of enzyme I have similar kinetics.

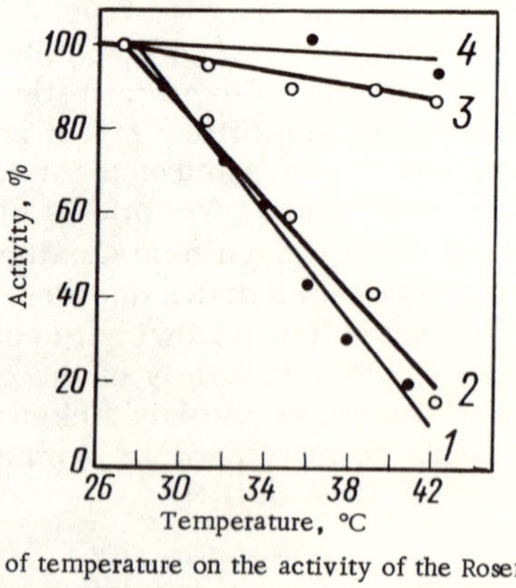

FIGURE 1. Effect of temperature on the activity of the Roseman system and the accumulation of glucose in ts 19 (1,2) and HfrC (3,4) cells:

2,3) activity of the Roseman system; 1,4) activity of the glucose accumulation system.

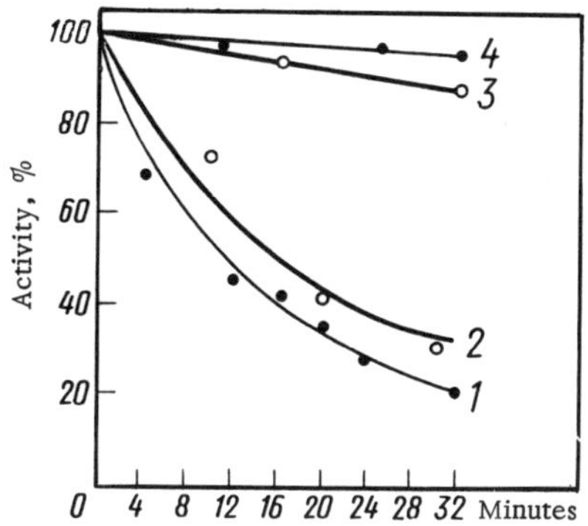

FIGURE 2. Effect of the time of incubation on the activity of the Roseman system and the accumulation of glucose in ts 19 (1, 2) and HfrC (3, 4) bacteria at 37°.

Synthesis of DNA, RNA and protein in heated and unheated ts 19 cells

The following procedure was used for determining the rate of biosynthesis of macromolecules in ts 19 bacteria. Bacteria incubated at 27° in a salt medium with added gluconate were washed and resuspended in minimal medium, after which they were divided in two batches. One batch was placed in an incubator at 27° while the other was heated at 37° to inactivate the enzyme I and was then placed at 27°. The rates of synthesis of DNA, RNA and protein were gauged from the incorporation or radioactive precursors in the acid insoluble fraction after 5 minutes of incubation (see G. I. Burd et al., 1971). The medium used for this contained gluconate as the only carbon source. This experiment enabled determination of the effects of the heat treatment on the activity of the Roseman system without interference from extraneous factors. The results of these experiments are given below.

	ts 19	HfrC
C^{14} threonine, %	75 ± 16(11)*	107**
C^{14} thymidine, %	108 ± 17(9)*	102
	40 ± 15(8)*	104
C^{14} uracil, %	52†	—

* The number of experiments is indicated in parentheses.
** Average of two experiments.
† 40-second tag, with 1-tryptophan (500 μ g/ml) added together with the uracil.

These results show convincingly that the denaturation of enzyme I at 37° markedly decreases the rate of RNA synthesis. Protein synthesis is less affected by inactivation of the Roseman system. The biosynthesis of DNA is not disturbed. It was evident from the controls that heating does not affect the rate of biosynthetic processes in cells of the original HfrC strain.

DISCUSSION AND CONCLUSIONS

The data obtained show clearly that mutational damage of the Roseman phosphotransferase system of E. coli lowers the rate of RNA and protein synthesis. This explains the slowing down of the growth of ctr⁻ bacteria on substrates which do not serve as phosphate acceptors in the phosphotransferase reaction. What is the mechanism of this effect? Most probably the inactivation of the enzymes of the Roseman system exerts only an indirect influence on the activity of the RNA-synthesizing system. At present it is difficult to explain the nature of the relationship between ctr⁻ mutations and the activity of the RNA-synthesizing systems. Since enzyme II of the phosphotransferase system is located in the bacterial membrane it can conceivably serve as a structural protein of the membrane, apart from its known functions. In such a case the mutational damage of one of the components of the phosphotransferase system may alter the conformation of the cytoplasmic membrane of the cell. This in turn may trigger a series of nonspecific changes affecting the transcription process in the bacterial cell. It should also be borne in mind that the membrane complex is the site of the synthesis of bacterial RNA (Rouvieri et al., 1969). Changes in the conformation of this complex may therefore exert a direct influence on the rate of RNA synthesis.

BIBLIOGRAPHY

Asensio, C., G. Avigad, and B. Horecker. — Archs. Biochem. Biophys. **103**, No.3 (1963), 299.

Burd, G.I., T.N. Bol'shakova, T.P. Saprykina, V.V. Klyucheva, and V.N. Gershanovich. — Molekulyarnaya Biologiya, Vol. 5 (1971), 384.

Burd, G.I., V.P. Shabolenko, and V.N. Gershanovich. — Molekulyarnaya Biologiya 1, No. 5 (1967), 713.

Burd, G.I., V.P. Shabolenko, I.V. Andreeva, V.V. Klyucheva, and V.N. Gershanovich. — Molekulyarnaya Biologiya 3, No.2 (1969), 256.

Gershanovich, V.N., N.V. Yurovitskaya, and G.I. Burd. — Molekulyarnaya Biologiya, Vol. 4 (1970), 534.

Kuback, H. — J. Biol. Chem. **243**, No.13 (1968), 3711.

Kundig, W., S. Ghosh, and S. Roseman. — Proc. Natn. Acad. Sci. U.S.A. **52**, No. 4 (1964), 1067.

Kundig, W., F. Kundig, B. Anderson, and S. Roseman. — J. Biol. Chem. **241**, No.13 (1966), 3243.

Rouvieri, J.S. Lederberg, P. Granboulau, and F. Gross. — J. Molec. Biol. **46**, No. 3 (1969), 413.

Sanno, Y., T. Wilson, and E. Lin. — Biochem. Biophys. Res. Commun. **32**, No. 2 (1968), 344.

Simoni, R., M. Levinthal, F. Kundig, B. Anderson, P. Hartman, and S. Roseman, — Proc. Natn. Acad. Sci. U.S.A. **58**, No. 5 (1967), 1963.

Tanaka, S. and E. Lin. — Proc. Natn. Acad. Sci. U.S.A. **57**, No. 4 (1967), 913.

Wang, J. and M. Morse. — J. Molec. Biol. **32**, No. 1 (1968), 59.

UDC 575.24:576.851.48

GENES CONTROLLING NUCLEOSIDE METABOLISM IN ESCHERICHIA COLI K12

V.V.SUKHODOLETS, T.I. CHUKANOVA, E.S. PUKHOVA,
Ya.V. FLYAKH and G.E. BAUMANIS
All-Union Research Institute of Genetics and Selection of Industrial Microorganisms,
Central Administration of the Microbiological Industry, Council of Ministers of the USSR,
Moscow

Reduction of the quantitative requirement for thymine is most pronounced in deoxyribonuclease (tlr) and deoxyribo-5-phosphate aldolase (thd-s) mutants and less so in purine phosphorylase (rib A) mutants. The genes tlr, thd-s and rib A are densely clustered and are located near the threonine gene associated with tdd-type mutations affecting the ability of the bacteria to convert thymine to thymidine and split the latter.

Genes controlling the metabolic conversions of nucleosides were discovered in the course of studies on thymine metabolism in bacteria auxotrophic for thymine (thy) (Kameneva et al., 1965; Sukhodolets and Alikhanyan, 1965; Alikhanian et al., 1966). Exogeneous thymine is assimilated by thy bacteria which convert it first to thymidine according to the reaction:

Thymine + deoxyribose-1-phosphate ⇄ thymidine + orthophosphate. This reaction, controlled by the enzyme thymidine phosphorylase, is normally biased toward the breakdown of thymidine (Breitman and Bradford, 1964); thymine auxotrophs free from secondary mutations therefore require high concentrations of thymine (about 20 μg/ml) in order to shift the equilibrium toward thymidine synthesis. The efficiency of the uptake of exogenous thymine by thy cells rises as a result of mutations blocking the breakdown of deoxyribose and causing the accumulation of the thymidine precursor deoxyribose-1-phosphate. Two types of such mutations are known — tlr and thd-s, affecting respectively the enzymes phosphodeoxyribomutase and deoxyribose-5-phosphate aldolase (Munch-Petersen, 1968; Lomax and Greenberg, 1968; Beacham et al., 1968). In the thy tlr mutants, the blocking of the conversion of deoxyribose-1-phosphate into deoxyribose-5-phosphate lowers the requirement for thymine to 0.5 — 1.0 μg/ml. The uptake of exogenous thymine is slightly less efficient in the thy thd-s mutants which are unable to perform the next catabolic stage — the breakdown of deoxyribose-5-phosphate into acetaldehyde and glyceraldehyde-3-phosphate; these mutants require about 2μg of thymine per ml. The thd-s mutants also differ from the tlr mutants in their sensitivity to deoxyribosides; the accumulation of deoxyribose-5-phosphate in the thd-s cells inhibits protein synthesis and stops cell growth. Mutations affecting purine nucleoside phosphorylase — an enzyme catalyzing the breakdown of purine ribosides and deoxyribosides — lower the requirement for thymine only slightly (to 5—10 μg/ml). Purine phosphorylase mutants were isolated from the riboside-sensitive thy bacteria as forms resistant to purine ribosides — rib A-r (Stenchuk and Sukhodolets, 1969). The lower requirement for thymine in these mutants may be due to the participation of purine phosphorylase in the conversion of exogenous thymine into the inert ribosylthymine. Purine phosphorylase and thymidine phosphorylase mutants can be selected from thd-s bacteria as organisms resistant to purine and pyrimidine deoxyribosides respectively. In this case there is a phenotypic reversion of the thd-s mutants with respect to the sensitivity to deoxyribosides; the occurrence of additional mutations affecting phosphorylases in the thd-s cells blocks earlier stages in the catabolism of deoxyribose and leads to an accumulation of a growth inhibitor — deoxyribose-5-phosphate.

thy bacteria with a mutation affecting thymidine phosphorylase (tdd) cannot assimilate exogenous thymine and require thymidine for growth (Mineeva and Il'ina, 1969).

All these mutations (tlr, thd-s, rib A-r, tdd) are located in the same area of the genetic map of E. coli, near the threonine gene, and possibly affect genes belonging to a single operon.

This article presents the results of our further genetic studies of the tdd and rib A-r mutations.

Figure 1 shows the position of 21 tdd mutations, four of which (Nos. 1, 9, 11, 12) are of the deletion type. All tdd mutations from No. 1 through No. 12 were obtained in thd-s strains as phenotypic reversions with respect to thymidine sensitivity; their position on the map was determined by Zlotnikov et al. (1969). Apart from these twelve mutants, carrying the thd-s mutation in their genome, our experiments with 3-factor conjugational crosses has enabled the mapping of nine tdd mutations (Nos. 20—23, 27, 28, 31, 33, 34), obtained by nitrosoguanidine treatment of the thymine auxotroph strain E. coli Hfr 3. OSO thy.

These nine thy mutants were selected as thymidine-dependent cells, i. e. cells incapable of utilizing thymine as a source of thymidylic acid. Thymidine phosphorylase mutants can also be selected on the basis of their inability to use thymidine as a carbon source (Dale and Greenberg, 1967). However, four of the nine thy tdd mutants obtained (Nos. 20, 27, 33, 34) proved capable of growth on thymidine or deoxyuridine as the carbon source. All these four mutations lie in a particular area of the genetic map — at the left end of the tdd locus. The tdd 5 (thd-s) mutant, previously mapped in the same part of the chromosome, similarly belongs to the group of tdd strains which retain the ability to catabolize thymidine. This was made clear from an analysis of the properties of the recombinants obtained by crossing the strains Hfr 3. OSO thy tdd 5 thd-s 5 × F^- P678 thy thd-s. All the thd-r recombinants inheriting the tdd 5 mutation proved capable of growth on thymidine as the carbon source. The tdd mutations 5, 20, 27, 33 and 34 could affect the enzyme thymidine phosphorylase (altering its activity in such a way that its catabolic function was maintained) and also another enzyme likewise involved in the conversion of thymine into thymidine (for example, an enzyme participating in the deoxyribosyl transferase reaction).

The first hypothesis is based on the fact that the tdd mutants 5, 20, 27, 33 and 34 appear to be leaky with respect to thymidine phosphorylase — an enzyme with a pronounced catabolic function. It is possible that the partial lowering of the activity of this enzyme affects mainly its anabolic function: the residual activity is not sufficient for the synthesis of thymidine whereas the breakdown of thymidine and the subsequent utilization of deoxyribose is still enough to enable cell growth.

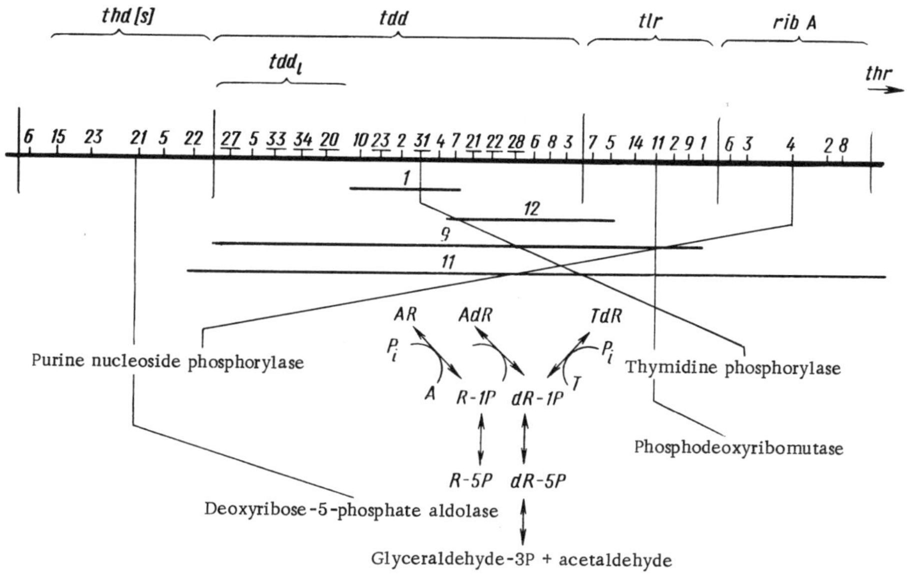

FIGURE 1. Genes, enzymes and reactions involved in the catabolism of deoxynucleosides in E. coli K12.

The position of the mutations in the ribA locus as well as that of mutations Nos. 20, 21, 22, 23, 27, 28, 31, 33 and 34 in the tdd locus were established in this work. The position of all other mutations, including those of the deletion type, is according to Zlotnikov et al. (1969). AR) adenosine; AdR) deoxyadenosine; TdR) thymidine; A) adenine; T) thymine; P_i) orthophosphate; R) 1(5) P-ribose-1(5) phosphate; dR) 1(5)P-deoxyribose-1(5) phosphate.

We were able to prove that the tdd mutants which retain the ability to break down thymidine can also synthesize this compound in conditions which enhance the anabolic function of thymidine phosphorylase. It became clear that the tdd mutants 5, 20, 27, 33 and 34 can utilize exogenous thymine for growth in the presence of deoxyadenosine in the medium (Figure 2). The results of the experiment, illustrated in Figure 2, show that the ability of the different mutants to grow on thymine in the presence of deoxyadenosine is correlated with their ability to grow on thymidine, i. e. the anabolic function of thymidine phosphorylase appears to be equally impaired in all mutants. Moreover, the growth of the tdd-leaky (l) bacteria on thymine in the presence of deoxyadenosine in this experiment actually depends on the activity of thymidine phosphorylase since the double mutants carrying mutations of the tdd$_l$ type 5, 20 etc. as well as a mutation affecting the purine nucleoside phosphorylase (rib A-r) lack the ability to assimilate thymine in the presence of deoxyadenosine (Figure 3). Thus, deoxyadenosine serves as a source of the thymidine precursor deoxyribose-1-phosphate which is split off by purine nucleoside phosphorylase. Since all the tdd_1 mutants were mapped at the left end of the tdd locus, we studied the complementation of tdd mutants in order to determine whether the group of tdd_1 mutants belongs to a separate cistron. This work was carried out with the diploid strain F' recA, containing an episome with the genetic markers thr^+ and leu^+ in the pertinent part of the chromosome of E. coli K12. At first we obtained three thy tdd (thd-s) mutants, selected among revertants with respect to thymidine sensitivity in the haploid strain F^- thy thd-s (thr, leu, his, arg). All three tdd mutants obtained were revertants; one of them could utilize thymine in the presence of deoxyadenosine in the medium. The three mutant strains F^- thy thd-s tdd were then used to obtain heterozygous F'-type diploids tdd thd-s/F' tdd^+ thd-r (by crossing with F' recA). These heterozygotes had the phenotype of thymine auxotrophs with a high quantitative requirement for thymine, i. e. the tdd and thd-s mutations are recessive (Sukhodolets and Pukhova, 1970).

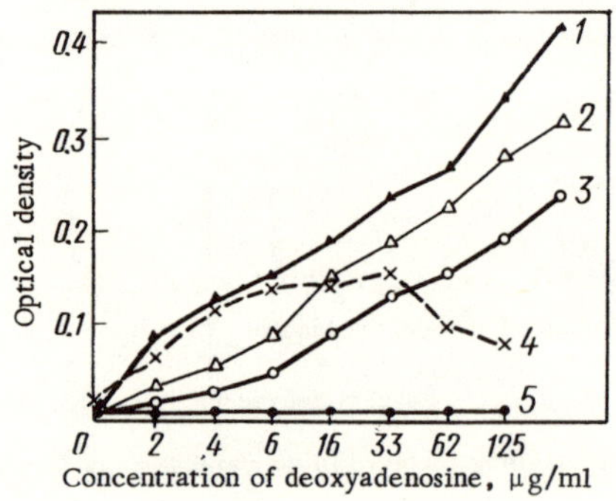

FIGURE 2. Growth of F^- tdd (thr, leu, thy, str-r) strains on a synthetic medium containing thymine (20 μg/ml) and different concentrations of deoxyadenosine:

1) tdd 34; 2) tdd 33; 3) tdd 20; 4) tdd 5 (thd-s); 5) tdd 22; inoculum — 10^6 cells/ml; duration of growth (on shaker, 37°) — 18 hours.

The heterozygous bacteria tdd thd-s/tdd^+ thd-r were converted to homozygotes in two stages. First of all, using a selective medium with a low concentration of thymine (5 μg/ml) we obtained diploids of the genotype tdd thd-s/tdd^+ thd-s; afterwards we obtained tdd thd-s/tdd thd-s homozygotes on a selective medium containing a high concentration of thymidine (200 μg/ml). F' strains bearing the tdd mutations in the episome were obtained in this manner. These strains were later used in crosses with a series of haploid rec^+-F^- tdd- and F^- tdd thd-s strains containing different tdd mutations of known position. Despite the presence of the rec^+ allele in the F^- tdd recipients, each cross yielded relatively stable diploid tdd/F' tdd clones which always had the tdd phenotype. If the tdd mutations in the episome and chromosome of these diploids belong to different sites on the tdd locus, plating of the corresponding strains on the selective medium with thymine would yield tdd+ haploid clones. In other words, the data obtained are not compatible with a complementation between tdd mutations in the trans position, i. e. all the tdd mutations apparently belong to a single cistron regardless of their phenotypic manifestation.

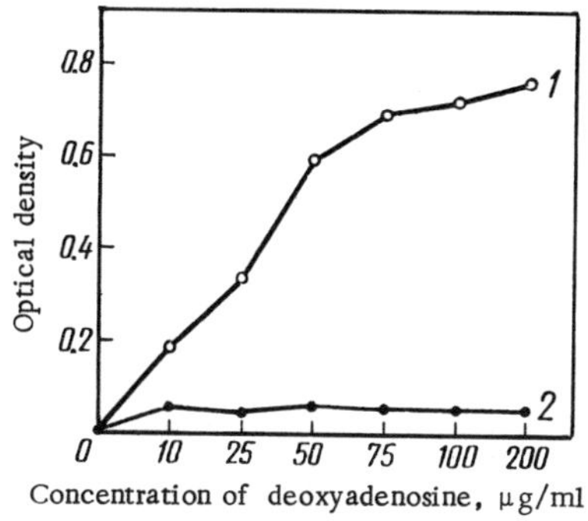

FIGURE 3. Growth of F⁻ tdd_l 34 thy (1) and F⁻ tdd_l 34 ribA-r2 thy (2) on a synthetic medium containing thymine (40 μg/ml) and different concentrations of deoxyadenosine.

inoculum — 10^6 cells/ml; duration of growth (on shaker, 37°) — 16 hours.

The hypothesis regarding the leaky nature of the tdd mutants catabolizing thymidine was confirmed by determinations of the activity of thymidine phosphorylase in vitro (see below).

Hfr 30 SO	Thymidine phosphorylase*	Purine nucleoside phosphorylase**
thy tdd_l 20	1.03	6.3
thy tdd_l 27	1.47	8.8
thy tdd 28	0.38	11.4
thy thd-s rib A-r 32	17.8	<0.5
thy thd-s rib A-r 1	16.0	<0.5
thy thd-s rib A-r 2	18.6	<0.5
thy+	5.7	9.0

* Determined as the quantity of deoxyribose-1-phosphate formed by the breakdown of thymidine (micromoles per hour per mg of protein) (Chukanova and Sukhodolets, 1969).

** Determined as the quantity of guanine formed by the breakdown of deoxyadenosine (micromoles per hour per mg of protein) (Tsuboi and Nudson, 1957).

The tdd-leaky (tdd_l) mutants show indeed a comparatively high residual activity of thymidine phosphorylase. It is apparently this residual activity of thymidine phosphorylase which enables the tdd_l mutants to grow on thymidine as the carbon source and on thymine and deoxyadenosine as sources of thymidylic acid.

It is evident from the above results that the activity of purine nucleoside phosphorylase is very low in the ribA-r mutants selected from the deoxyadenosine-resistant revertants of the strain Hfr 3. OSO thy thd-s. Figure 1 shows the position of the ribA locus in relation to the other genes involved in the catabolism of nucleosides, as well as the position of five different ribA-r mutations within the locus. The position of the ribA locus on the map was established from the results of reciprocal conjugation crosses between the ribA-r mutants and the tdd 9 deletion-type mutant. Our data on the position of the ribA locus confirm those of Ahmad and Pritchard (1969).

As noted above, the tdd_1 mutants require purine phosphorylase for the synthesis of thymidine from thymine and deoxyadenosine. It is known that the addition of deoxyadenosine to the medium enhances the uptake of exogenous thymine (Boyce and Setlow, 1962) and lowers the quantitative requirement for thymine in thy mutants. This effect is completely eliminated by the ribA-r mutation: the growth of the strain Hfr 3. OSO thy ribA-r (obtained in our laboratory by the conjugational cross Hfr 3. OSO thy thd-s 17 × Hfr. 3. OSO thy thd-s5 ribA-r 32) on a thymine-containing medium was even inhibited to some extent by deoxyadenosine (Figure 4). Thus, the deoxyribosyl transferase activity (the transfer of deoxyribose from deoxyadenosine to thymine) apparently results from the combined action of two enzymes, purine nucleoside phosphorylase and thymidine phosphorylase — a conclusion corresponding to the data of other workers (Imada and Igarashi, 1967).

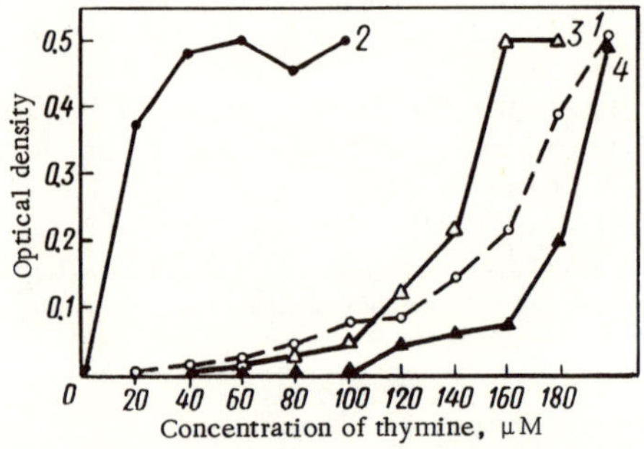

FIGURE 4. Growth of Hfr 3.OSO bacterial strains at different concentrations of thymine in the medium:

1) thy; 2) thy in the presence of deoxyadenosine (500 μM); 3) thy ribA-r; 4) thy ribA-r in the presence of deoxyadenosine; inoculum — 10^6 cells/ml; duration of growth (on shaker, 37°) — 14 hours.

BIBLIOGRAPHY

Ahmad, S.I. and R.H. Pritchard. — Molec. and Gen. Genet., Vol. 104 (1969), 351.

Alikhanian, S.I., T.S. Iljina, E.S. Kaliaeva, S.V. Kameneva, and V.V. Sukhodolec. — Genet. Res., Vol. 8 (1966), 83.

Beacham, I.R., A. Eisenstark, P.T. Barth, and R.H. Pritchard. — Molec. and Gen. Genet., Vol.102, p.112.

Boyce, R.P. and R.B. Setlow. — Biochim. biophys. Acta, Vol. 61 (1962), 618.

Breitman, R.R. and R.M. Bradford. — Biochem. Biophys. Res. Commun., Vol.17 (1964), 786.

Chukanova, T.I. and V.V. Sukhodolets. — Molekulyarnaya Biologiya 3, No. 4 (1969), 584.
Dale, B. and G.R. Greenberg. — J. Bact., Vol. 94 (1967), 778.
Imada, A. and S. Igarashi. — J. Bact., Vol. 94 (1967), 1551.
Kameneva, S.V., E.S. Kalyaeva, and S.I. Alikhanyan. — Genetika, No.1 (1965), 100.
Lomax, M.S. and G.R. Greenberg. — J. Bact., Vol. 96 (1968), 501.
Mineeva, L.A. and T.S. Il'ina. — Biokhimiya 34, No. 1 (1969), 21.
Munch-Peterson, A. — Biochim. biophys. Acta, Vol. 161 (1968), 279.
Stenchuk, N.N. and V.V. Sukhodolets. — Genetika 5, No.11 (1969), 106.
Sukhodolets, V.V. and S.I. Alikhanyan. — Genetika, No.2 (1965), 47.
Sukhodolets, V.V. and E.S. Pukhova. — Genetika 6, No.4 (1970), 190.
Tsuboi, K. K. and P.B. Nudson. — J. Biol. Chem., Vol. 224 (1957), 879.
Zlotnikov, K.M., V.V. Sukhodolets, and G.E. Baumanis. — Genetika 5, No. 12 (1969), 114.

UDC 575.24:576.858.9

THE EFFECT OF MUTAGENS ON AMBER MUTANTS OF THE BACTERIOPHAGE T4B

E.S.PIRUZYAN
All-Union Research Institute of Genetics and Selection of Industrial Microorganisms, Central Administration of the Microbiological Industry, Council of Ministers of the USSR, Moscow

The effect of the base analogs 2-aminopurine, 5-bromouracil and a number of their derivatives on the frequency of reversions in two amber mutants of the phage T4B was examined. It was found that the level of mutagenic activity of the base analog derivatives depends on the position of the radicals on the ring. The intragenic specificity of the mutation process was studied in 15 amber mutants of the T4B phage, all of them mapped in gene 43. It was established that the frequency of induced and spontaneous reversions varies among the different amber mutants.

This paper deals with the mechanism of chemical mutagenesis in the case of amber mutants of the bacteriophage T4B. The mutagenic effect was gauged from the frequency of induced back mutations (Alikhanyan et al., 1967). It is known that the amber mutations are point mutations involving the formation of the nonsense triplet ATC instead of the original triplet in the DNA. In all cases of a mutagenic effect, therefore, the chemical agent reacts with the ATC triplet. This fact provides a clue as to the chemical reaction involved and consequently a convenient model for studies of the mechanism of mutagenesis. Back mutations can arise in amber mutants under the influence of mutagens capable of bringing about the transition T $\rightarrow$ C. Consequently, base analogs capable of inducing transitions in either direction must cause reversions in amber mutants.

E. G. Mugnetsyan and I examined the effect of 2-aminopurine and 14 of its derivatives, as well as that of 5-bromouracil and 24 of its derivatives on two amber mutants of the phage T4. The base analog derivatives used in this work were synthesized at the Moscow Chemical-Pharmaceutical Institute.

Examination of these compounds revealed a close relationship between the chemical structure of the compound and the mutagenic and inactivating effects. Thus, the activity depends on the presence of particular radicals in specific positions on the ring.

The compounds tested included also orotic acid which serves as a precursor of base analogs, as well as some of its derivatives, three of which (5-aminoorotic acid, 5-bromoorotic acid, and N^3 allyl orotic acid) proved mutagenic.

The study of the intragenic specifity of the effect of mutagens provides another approach to the mechanism of mutagenesis.

It was shown that alleles of a given gene may differ from one another in several ways. We examined the comparative mutability of different alleles of the same gene on the basis of the map drawn up by V. Z. Pogosov and S. I. Alikhanyan (1969) for gene 43 which controls the synthesis of phage-induced DNA polymerase. The map of gene 43, constructed by these authors on the basis of mapping 44 mutations, shows that the length of the gene equals 6.4 recombination units. This enabled us to use mutants spread throughout the length of gene 43 (see figure).

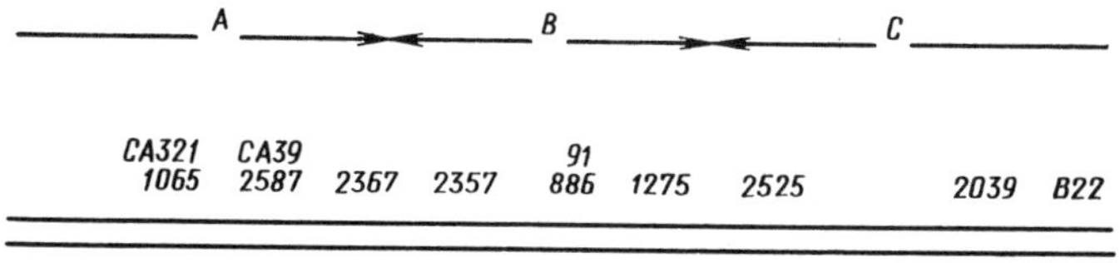

Position of the amber mutants examined on the map of gene 43 of bacteriophage T4B.

The mutagenic effect of nitrous acid and 2,6-diaminopurine was determined on 12 mutants mapped in different areas (A, B, C) of the gene. Also selected were three pairs of mutants localized at a single site. Determinations of the level of spontaneous reversions in mutants examined showed differences from one mutant to another: thus, the proportion of revertants varies within 4 orders of magnitude (from 0.42×10^7 to 2198×10^7). It was assumed that the rates of induced mutation similarly vary from one site to another.

The set of amber mutants of gene 43, being scattered throughout the gene, represents a convenient model for the study of some aspects of intragenic mutation: it provides data on the mutational lability of different parts of one gene of phage T4 and at the same time the use of alleles of a single gene reduces all the tested mutants to a common denominator in terms of their chemical accessibility to the mutagen in vitro, in view of the fact that one of the possible causes of the differing mutability of phage genes is known to be their different position in the packed DNA in the phage head. Since all the amber mutants examined were obtained by the use of hydroxylamine, the transitions leading to the formation of the ATC triplet could originate from two codons — for glutamine (GTC) and for tryptophan (ACC).

Preliminary tests of the amber mutants used on three Su^+ strains of E. coli (Table 1) showed that some of them do not grow on certain strains containing a suppressor gene: thus, the mutant 1275 does not grow on the Su^+-3 strain which translates tyrosine, while the mutants 91, CA39, 2587 and CA321 do not grow at 42° in the E. coli C600 strain which translates glutamine at the point of mutation in the polypeptide chain. It is assumed, therefore, that in the absence of a suppression of the amber mutation on the E. coli strain 600 Su^+-2 glutamine is unacceptable, which means that the amber codon induced in these mutants by hydroxylamine replaces the triplet coding for tryptophan. Thus, in mutants incapable of growth on certain Su^+ strains the corresponding insertions of amino acids are unacceptable.

TABLE 1. Ability of amber mutants of phage T4B to grow on Su^+ strains of E. coli

Amber mutants of phage T4B	CR 63 27	(Su+-1) 42	C 600 27	(Su+-2) 42	CA 265 27	(Su+-3) 42
CA321	+	+	+	−	+	+
1065	+	+	+	+	+	+
CA39	+	+	+	−	+	+
2587	+	+	+	−	+	+
91	+	+	+	−	+	+
886	+	+	+	+	+	−
2364	+	+	+	+	+	+
B22	+	+	+	+	+	+
2357	+	+	+	+	+	+
2525	+	+	+	+	+	+
2039	+	+	+	+	+	+
1275	+	+	+	+	−	−

Note: "+ " indicates presence of growth, "− " lack of growth.

Data on the mutagenic effect of 0.1 M nitrous acid on amber mutants of gene 43 are given in Table 2. As in the case of 2,6-diaminopurine, nitrous acid sharply increases the frequency of back mutations in all the mutants examined above the spontaneous background. This is explained by the fact that both mutagens cause AT — GC transitions.

TABLE 2. Frequency of reversions of gene 43 per 10^7 phage particles in amber mutants of phage T4B caused by nitrous acid

Duration of treatment, min	Amber mutants											
	91 (496)	886 (0.42)	CA39 (3.5)	2587 (0.75)	CA321 (4.2)	1065 (2.6)	2364 (9.4)	B22 (2.6)	1275 (5.5)	2039 (15.5)	2525 (5.1)	2357 (2540)
5	2.3	72.5	33.0	22.4	29.4	18.0	12.6	144	18.6	39.2	15.5	11.2
10	49.6	1098.9	337.7	484.4	574.7	185.3	208.8	3180	90.9	595.5	673.3	69.1
15	315.3	1492.7	836.0	640.0	865.8	2237.7	574.4	9259	376.2	1202.8	2633.9	461.4

Note: The frequency of spontaneous reversions is given in parentheses.

However, the mutability, i. e. the ability to cause reversions, differed from one mutant to another. The first six amber mutants listed in the table were previously mapped at a single site. Since the resolving capacity of crosses in genetic mapping is 0.02 per cent recombination, corresponding in physical terms to 2 — 5 nucleotide pairs, closely clustered mutations can be mapped at the same site. Nevertheless, these two mutants may differ in mutability.

The capacity for reversions on treatment with nitrous acid is five times greater in the mutant 886 than in the mutant 91, although both lie at the same site. Moreover, the mutant 91 differs from 886 in being unable to grow on the Su^+-2 strain, i. e. the original triplet in its DNA probably coded for tryptophan. Thus, the mutations 91 and 886, although situated close to one another in the phage genome, are not identical. The lower mutability of site 91 results from the unacceptability of certain amino acids in the formation of revertants. According to A. I. Maisuryan and N. D. Lomovskaya (1968), the acceptability or unacceptability of amino acid substitutions depends largely on the amino acid situated at the site of the original lesion, i. e. the direct mutations. Tryptophan is a highly reactive amino acid owing to the presence of an indole ring, and its substitution with another amino acid may prove ineffective. Indeed, all four mutants incapable of growth on the Su^+-2 strain show a lesser capacity for nitrous acid-induced reversions in comparison with the other mutants. The mutant 1275, which does not grow on the Su^+-3 strain, similarly has a lower mutability.

Two mutants mapped at another site show an equal capacity for reversions and an equal ability for growth in different Su^+ strains. Consequently, they can be regarded as identical alleles.

TABLE 3. Frequency of reversions of gene 43 per 10^7 phage particles in amber mutants of T4B induced by 2,6-diaminopurine

Duration of treatment, min	Amber mutants						
	886 (0.73)	CA321 (3.0)	2587 (1.3)	1065 (4.3)	2364 (29.5)	B22 (0.9)	2525 (16.4)
30	5.4	11.4	5.7	12.6	3.3	36.5	23.2
60	7.5	22.3	8.3	20.55	4.3	53.8	32.8
90	12.7	31.9	11.1	34.3	4.8	55.5	42.2

Note: The frequency of spontaneous reversions is indicated in parentheses.

Comparison of the mutants B22 and 2364, which are mapped at different sites, shows that the frequency of nitrous acid-induced mutations in the mutant B22 is 16.5 times greater than in the mutant 2364. Similar results were obtained with 500 μg of 2,6-diaminopurine per ml (Table 3) — the mutation frequency in the mutant B22 is 11.5 times greater. Since the two mutants do not differ in their growth on the three suppressor strains used, there appears to be another factor affecting the mutability of the site. This factor may be the adjacent bases or the base sequence, since the mutations B22 and 2364 lie in different areas of the map rather far apart (about 5 recombination units from one another). This is only a preliminary conclusion, however, since the growth of the mutants has so far been studied on only three suppressor strains of E. coli.

Thus, the difference in the mutability of alleles of a single gene may result from a number of causes. Given equal exposure to the action of mutagen, the mutability may depend largely on the acceptability or unacceptability of an amino acid at the point of mutation. It is also possible that the mutability of the given site is affected by the adjacent nitrogen bases or groups of bases.

BIBLIOGRAPHY

Alikhanyan, S.I., E.S. Piruzyan, and N.M. Mkrtumyan. — Genetika, No.9 (1967), 76.
Maisuryan, A.N. and N.D. Lomovskaya. — Molekulyarnaya Biologiya 2, No. 3 (1968), 389.
Pogosov, V.Z. and S.I. Alikhanyan. — Genetika, 5, No. 6 (1969), 127.

UDC 575.24:576.858.9

THE GENE-SPECIFIC SUPPRESSOR suA IN THE GENOME OF THE PHAGE T4

V.N. KRYLOV and T.G. PLOTNIKOVA
All-Union Institute of Genetics and Selection of Industrial Microorganisms,
Central Administration of the Microbiological Industry,
Council of Ministers of the USSR, Moscow

A suppressor mutation suA of phage T4B was discovered, partly suppressing the phenotypic manifestation of an am mutation (am 765) in gene 46) of the phage (phage DNA ase). SuA is a gene-specific suppressor which acts with the same efficiency on am mutations in gene 46 regardless of their position (at 6 different sites on gene 46) but does not suppress am mutations in other genes.

INTRODUCTION

It has been calculated that the genome of bacteriophage T4 can code for about 200 polypeptides. However, not more than 80 genes of this phage have been identified so far. It is known, on the other hand, that infection of bacteria with the phage causes the appearance of new phage-specific enzymes the loci of which have not yet been established on the genetic map of the phage. An example of these is the group of nucleases. It is furthermore known that genes 46 and 47 of phage T4 control some nuclease activity, but the biochemistry of this enzyme is still obscure. If it is accepted that new genes can arise in the course of evolution by the duplication of preceding genes, it can be assumed that the various genes controlling the different nucleases are descended from a single gene. If this is so, any member of the group of these nuclease genes may be able to function instead of another nuclease gene as a result of a "successful" mutation. This might be detected in genetics experiments by investigating the phenomenon of gene-specific suppression. Previously "latent" genes might possibly be revealed in this manner. On the basis of these general considerations we tried to select mutations capable of the specific suppression of amber mutations in gene 46.

MATERIALS AND METHODS

The bacterial strains used were E. coli B and E. coli CR63 which are respectively restrictive and permissive to amber mutants. The phage strains were amber mutants with respect to various genes and the wild type phage. The amber mutants were obtained from Gordeev (Moscow

State University). The medium used was Hottinger broth (containing 100 mg% of amine nitrogen) in 1.2 or 0.7% agar. The methods of work with the phage were as described in the monograph by Adams (1959).

RESULTS AND DISCUSSION

The am 765 revertants (am 765 being an amber mutation in gene 46 of phage T4B) form several types of plaque. One of the revertant types (R13) forms small plaques on E. coli B (Figure 1).

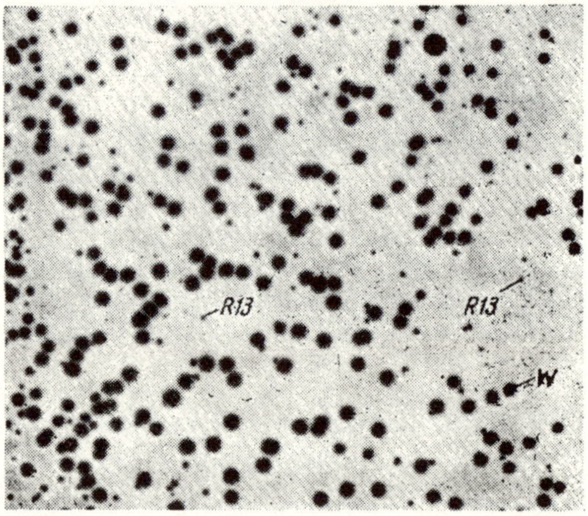

FIGURE 1. Wild-type (w) and R13 mutant plaques on a culture of E.coli B cells.

The genetic structure of phage R13 was established in crosses with the wild type phage. The progeny of this cross included am 765 mutants. Thus, the phage R13 contains the original mutation am 765 and a suppressor mutation. We designated this suppressor mutation suA. The phage bearing the suA mutation was isolated in a pure form. Its plaques on E. coli CR63 are indistinguishable from those of the wild type phage, whereas the plaques obtained on E. coli B are somewhat more opaque than those of the wild type phage (although this trait is not constantly reproducible).

Several crosses were made in order to determine the position of the suA mutation on the genetic map of phage T4B, namely ++ × am 765 suA; r I am 765 + × ++suA; am 765 + × + suA; rI am 765 + × + am 765 suA, as well as several crosses between standard amber mutants in different genes located in the same region of the genetic map. It was concluded on the basis of these crosses that the suA mutation lies between genes 49 and 55. However, crosses of multiple mutants of the type am 765 am N suA × am 765, where N is a mutation in any gene of phage T4 and subsequent determination of the proportion of R13 plaques yielded new and more reliable data. According to these findings the suA mutation maps near genes 33 and 34.

In order to determine the specificity of action of the suA suppressor, i.e. whether it can suppress amber mutations also in other genes of the phage, we crossed phage carrying the suA suppressor with various amber mutants located in other genes: 10, 18, 20, 23, 26, 30, 42, 43, 49, 47, 63. None of these crosses yielded progeny forming R13-type plaques.

On the other hand, crosses with amber mutants located at different sites of the same gene invariably gave rise to phage forming R13-type plaques (see Figure 2 showing the map prepared by us on the basis of preliminary tests). Thus, the suA suppressor is clearly gene specific.

In principle, suppression can be brought about in two ways — either by inactivating the product of the suA gene (inhibiting the development of the phage and normally inactivated by the product of gene 46), or by modifying this product. To answer this question we performed the following experiment. A suppressor-bearing pseudorevertant was isolated from the am752 mutant, located at a different site on the genetic map from that of am 765. This suppressor was obtained in a pure form (by crosses with the wild type phage) and was then introduced into the genome of phage am 765. The suppressor was designated suB. Then the cross am 765 suA× am 765 suB was made. If the independently selected suppressors suA and suB map at the same site, this might indicate that suppression is brought about in this particular case by a modification of the product of gene suA. On the other hand, if the suA and suB mutations are located at different sites of the suppressor gene, the suppression probably results from inactivation of the product of the suppressor gene. The degree of linkage between suA and suB was assessed from the yield of the am 765 amber mutant in the progeny of the cross. Not a single amber mutation was found in a total of 800 plaques examined. We interpret this result as an indication that here the observed suppression probably results from a modification of the product of the suA gene.

We determined the effect of various conditions on the harvest of phages bearing the suppressor mutation in order to obtain some preliminary physiological data on this mutation.

On E. coli CR63 the duration of the latent period, the adsorption rate and the harvest of phage am 765 suA do not differ from those of the wild type phage. When grown on E. coli B, however, the suppressed variety of the phage yields a reduced harvest amounting to about 25—40% of the harvest of the wild type phage. This means that the suppression of the am 765 mutation is incomplete, though highly efficient.

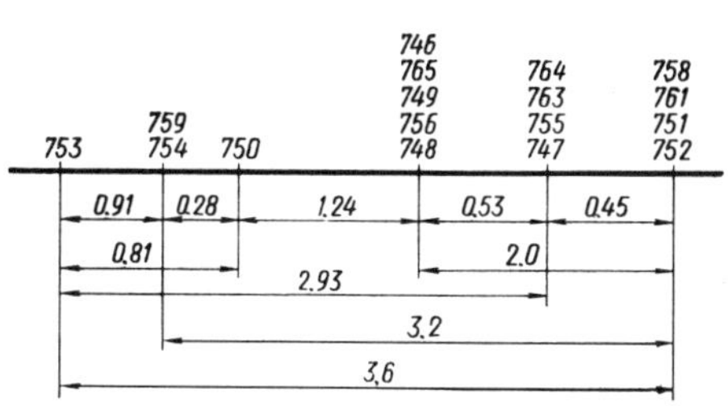

FIGURE 2. Location of amber mutations in gene 46.

It was desirable to determine whether the suA suppressor acts catalytically or stoichiometrically. It is known that gene 46 controls a catalytic function (an enzyme). If the suA suppressor compensates for the function of gene 46, its own function must also be catalytic. Snustad (1968) has proposed a special test for determining the manner in which genes operate. This test can be outlined as follows. The restrictive host E. coli B is infected with a mixture of the wild type phage and the mutant to be tested. The multiplicity of infection is 0.5 for the wild type phage and from 0 to 15–30 for the amber mutant. If the gene controls a catalytic function, a marked increase in the multiplicity of infection of the cells with the amber mutant will cause only a slight decrease in the harvest from the double-infected cells. If the gene controls a stoichiometric function, the harvest would drop more sharply and to lower values than in the previous case. Figure 3 shows the results of such an experiment with the wild type phage and the am 765 suA mutant. A mixed infection of the cells with the wild type phage and am 765 causes a response which is typical of a catalytic function, i. e. the harvest decreases by only half when the multiplicity of infection is increased to 15. On the other hand, mixed infection of the cells with am 765 suA and the amber mutant leads to what appears to be a typical stoichiometric response. Nevertheless, this experiment does not constitute proof of a stoichiometric function of the suA suppressor since the observed incomplete suppression may be due to the formation of a minute amount of the active enzyme produced by the gene suA.

The suA suppressor can operate by two mechanisms assuming that the suppression results from a modification of the product of this gene.

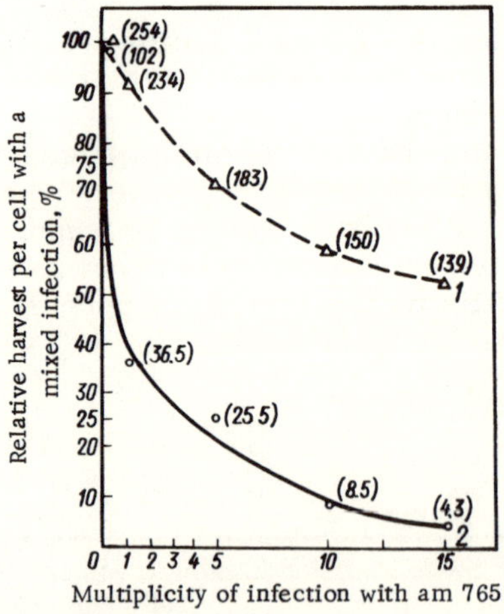

FIGURE 3. Results of the mixed infection of E. coli B cells with different phages:

1) am$^+$ × am 765; 2) am 765 suA × am 765; infection with am$^+$ and am 765 suA was at a multiplicity of 0.5; the figures in parentheses indicate the absolute harvest per cell.

A substitutive suppression would mean that the suA gene of phage T4 normally controls one of the nucleases of which the genetic locus has not yet been mapped. Such a nuclease was described by Kutter and Wiberg (1968). This nuclease is responsible for the preliminary breakdown of bacterial host DNA into large, acid-insoluble fragments. If the suA mutation alters the specificity of this enzyme so that it breaks down the host DNA directly into acid-soluble fragments, this may explain the gene-specific suppression in this particular case.

Another explanation of the suppressor effect can also be given. It is known that the amber mutations in gene 46 are incomplete in the sense that the mutant phage, when grown in restrictive conditions, yields a minimal harvest of 6—15 phage particles per infected cell. One possible explanation of this is that with some suA mutations, certain nucleases (of bacterial or phage origin) can partly perform the function carried out by the product of gene 46. An increase of the efficiency of the nucleases can be brought about either by a structural change (for example at the quaternary level by an interaction with the product of the suA gene), or by an increase in the affinity between the nucleases and the particular areas of the cytoplasmic membrane where they usually operate. In the latter case the suA mutation could alter the structure of this area by modifying some product included in the cytoplasmic membrane and controlled by the suA gene.

Our data are not yet sufficient to establish the mechanism of suppression. There is no doubt, however, that the study of gene-specific phage suppressors enables the discovery of new, hitherto unidentified genes.

CONCLUSIONS

1. The genome of phage T4 was found to contain a suppressor, suA, which specifically suppresses the phenotypic manifestation of amber mutations in gene 46.

2. The efficiency of the suA suppressor is quite high (the phage carrying a mutation in gene 46 and a suppressor mutation yields a harvest corresponding to 25—40% of that of the wild type phage) but is not complete.

3. The suppression apparently results from a modification of the product of the suA gene.

4. The suA suppressor was first mapped in phage T4B between genes 55 and 49. Subsequent data showed that suA maps between genes 33 and 34.

BIBLIOGRAPHY

Adams, Mark. Bacteriophages. — (Interscience) Wiley, 1959.
Kutter, E.M. and J.S. Wiberg. — J. Molec. Biol., Vol. 38 (1968), 395.
Snustad, P. — Virology, Vol. 35 (1968), 550.

UDC 575.24

THE MECHANISM OF THE MUTAGENIC ACTION OF HYDROXYLAMINE

E.I. BUDOVSKII and E.D. SVERDLOV
Institute of the Chemistry of Natural Compounds,
Academy of Sciences of the USSR, Moscow

A study was made of the mechanisms and kinetics of the reactions of nucleotide residues with hydroxylamine (HA) and O-methylhydroxylamine (OMHA), the chemical and physical properties of nucleotide residues modified by HA and OMHA and finally the effect of the conditions of the reaction on the rates of its individual stages and the proportions between the different types of modified residues.

Thorough knowledge of the whole sequence of processes from the application of the mutagen to the manifestation of the modified genetic information is of crucial value in any study of induced mutagenesis. The interpretation of genetic data is considerably simplified if data on the chemical and functional properties of the products of the reaction between the mutagen and DNA and on the mechanisms and kinetics of their formation are available. Regrettably, the chemical aspects of the action of many mutagens are still obscure and so it is impossible to correlate the chemical and functional changes in the genetic material. Here we shall discuss our own findings and those published in the literature on the reaction of hydroxylamine (HA) and O-methylhydroxylamine (OMHA) with nucleic acid residues and draw some conclusions as to the nature of their mutagenic effect.

It is well known that HA and OMHA induce both transitions of the C → T type and inactivation of the genetic material (Bautz-Freese and Freese, 1964; Phillips and Brown, 1968). According to the data of Phillips et al. (1965) and Wilson and Caicuts (1966), both these effects are brought about at the polynucleotide level by the modified cytidine residues. There are at least two types of these, one causing erroneous incorporation during template synthesis and the other blocking template synthesis. Calculations made on the basis of the data of Phillips et al. (1965) show that the residues responsible for the errors are formed by a single-stage chemical reaction while those leading to inactivation result from a two-stage reaction.

According to the conversion scheme given below, the reaction of HA and OMHA with cytidine yields residues of type III in two stages; conceivably these compounds are totally or largely devoid of functional activity and cause inactivation of the template (Figure 1). Residues of types II and IV are formed in one stage. In view of the great instability of type II it can be assumed that residues of type IV retain their functional activity in the template. This is confirmed by the finding that the synthesis of poly-G in theRNA polymerase system of E. coli is the same on poly-(C, IV) as on the normal poly-C.

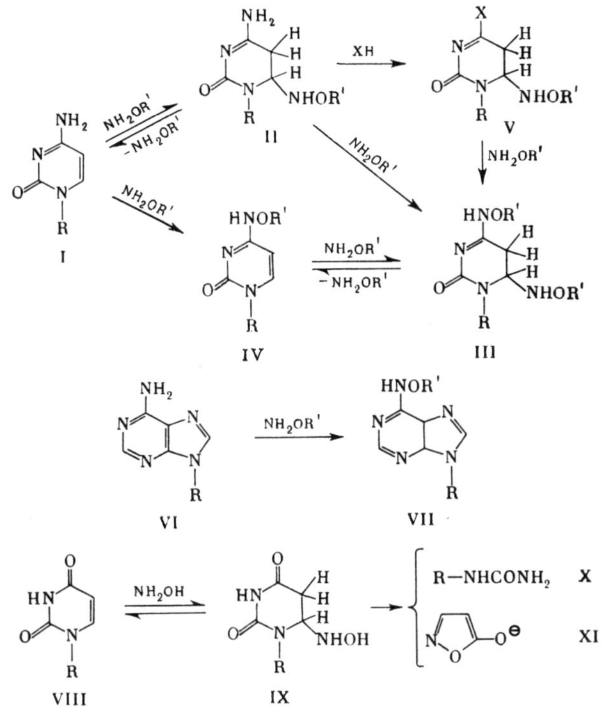

FIGURE 1. Scheme of the conversions of nucleoside residues induced by hydroxylamine and O-methylhydroxylamine.

Since both guanyl and adenyl residues can be incorporated by the RNA polymerase system on poly-C modified by HA or OMHA (Phillips et al., 1965; Wilson and Caicuts, 1966), it is evident that the type IV residues, while retaining their functional activity, can also show an altered functional specificity i. e. can lead to transitions.

These assumptions are supported by two findings: first, the reaction of HA or OMHA with 5-substituted cytosine nuclei yields type IV nuclei only (Janion and Shugar, 1967); secondly, the mutagenic effect of HA and OMHA is maximal in T-even phages, which are distinguished by the presence of 5-oxy-methylcytosine (Freese and Strack, 1962).

The type III units differ markedly in structure from the natural nucleoside residues, whereas the structure of IV in the imino (IVb) and amino (IVa) forms is analogous to that of thymine (uracil) and cytosine respectively.

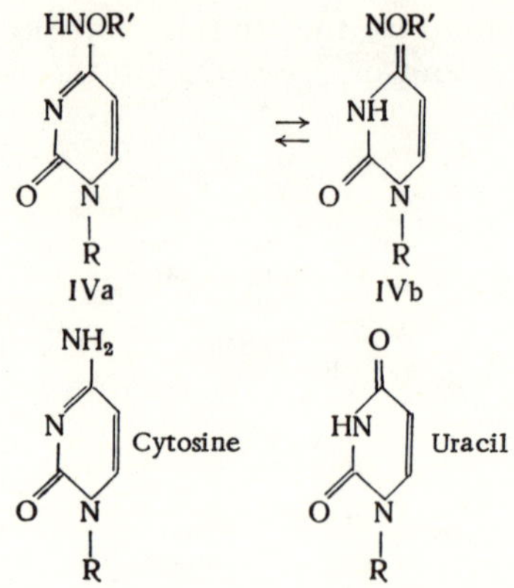

In contrast to the case of cytidine, the tautomeric equilibrium IVa ⇄ IVb is shifted strongly towards the imino form (Brown, Hewlins and Shell., 1968). This constitutes further proof that the type IV residue can function both as a cytidine analog and as a thymine (uridine) analog, i. e. it shows an amphoteric template specificity.

The above data concern mainly the functional activity and specificity of the modified cytidine residues within the template. In order to test the validity of these conclusions with respect to the precursors, we synthesized the nucleoside triphosphates III and IV (R' = H and CH_3) and examined their ability to replace the normal precursors in the RNA polymerase system of E. coli. It became clear that triphosphate III cannot replace any of the precursors, whereas triphosphate IV can be incorporated both instead of cytidine triphosphate and instead of uridine triphosphates. Thus, the type IV residues possess an amphoteric functional specificity not only within the template but also as precursors.

This conclusion as regards the amphoteric properties of type IV residues finds support in the work of Tessman et al. (1965), who examined the mutagenesis of phase S13 in vivo. Transitions of all types were obtained — a finding which can be explained only by the formation of amphoteric precursors in the bacterial cell and their subsequent incorporation into the phage DNA (Figure 2).

The relationship between the mutagenic and inactivating effects of HA and OMHA on DNA thus clearly depends to a large extent on the ratio between IV and III residues. This ratio depends on the conditions of the experiment and can easily be controlled within a wide range by changes in the pH, temperature, concentration of the reagent, etc. (Kochetkov and Budowsky, 1969). These correlations were studied on monomers and should not, of course, be indiscriminately applied to polynucleotides and nucleoproteins. It can be assumed, however, that data obtained with monomers can be used at the polymer level within the range of conditions in which the higher structures of polynucleotides and nucleoproteins show no marked changes.

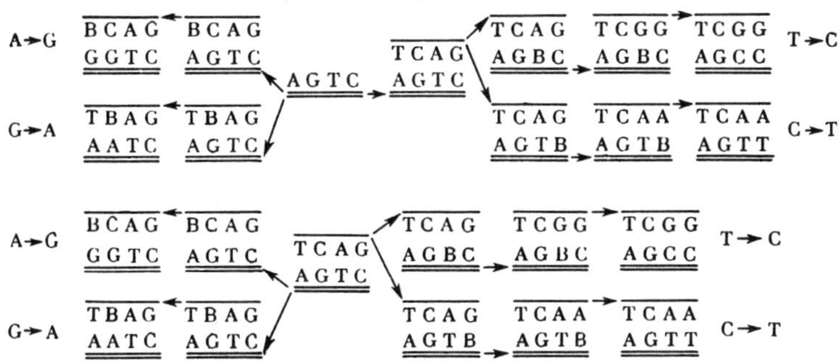

FIGURE 2. Possible substitutions resulting from the incorporation of a precursor (B) with an amphoteric functional specificity into the polynucleotide.

An experimental verification of these conclusions is being carried out currently in collaboration with A. S. Kriviskii and T. P. Shcherban. When phage MS2 is exposed to OMHA, which does not affect the uridine residues, the effect of pH on the inactivation rate corresponds to the effect of pH on the rate of the modification of the cytidine residues (Budovskii et al., 1971). The mutagenic effect of OMHA on phage MS2 is enhanced by a decrease in the pH which is correlated with a change in the IV/III ratio (Figure 3).

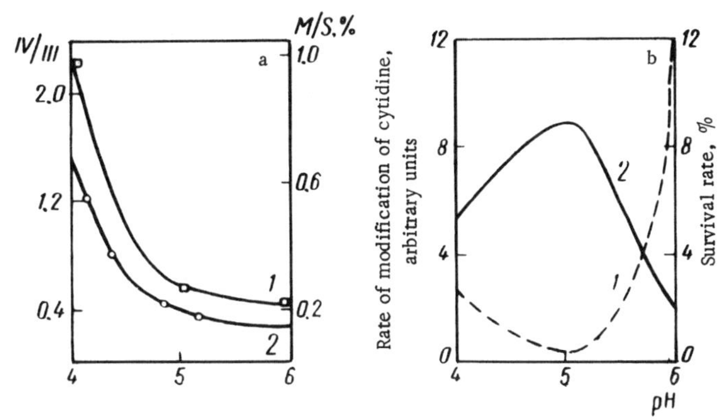

FIGURE 3. Correlation between the chemical and functional effects of O-methylhydroxylamine on cytidine and the phage MS2:

a: 1 — M/S, 2 — IV/III; b: 1) survival rate, 2) rate of cytidine modification; IV/III) ratio between cytidine modification products (see Figure 1); M/S) ratio between mutants and non-activated phage particles.

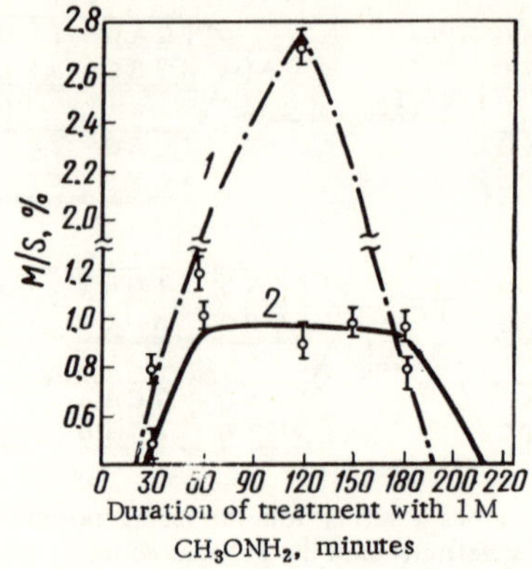

FIGURE 4. The effect of Versene on the course of O-methylhydroxylamine-induced mutagenesis of phage MS2:

1) water; 2) Versene ($1.5 \cdot 10^{-4}$M). Other designations as in Figure 3.

Versene exerts a considerable influence on the kinetics of mutagenesis in the phage (Figure 4). Since Versene does not affect the kinetics of the chemical reaction in the case of monomers, it can be assumed that the chelating of metal ions alters the higher structure of the phage nucleoprotein and thus changes the reactivity of some of the nucleoside residues.

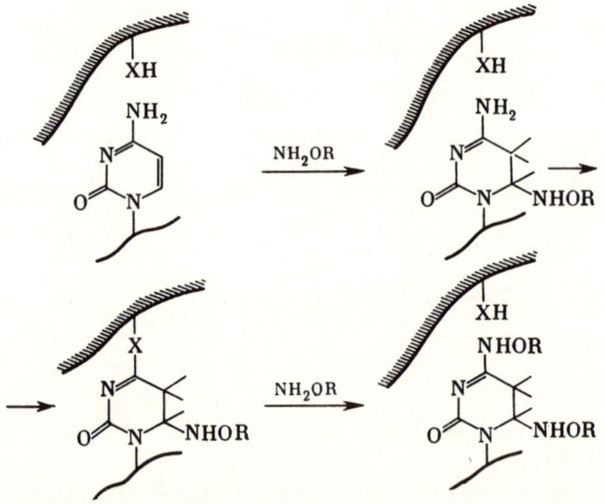

FIGURE 5. Scheme of the formation and rupture of covalent bonds between polynucleotide and protein as a result of the action of O-methylhydroxylamine on nucleoproteins.

The interaction between polynucleotide and protein in nucleoproteins may affect not only the rate but also the mechanism of the reaction. When phage S_d is treated with OMHA, for example, type III residues are practically the only end product of the modification as a result of the shielding of the cytosine nucleus by the protein (Sklyadneva et al., 1970). Moreover, in the process of modification changes in the properties of the phage particle occur indicating the reversible formation of covalent bonds between polynucleotide and protein (Tikhonenko et al., 1971). The mechanism of this phenomenon can be outlined as follows. The addition of OMHA at the $C_{(5)} = C_{(6)}$ double bonds greatly facilitates the substitution of the amino group of residue II with the nucleophilic group of the amino acid residue, which leads to a covalent bond between the phage DNA and the protein. The further action of OMHA dislodges the amino acid residue, creating a type III residue and splitting the polynucleotide-protein bond (Figure 5). Conceivably, the formation of type V residues is likewise accompanied by inactivation of the phage particle.

Apart from their reaction with the cytosine nucleus, HA and OMHA can also substitute the amino group of the adenine nucleus. This reaction has a maximum velocity at pH 5. Its velocity is 50 to 100 times lower than the overall rate of modification of the cytosine nucleus (Budowsky et al., 1969, 1971).

This difference is much smaller if one considers only the rate of substitution of amino groups in adenine and cytosine, and especially in 5-oxymethylcytosine (Brown and Hewlins, 1968). In other words, the mutagenesis of T-even phages involves not only the modification of 5-oxymethylcytosine units but also an intensive modification of adenosine units. No data are available so far on the genetic consequences of the modification of the adenine nucleus, but the possibility of such consequences and in particular the lability of the $N_{(6)}$-oxyadenosine unit should be borne in mind.

Thus, the treatment of genetic material with OMHA triggers a complex of interlinked processes, associated not only with chemical changes in the polynucleotides but also with modifications of the higher structures of the polynucleotides and nucleoproteins. For this reason a conclusive, unequivocal interpretation of the observed genetic effects is not possible at present. One of these effects is the existence of a peak in the ratio between the number of mutations and the surviving phage, e. g. in MS2 phage treated with OMHA. The presence of such a peak probably reflects changes in the higher structure of the nucleoprotein in the course of mutagenesis.

HA reacts not only with cytidine and adenosine residues but also with uridine residues; the rate of the formation of type IX residues is optimal at pH 8 (Domkin et al., 1970). One might assume that the IX residues are devoid of functional activity, as is the case with the III residues. In order to test this assumption and determine the effect of the conditions of storage on the modified genetic material we examined, together with N. N. Pashneva, the inactivation of phage MS2 on exposure to HA.

At a pH higher than 7, the treatment of RNA or RNP with HA causes mainly a modification of the uridine residues and the formation of type IX residues; these readily revert to the original compounds in the absence of HA at pH 4—7, whereas in a more alkaline medium (at about pH 10) a proportion of them is converted into ribosylurea residues (Domkin et al., 1970). Thus, phage inactivated by HA at pH 8 is largely reactivated by a subsequent incubation without HA, the degree of reactivation being higher at

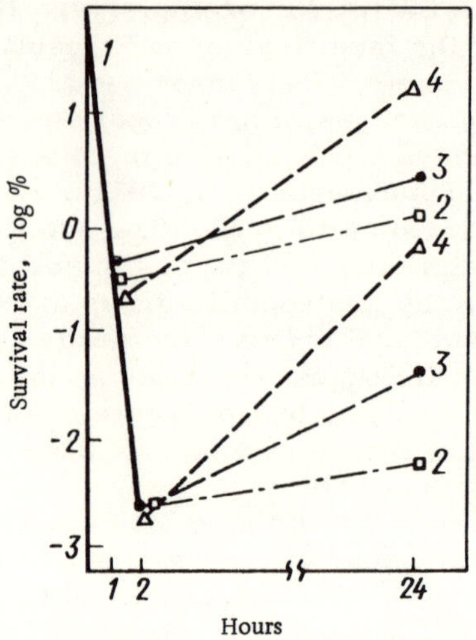

FIGURE 6. Infectivity of phage MS2 following treatment with 1 M hydroxylamine (pH 8.0, 30°).

The infectivity was determined immediately after the removal of hydroxylamine (1) and after a 24-hour incubation without hydroxylamine at pH 4.3 (2); 7 (3); 10.0 (4).

pH 4.3 than at pH 7. As shown in Figure 6, the experimental data agree well with the theoretical assumptions.

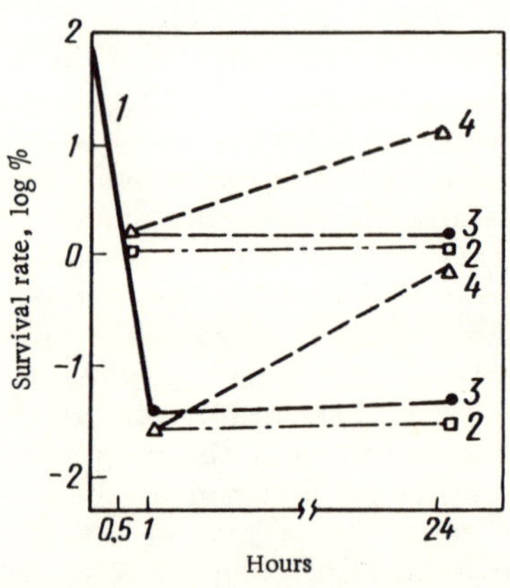

FIGURE 7. Infectivity of phage MS2 following treatment with 1M hydroxylamine (pH 5.0; 30°).

Key as in Figure 6.

Treatment of the phage with HA at pH 5 modifies mainly the cytidine residues. Since the modified cytidine residues are stable at about pH 7, little reactivation of the phage in the absence of HA can be expected in these conditions, except as a result of the reversion of uridine residues. At a pH less than 5 the type III residues are converted into type IV residues and this is necessarily accompanied by a marked reactivation of the phage. As shown in Figure 7, the experimental data and theoretical considerations are in good agreement here as well.

The detailed study of the mechanisms and kinetics of the interaction of nucleic bases with HA and OMHA and the properties of the modified nucleoside residues will thus enable the rational use of HA and OMHA in functional research and lead to a better understanding of the observed effects, especially in the field of genetics.

BIBLIOGRAPHY

Bautz-Freese, E. and E. Freese. — Proc. Natn. Acad. Sci. U.S.A., Vol. 52 (1964), 1289.

Brown, D.M. and M. J. Hewlins. — J. Chem. Soc., (c), p.1922, 1968.

Brown, D.M., M. J. E. Hewlins, and P. Shell. — J. Chem. Soc. (c), p.1925, 1968.

Budovskii, E.I., A.S. Kriviskii, and T.P. Shcherban. — Genetika, Vol. 7 (1971), 120.

Budowsky, E.I., E.D. Sverdlov, and G.S. Monastyrskaya. — J. Molec. Biol., Vol. 44 (1969), 205.

Budowsky, E.I., E.D. Sverdlov, and G.S. Monastyrskaya. — Biochim. biophys. Acta, Vol. 246 (1971), 320.

Domkin, V.D., A.T. Pogorelov, V.N. Shibaev, E.I. Budovskii, and N.K. Kochetkov. — Doklady AN SSSR, Vol. 193 (1970), 97.

Freese, E. and H. Strack. — Proc. Natn. Acad. Sci. U.S.A., Vol. 48 (1962), 1796.

Janion, C. and D. Shugar. — Acta biochim. pol., Vol. 14 (1967), 293.

Kochetkov, N.K. and E.I. Budowsky. — In: "Progress in Nucleic Acid Research and Molecular Biology," Acad. Press, Vol. 9 (1969), 403.

Phillips, J.H., D.M. Brown, R. Adman, and L. Grossman. — J. Molec. Biol., Vol. 12 (1965), 816.

Phillips, J. H. and D. M. Brown. — In: "Progress in Nucleic Acid Research and Molecular Biology," Acad. Press, Vol. 7 (1968), 349.

Sklyadneva, V.B., N.P. Kiseleva, E. I. Budovskii, and T.I. Tikhonenko. — Molekulyarnaya Biologiya, Vol. 4 (1970), 110.

Tessman, I., H. Iohsiwa, and S. Kumar. — Science, Vol. 148 (1965), 507.

Tikhonenko, T.I., E.I. Budowsky, V.B. Sklyadneva, and I.A. Khromov. — J. Molec. Biol., Vol. 55 (1971), 535.

Wilson, R.G. and M. J. Caicuts. — J. Biol. Chem., Vol. 241 (1966), 1725.

UDC 575.24

THE MECHANISM OF ACTION OF DIAZOKETONES

L.A.PLUGINA and N.G. SHUPPE
Institute of Chemical Physics, Academy of Sciences of the USSR,
Moscow

Diazoketones inhibit the synthesis of DNA and affect to a lesser extent the synthesis of RNA and proteins. The inhibition of DNA synthesis is reversible; DNA synthesis returns to normal after removal of the diazoketone. Protein synthesis is a necessary condition for the restoration of DNA synthesis. It appears that diazoketones inactivate one of the enzymes of the DNA synthesizing system, thus stopping DNA synthesis.

Diazoketones ($RCOCHN_2$) are biologically active in a number of different ways. The mutagenic effect of 1,4-bis-diazoacetylbutane (BDB) has been demonstrated in *Drosophila* (Rapoport, 1960), higher plants (Zoz, 1966), and various species of actinomycetes (Gumanov et al., 1966). 1,2-Bis-diazoacetylethane was found to be a potent anticancer agent with a wide spectrum of action (Emanuel' et al., 1968). Diazoketones belong to the category of cytostatic agents on the basis of their effect on the viability of treated objects.

The molecular mechanism of the biological action of diazoketones is still unknown. Our experiments on *Escherichia coli* cells have shown that diazoketones inhibit the biosynthesis of DNA, RNA and proteins (Figure 1).

Similar results were obtained with cultures of Ehrlich ascites tumor cells.

Diazoketones inhibit in particular the synthesis of DNA; the synthesis of RNA and proteins is affected to a lesser extent. All the diazoketones used in our experiments on bacteria caused predominantly inhibition of DNA synthesis.

The inhibition of DNA synthesis by diazoketones is reversible. If diazoketone-treated cells, in which DNA synthesis has been inhibited, are washed free of the inhibitor and incubated again, DNA synthesis returns to normal (Figure 2).

Protein synthesis is a necessary condition for the restoration of DNA synthesis. DNA synthesis is not resumed after removal of the diazoketone from the system if chloramphenicol is present (Figure 3). This suggests that diazoketones inactivate some enzyme of the DNA-synthesizing system. According to Kozlov et al. (1967), diazocarbonyl compounds can inactivate enzymes by esterifying the carboxyl group of the active center. We were able to show by equilibrium dialysis that diazoketones bind bacterial proteins but not DNA.

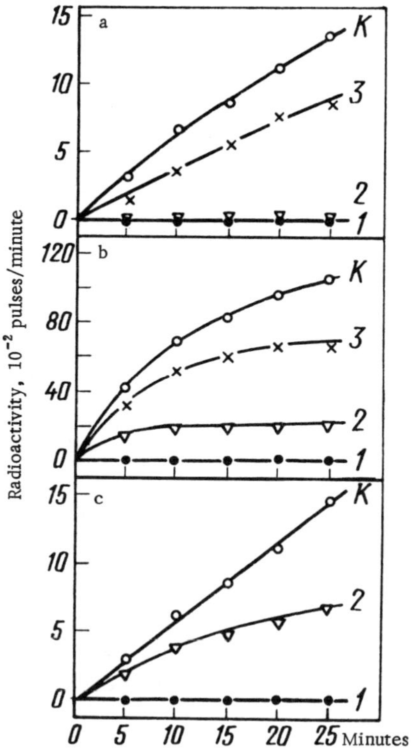

FIGURE 1. Effect of BDB on the synthesis of DNA (a), RNA (b), and proteins (c) in E.coli.

Concentration of BDB (in mM): 1) 100, 2) 25, 3) 2.5, K) control.

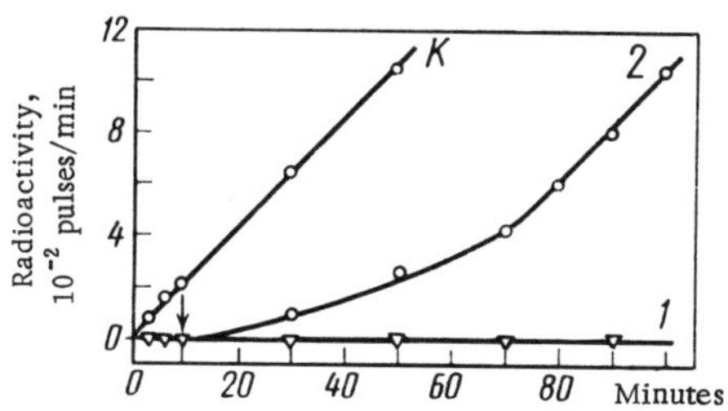

FIGURE 2. Restoration of DNA synthesis in E. coli after removal of the inhibitor:

1) 25 mM BDB, 2) washed cells; K) control; the arrow indicates the moment of washing.

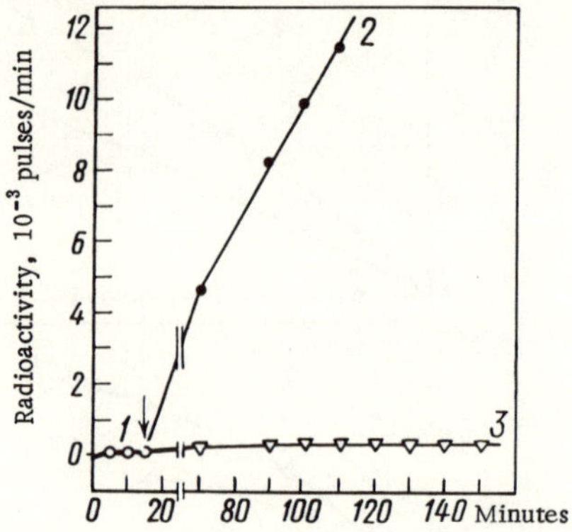

FIGURE 3. Effect of chloramphenicol on the restoration of DNA synthesis in E. coli after removal of the inhibitor:

1) 15 mM BDB; 2) washed cells; 3) washed cells with 100 μg/ml chloramphenicol; the arrow indicates the moment of washing.

If E. coli cells in which the DNA has been previously labeled with ^{14}C-thymidine are infected with phage T4, the label does not disappear from the acid-insoluble fraction. Experiments performed with mutant phage in which DNA polymerase is defective have shown that treatment with diazoketone does not inactivate nucleases but destroys bacterial DNA (Figure 4).

The retention of the thymidine label in this case indicates that in the presence of diazoketones phage DNA is synthesized from the deoxyribose monophosphates formed by the breakdown of bacterial DNA. In other words, diazoketones do not affect the enzymes participating in DNA synthesis after the stage of reduction of ribose derivatives to deoxyribose derivatives.

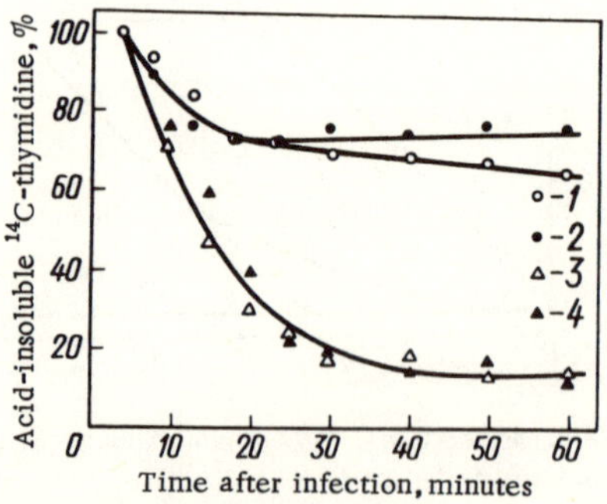

FIGURE 4. Effect of BDB on the distribution of labeled thymidine during the infection of E. coli with wild type and mutant phage:

1) phage T4B; 2) phage T4B, 25 mM BDB; 3) phage 1141; 4) phage 1141, 25 mM BDB.

Since diazoketones inhibit mainly the synthesis of DNA, blockage at an earlier stage, when DNA and RNA have common precursors, is scarcely possible. It appears that the formation of deoxyribose monophosphates from bacterial DNA during phage infection provides an alternative pathway bypassing the blockage of DNA synthesis at the reduction stage.

The presence of labeled 5-hydroxymethylcytosine in the DNA hydrolysate confirms that the retention of the label in the acid-insoluble fraction is due to DNA synthesis by the phage.

According to a preliminary analysis of the pool of nucleic acid precursors in E. coli cells in which DNA synthesis has been inhibited by 1,4-bis-diazoacetylbutane, the amount of UDP-glucose in the experiment is four times greater than in the control. This accumulation of diphosphates could result from the inactivation of ribonucleotide reductase, since reduction takes place at this level.

These experimental data present strong evidence that the essence of the biological activity of diazoketones is inhibition of DNA synthesis as a result of inactivation of the ribonucleotide reductase system.

BIBLIOGRAPHY

Emanuel', N.N., L.L. Gumanov, N.P. Konovalova, R.F. D'yachkovskaya, G.N. Bogdanov, L.S. Vasil'eva, L.A. Plugina, and Zh. I. Orlova. — Doklady AN SSSR, Vol. 183 (1968), 724.

Gumanov, L.L., S.D. Kononova, and N.P. Norenko. — In: "Supermutageny," p.49. Moscow, "Nauka," 1966.

Kozlov, L.V., L.M. Ginodman, and V.P. Orekhovich. — Biokhimiya, Vol. 32 (1967), 1011.

Rapoport, I.A. — Doklady AN SSSR, Vol. 130 (1960), 1134.

Zoz, N.N. — In: "Supermutageny," p.93. Moscow, "Nauka," 1966.

UDC 576.36

INTRACHROMOSOMAL PROCESSES AND THE FORMATION OF CHROMOSOMAL ABERRATIONS

A.P. AKIF'EV
Institute of General Genetics, Academy of Sciences of the USSR, Moscow

Data concerning the relationship between the formation of structural mutations of chromosomes and normal intracellular genetic processes are discussed. Experimental evidence that premutational potential changes in chromosomes are distinct from primary molecular lesions of DNA is presented. It is suggested that the transition of potential changes to a detectable form (chromosomal aberrations) is brought about by a genetic process checking the nucleotide sequence of DNA.

Most of the information available on the various aspects of the structure and function of the gene has been derived from the analysis of spontaneous and induced mutations. Yet one must admit that the vast amount of data relating to the structural mutations of chromosomes has up to now contributed little towards an understanding of the specific organization of the genonema of eukaryotes. This might logically be interpreted as an indication that the analysis of structural mutations cannot be regarded as a method of research into the structure of chromosomes. Such a conclusion is, however, quite premature. Cytological experiments designed in particular to determine the number of subunits (Makarov et al., 1968) clearly indicate that the chromosome of eukaryotes is single stranded, and this conclusion agrees with recent data on the renaturation kinetics of denatured DNA (Laird, 1971). The same experiments also demonstrated quite clearly that the chromosome (the chromatid in S and G_2) functions as a single entity during the development of structural mutations.

We suggest in view of these facts that while the cytogenetic approach does not provide a detailed picture of the entire sequence of molecular events accompanying the appearance of chromosomal breaks, it may help to establish some general principles of a new approach to the mechanism of the development of structural mutations and the processes determining these mutations.

Only a few authors still uphold the original version of the "breakage-reunion" concept. However, its alternative, the exchange hypothesis, has little value except in the sense that it treats exchange as a general genetic mechanism responsible for the appearance of chromosomal mutations. Modern hypotheses on the origin of chromosomal aberrations essentially regard these phenomena as associated with regular intrachromosomal genetic processes. Brewen and Peacock (1969) showed recently that the reunions of DNA polynucleotide strands during isolocus fusions are not random. This means that chromosomal aberrations are created by a mechanism which recognizes the complementary DNA strands. Of particular interest in this connection are the experimental studies of chromosomal breaks induced by nucleotide antimetabolite inhibitors outside the DNA replication stage (Aingorn and Akif'ev, 1970). These agents do not react directly with DNA but cause a deficiency of certain precursors, which in turn inhibits DNA synthesis. On these grounds the cytogenetic effect of the inhibitors outside the S stage can be interpreted as evidence of intrachromosomal processes involving a temporary disruption of the continuity of the DNA molecule.

The essential and permanent component of the chromosome of higher animals and plants is the giant DNA molecule which attains a length of several centimeters. The main feature of the genetic organization of the chromosome of eukaryotes is the presence of a discrete structure composed of numerous polymeric functional-genetic units or chromomeres which are apparently the units of replication. The autonomy of the chromomeres is well illustrated in cases of selective gene amplification, when the polycystronic nucleolus-forming locus or chromomere undergoes multiple replication during the prophase of meiosis.

Various examples of radical reorganization of chromosomal structure occur under natural conditions. For example, diminution of the chromatin — a phenomenon which appears to be more widespread than was assumed previously — may involve the removal of considerable parts of the genonema not only in the distal parts of the chromosome but also proximally (i. e. near the centromeres) (Beermann, 1966). From a strictly phenomenological viewpoint this process closely resembles the events following the action of radiation and alkylating agents on the cell. However, the process of chromatin diminution differs sharply from spontaneous mutations of chromosomes in its regularity and its strict association with specific stages of cleavage in ascarids and cyclopoids. These peculiarities of the diminution process suggest that the change in the chromosomal structure is brought about by a particular intracellular enzymatic mechanism functioning during one of the stages of manifestation of the genetic code of the species. It is indeed regrettable that the process of chromatin diminution, though discovered long ago, has not yet been subjected to experimental genetic analysis.

Recently many scientists working in the field of mutations have inclined towards the view that aberrations arise in close connection with DNA repair or genetic recombination. Mutagens induce potential lesions in the chromosomes which turn into definitive aberrations during certain periods of the mitotic cycle (Dubinin and Akif'ev, 1970). Any molecular interpretation of the potential change assumes its specific association with a particular genetic process. In other words, if the molecular basis of the potential mutation can be discovered it will be possible to establish with a high degree of certainty the intrachromosomal event causing its conversion into a definitive mutation. Analysis of the strictly mutational aspect of the processes leading to aberrations may therefore shed light on the intrachromosomal processes which normally occur in the cell and are dictated by the necessity of reorganizing the genonema.

Let us examine the recent evidence regarding the nature and manifestation of potential changes. At present there are insufficient grounds for assuming that chromosomal aberrations can arise on the basis of erroneous repair of primary molecular lesions of DNA. DNA repair in mammalian cells can take place throughout interphase and even during mitosis itself (Stich and San, 1970), whereas aberrations appear, as a rule, at strictly defined moments of the mitotic cycle depending on the nature of the mutagen used. Aberrations caused by delayed-action mutagens appear only after the treated cells have completed the phase of DNA synthesis. If the lesion is inflicted during the G_2 phase, the aberrations appear in the next mitotic cycle. It has been shown experimentally that there is no relationship between the formation of aberrations and the repair synthesis of DNA (Wolff and Scott, 1969). The fundamental principle of the repair hypothesis of chromosomal aberrations — the equal probability of cleavage of the normal and damaged DNA strands by endonuclease — is no longer tenable in view of recent evidence (Paribok and Tomilin, 1970; Carrier and Setlow, 1970) which proves that endonuclease attacks the damaged strand only. On the other hand, one should not go so far as to assume that there is no relationship whatsoever between primary molecular lesions and chromosomal aberrations. To do so would be tantamount to saying that mutations can appear without a mutagenic effect. Moreover, the data of Wolff and Scott (1969), who found no difference in chromosome mutations between normal human cells and those of a patient suffering from xeroderma pigmentosum, i. e. mutant at one of the initial stages of dark repair, are not sufficient for such a far-reaching conclusion since there is no doubt that the efficiency of the repair of primary molecular lesions in DNA is a major factor in the final yield of chromosomal mutations. Preliminary data on the different quantitative level of aberrations in normal humans and patients suffering from xeroderma pigmentosum have already been published (Mekhel'son, 1970).

The central stage of mutagenesis, defined as the period between the infliction of the primary lesion in the DNA and the appearance of visible structural disturbances, has been studied in detail during the last five years in the laboratory of N. P. Dubinin. It was found that this stage, during which the mutations exist in a potential form, can last for days, months and even years. N. P. Dubinin and L. G. Dubinina (1968) have examined in detail the kinetics of chromosomal aberrations in the cells of stored dry seeds of *Crepis capillaris* treated with the alkylating agents ethylenimine and triethylenemelamine. These experiments showed that the yield of structural mutations varies in the course of storage; in a number of independent tests the mutability level was observed to fluctuate regularly, alternating between peaks and troughs. These findings gave reason to assume that the "probability of manifestation" (Dubinin and Tarasov, 1966), measuring the transition to definitive mutations, varies with the duration of the potential change. It must be stressed that such a conclusion could be drawn only from the results of experiments performed on synchronous cell systems. The undulatory course of chromosomal mutability was subsequently confirmed in the same laboratory on another synchronous population — rat hepatocytes (Movsesyan and Akif'ev, 1969). The repair of primary molecular lesions of DNA in eukaryote cells takes place practically within 24—48 hours (Harikawa et al., 1968), whereas the undulating fluctuations in the mutability of rat hepatocytes last 25 days at least. This comparison similarly suggests that primary molecular lesions in the chromosomes are directly followed by properly premutational potential changes which may develop into definitive mutations during particular stages of the cell cycle, the probability of this fluctuating quite regularly according to a pattern determined by the object and mutagen dose. The undulatory kinetics of the yield of structural mutations of chromosomes in synchronous cell populations reflects the phenomenological image of the state of the premutational potential changes, i.e. the probability of their manifestation. L. E. Nemirovskii, V. V. Klimenko and A. P. Akif'ev have shown experimentally that a shift in the wave of the aftereffect of alkylating mutagens in liver cells can be achieved in the G_0 phase by means of inducers or depressors (phenobarbital, tryptophan and caffeine) causing a partial recoding of the genome and an intensification of transcription. Although the reasons for the undulatory fluctuations in chromosomal mutability remain largely unknown, the very existence of this phenomenon indicates clearly that the aberrations do not arise directly from primary lesions. N. P. Dubinin and his colleagues have established that the prolongation of the DNA synthesis phase in germinating seeds of C. capillaris achieved by the use of 5-aminouracil — a thymidylic acid inhibitor — practically eliminates the effect of the alkylating mutagen ethylenimine and reduces the number of aberrations to a negligible level. At first they interpreted this finding as an indication of repair during the S phase. This assumption, however, proved erroneous since repeated exposure to 5-aminouracil during the G_2 prophase restored the cytogenetic effect of ethylenimine. In other words, the "repair" does not lead to a true restoration of the original structure of the DNA but converts the potential changes to a latent form, so that if the S phase is greatly prolonged the probability of manifestation of the potential change is almost zero. Simultaneously R. A. Stepanenko and also K. Ergashev and V. V. Yurchenko, using another inhibitor of DNA synthesis — 5-fluorodeoxyuridine (FUDR) — at the G_2 stage of the second nuclear cycle, showed that some of the latent potential changes induced by ionizing radiation and alkylating agents persist during two cycles of DNA replication. Dubinin and his colleagues (1971) demonstrated autoradiographically that new aberrations appear with the participation of chromatids containing a template damaged by the mutagen, as well as in chromatids entirely synthesized after exposure to the mutagen. This constitutes a remarkable similarity between the self-replicating gene instability described by Auerbach (1946) and the chromosomal mutability of somatic cells persisting throughout several cell generations. The phenomenon of self-replicating chromosomal instability was shown much earlier in experiments performed on fish embryos (Dubinin and Tarasov, 1969) revealing the appearance of chromosomal breakages in tens of cell generations (in the gastrula stage) after exposure of the spermatozoa to radiation or thio-TEPA. It is obvious that two damaged templates present at that time in the embryo (in two cells, if sister chromatid exchanges are not taken into account) cannot cause the high level of chromosomal aberrations observed (approximately one per cell). Although there is some discrepancy between the data obtained with mammalian and fish embryo cells and with plant cells, at least in the first case there is no doubt that replication of premutational potential changes occurs leading to chromosomal aberrations. This leads to the quite plausible assumption that the molecular basis of premutational potential changes may be changes in the nucleotide sequence of the DNA, since there is no indication at present that anything else can be replicated in DNA. Once again, we are not speaking of a survival of primary molecular lesions in templates originally exposed to the mutagen. If the chromosomal breaks and exchanges arise on the basis of an altered nucleotide sequence, there must be an intrachromosomal process which checks this sequence, i. e. the primary structure of the DNA. The only known process of this kind is genetic recombination, which involves the pairing of complementary strands and the correction of molecular heterozygotes. Hence the search for parallels between specific stages of genetic recombination and the appearance of structural mutations of chromosomes. The analysis of chromosomal mutations and of sister chromatid exchanges in the cells of a desynaptic mutant of barley has so far failed to give clear-cut results (Evans and Swetlinsky, 1970). This should not be taken as grounds for a pessimistic evaluation of the above approach to the mechanism of aberrations, however, since the meiotic and mitotic crossovers are not necessarily identical at all stages (for example, the synaptonemal complex is lacking in the second case). On the other hand, it is not known which particular defect leads to the decrease in the probability of crossing over in the mutant examined. Moreover, the same authors report a positive correlation between chromosomal mutations and the frequency of chiasmata in a desynaptic human mutant. This trend of research is only just beginning, and we believe that the genetic analysis of the mechanism of formation of structural mutations will be no less fruitful than the study of repair and recombination mutants in bacteria and yeast.

We have already mentioned that the cytogenetic effects of inhibitors outside the S phase of the mitotic cycle suggest the existence of particular genetic processes requiring an additional (extrareplicative) synthesis of DNA. Of particular interest in this connection is the modifying effect of inhibitors during the G_2 prophase in cells previously treated with a mutagen. Under such conditions the number of aberrations (simple and isolocus breaks) usually rises sharply. As we pointed out previously, this proves the existence of latent potential changes, although the manifestation of these potential changes requires the inhibition of the additional DNA synthesis. Thus, under normal conditions, an intrachromosomal process occurs during the G_2 prophase in which premutational potential changes are discerned, i. e. the nucleotide sequence of DNA is checked at this time, in accordance with the concept outlined above. The sensitivity of this hypothetical process to inhibitors of DNA synthesis can likewise be interpreted as an indication of its similarity to crossing over.

Generally speaking, preparation for mitosis may be assumed to involve the gathering together of particular subunits (replicons) which are relatively autonomous during the phase of DNA synthesis. The sister chromatid exchanges and the "isolabeling" phenomenon (Peacock, 1963) probably reflect this assembly of the mitotic chromosome. The assembly involves the repair of quite large gaps or even the synthesis of entire areas, as is assumed with respect to the correction of heteroduplexes (Fincham and Holliday, 1970). If premutational potential changes are previously induced in the chromosome, there will naturally be a sharp increase in the number of defects in the secondary structure of DNA caused during the check of its primary structure; in this case inhibition of DNA synthesis will prolong the duration of single-strand areas and thus increase the yield of structural mutations under certain conditions. From this point of view it might be expected that intensification of the additional synthesis of DNA would decrease the number of aberrations. This indeed appears to be so. Together with K. Ergashev we were able to show recently that the addition of excess thymidine during the G_2 prophase markedly lowers the yield of chromosomal aberrations induced by nitrogen mustard in the G_1 phase of cells of C. capillaris. This fact in itself proves that at least with respect to chemical mutagenesis aberrations are definitively formed during the period of premitotic preparation of the chromosome.

An increasing array of recent evidence obtained by strictly genetic or cytogenetic procedures proves the existence of hitherto unknown genetic processes which have no precedent in molecular biology. For example, the genetic analysis of mutable genes in higher plants (Fincham, 1970) shows that the appearance of new alleles in this case cannot be attributed to strictly mutational events but represents the unmasking of preexisting genetic information.

We believe that the molecular genetic analysis of the mechanism of structural mutations of chromosomes will shed light on many important aspects of the functional organization and evolution of the genetic apparatus of eukaryotes.

BIBLIOGRAPHY

Aingorn, E.D. and A.P. Akif'ev. — Proceedings of the Scientific Conference of the 2nd Medical Institute, 3. Moscow, 1970. (Russian)

Auerbach, Ch. — Proc. R. Soc. Edinb. B, Vol. 62 (1946), 11.

Beermann, W. — Genetics 54, No.2 (1966), 567.

Brewen, J.G. and W. J. Peacock. — Proc. Natn. Acad. Sci. U.S.A. 62, No. 2 (1969), 389.

Carrier, W.L. and R.B. Setlow. — J. Bact. 102, No. 1 (1970), 178.

Dubinin, N.P. and A.P. Akif'ev. — Usp. Sovrem. Biol. 69, No. 2 (1970), 272.

Dubinin, N.P., A.P. Akif'ev, and T.A. Vishtynetskaya. — Genetika, Vol.7 (1971), 5.

Dubinin, N.P. and L.G. Dubinina. — Genetika 7, No. 9 (1968), 3.

Dubinin, N.P. and V.A. Tarasov. — Usp. Sovrem. Genetiki, Vol.2 (1969), 3.

Fincham, J.R.S. — Proc. R. Soc. 176, No. 1044 (1970), 295.

Fincham, J.R.S. and R. Holliday. — Molec. and Gen.Genet.109, No. 4 (1970), 309.

Harikawa, M., O. Nikaidi, and T. Sugachara. — Nature 218, No. 5140 (1968), 489.

Laird, C. — Chromosoma 32, No.4 (1971), 378.

Makarov, V.B., A.P.Akif'ev, and R.A.Stepanenko. — Doklady AN SSSR 182, No.4 (1968), 949.

Mekhel'son, V.M. Trudy Inst. Cytol. Histol. Embryol. Acad. Sci., USSR. Moscow—Leningrad, "Nauka," 1970.

Movsesyan, M.V. and A.P.Akif'ev. — Doklady AN SSSR 188, No.3 (1969), 685.

Paribok, V.P. and I.V.Tomilin. — Doklady AN SSSR 195, No.2 (1970), 489.

Peacock, W.J. — Proc. Natn. Acad. Sci. U.S.A. 49, No.6 (1963), 753.

Stich, U.F. and R. San. — Mutation Res. 10, No. 4 (1970), 389.

Wolff, S. and D. Scott. — Expl. Cell. Res. 55, No. 1 (1969), 9.

UDC 576.36

THE INTRACHROMOSOMAL AND INTERCHROMOSOMAL DISTRIBUTION OF CHROMATID ABERRATIONS INDUCED BY CHEMICAL AGENTS: AN ANALYSIS OF "SYNTHETIC" KARYOTYPES IN VICIA FABA

R.RIEGER and A. MICHAELIS
Central Institute of Genetics and Crop Plants
Academy of Sciences of the German Democratic Republic
Gatersleben

The distribution of chromatid aberrations induced by ethanol (E) and maleic hydrazide (MH) is not random. The aberrations are concentrated in the heterochromatic areas of the chromosome although the particular chromosomal segments most frequently affected by aberrations vary according to the mutagen used.

The standard karyotype of V. faba (2n=12) is widely used for analyses of the intra- and interchromosomal distribution of chromatid aberrations induced by ionizing radiation or chemical agents. A mutagen-specific and nonrandom distribution of aberrations was established both within and among chromosomes (Evans, 1962; Kihlman, 1966; Rieger and Michaelis, 1967). One shortcoming of the standard karyotype of V. faba as a model for the study of the distribution of chromatid aberrations is the morphological similarity between the ten small chromosomes which greatly hampers analysis (Michaelis and Rieger, 1968).

For this reason the small chromosomes are treated as a single group in the majority of works dealing with the distribution of aberrations. In order to facilitate the identification of the small chromosomes and enable a thorough study of the distribution of chromatid aberrations both within and between the individual chromosomes we rearranged the standard karyotype and created several new chromosome sets or "synthetic" karyotypes in which all the chromosomes are easily distinguishable during the metaphase of mitosis. Some of these karyotypes were used for a study of the intra- and interchromosomal distribution of chromatid aberrations caused by ethanol (E) and maleic hydrazide (MH) and also of the effect of translocations on the participation of different chromosomal segments in chromatid aberrations induced by E. A detailed description of the conditions of treatment and the preparation of the material has been published earlier (Michaelis and Rieger, 1968).

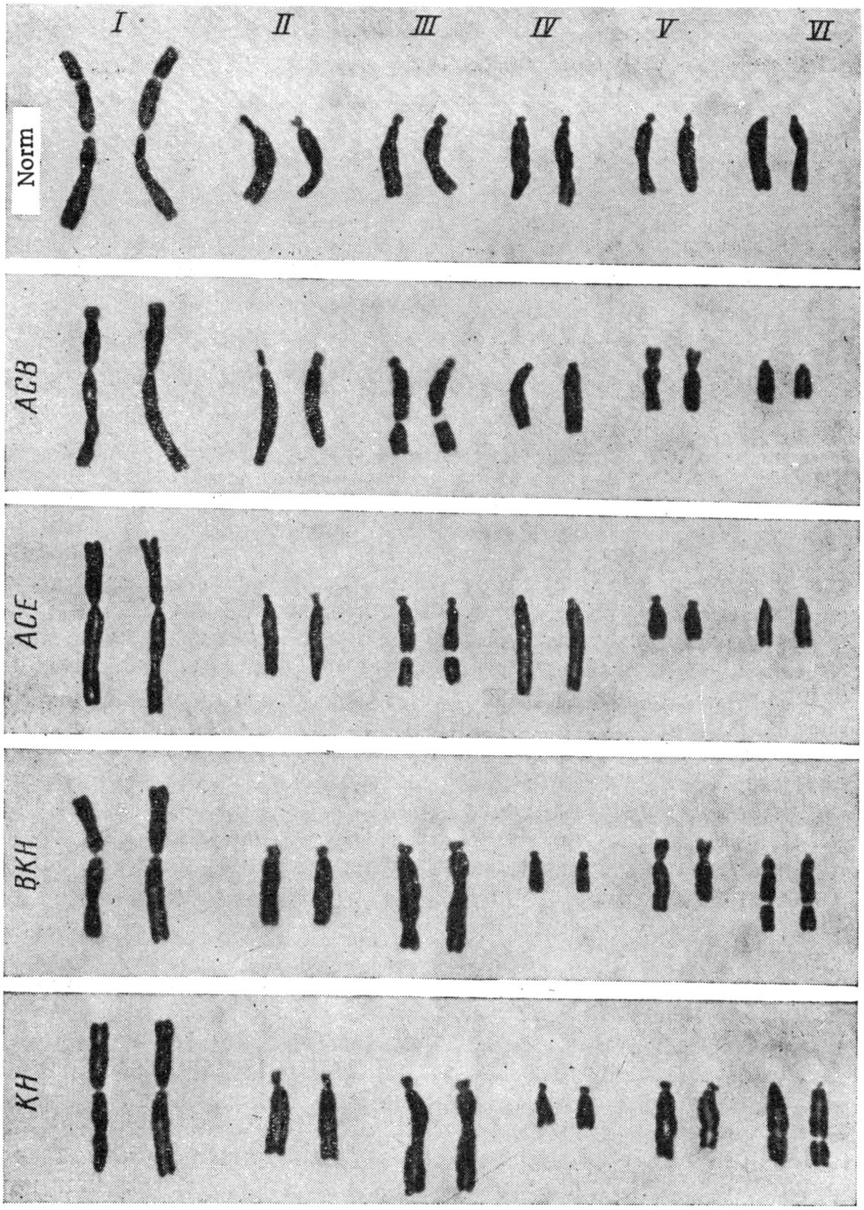

FIGURE 1. Karyograms of the normal and new karyotypes of Vicia faba.

The lettering (ACB, ACE, BKH, and KH) refers to the structural changes characterizing the new karyotypes.
A = TI − III_i, B = JV_i, C = T_1 I − VI_i, E = TIV − V_i, H = TIII − IV_i, K = T_2I−VI.

NEW KARYOTYPES

Figure 1 shows karyograms of new chromosome complements of V. faba, which differ from the standard karyotype by the presence of structurally altered chromosomes (marker chromosomes formed by interchromosomal exchanges and inversions).

For the mapping of the chromatid aberrations in the metaphase chromosomes, the standard and new karyotypes were divided into 20 segments, indicated in Figure 2. The first chromosome (I) of the standard chromosome complement (the satellite chromosome) contains eight segments (two of which have a primary and a secondary constriction), whereas each of the five small chromosomes of the chromosomal set usually treated as one (II—VI) is split into four segments of equal length. Since the two arms of chromosome I and the small chromosomes differ somewhat in length (Michaelis and Rieger, 1959), the actual size of the segments varies from one chromosome to another. The new karyotypes were divided into segments according to the same principle. Figure 2 shows the marker chromosomes of the new karyotypes and the position of the translocated segments.

RESULTS AND CONCLUSIONS

The agents used for the induction of aberrations — E and MH (1,2-dihydro-3,6-pyridazinedione) — cause exclusively delayed-type rearrangements (Michaelis et al., 1959; Darlington and McLeish, 1951; McLeish, 1953, 1954; Evans and Scott, 1964).

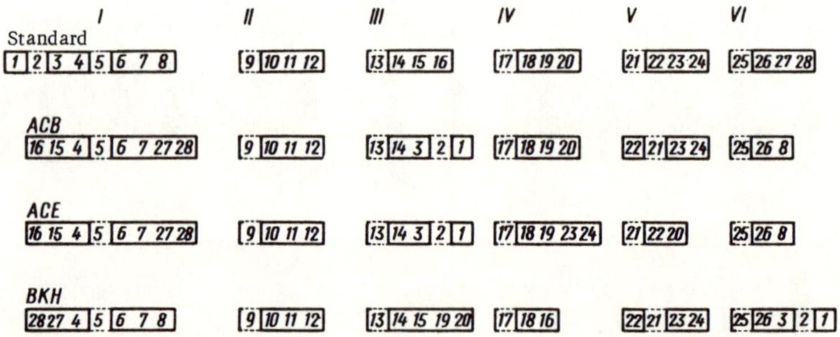

FIGURE 2. Diagram of the standard and new karyotypes of Vicia faba.

The sequential arrangement of the segments shows the structural changes which have taken place in the new karyotypes. The dotted lines indicate the position of the centromeres and the area of the nucleolar organizer (segment 2).

It was found that the formation of aberrations is enhanced by a rise in the temperature and requires the synthesis of chromosomal DNA. The rate of formation of aberrations decreases under conditions of oxygen starvation and on exposure to respiratory poisons or agents inhibiting oxidative phosphorylation (Kihlman, 1956; Rieger and Michaelis, 1960b). The following types of aberrations were distinguished: isolocus breaks (i), duplicative deletions (dd), intercalary deletions (d), and chromatid translocations (t) (Figure 3). In standard karyotypes treated with MH, aberrations were much more frequent in the two satellite chromosomes than in the group of small chromosomes (McLeish, 1953); the aberrations induced by E show the inverse picture (Rieger and Michaelis, 1960a).

The interchromosomal distribution of chromatid aberrations induced by E and MH

The interchromosomal distribution of chromatid aberrations observed after treatment of the new karyotype ACB (Figures 1 and 2) with E and MH was found to be nonrandom and specific to each agent (Michaelis and Rieger, 1968). Aberrations of the i and t types appeared mostly in chromosome IV after treatment with E and in chromosome III (the satellite chromosome in this karyotype) in the case of MH. In the presence of both agents, the majority of dd and d aberrations appeared in chromosome V. This pattern of the interchromosomal distribution of aberrations did not depend on the duration of the restoration period after treatment and persisted in the case of previous treatment with EDTA, which serves as sensitizer with respect to both mutagens (Michaelis et al., 1962; Rieger et al., 1963). These results lead to the following conclusions: a) chromosome IV (in the case of treatment with E) and chromosome III (in the case of MH) show a marked specificity with respect to the mutagen as regards the induction of chromatid aberrations; b) chromosome V is distinguished by a relatively high yield of aberrations after treatment with E or MH. The latter chromosome shows no specificity with respect to the agent used, but displays mostly duplicative deletions (dd) and intercalary deletions (d), i. e. a specific aggregate of two definite types of inner exchange. To our knowledge, such a predominant appearance of a specific type of aberration in a particular chromosome has not yet been described in any object.

FIGURE 3. Types of chromatid aberrations observed after exposure of V. faba with a standard chromosome set to MH or E:

i — isolocus break; dd — duplicative deletion; d — intercalary deletion; t — chromatid translocation (magnification 4,000 ×).

In standard karyotypes treated with MH, the ratio of chromatid aberrations between the large chromosome (I) and the small chromosomes (II—VI) was about 1:0.3 (McLeish, 1953), and there was a predominant concentration of the aberrations in a particular area (the heterochromatin) of the nucleolar arm of chromosome I. Owing to the translocation of this area together with the distal part of the chromosome, including the secondary constriction and the satellite, in the ACB genotype this segment is contained within chromosome III. In ACB karyotypes treated with MH, the ratio of chromatid aberrations between chromosome I and the other (II—VI) chromosomes was found to be 1:7.3 (compared with 1:0.3 in the standard karyotype). Thus, the transfer of part of the nucleolar arm from chromosome I to chromosome II is accompanied by a shift in the interchromosomal distribution of the aberrations (Michaelis and Rieger, 1968).

The intrachromosomal distribution of chromatid aberrations induced by E and MH

The nonrandom, mutagen-specific distribution of chromatid aberrations induced by E and MH between the different chromosomes is linked with the concentration of the aberrations in specific chromosomal segments. As demonstrated in Figure 4, as regards the agent-specific distribution of aberrations, these are most frequent in segments 14 and 19 of karyotype ACB (Rieger and Michaelis, 1970). In chromosome IV, which is structurally identical to the corresponding chromosome of the standard karyotype, the number of E-induced chromatid aberrations is greatest in segment 19, whereas the MH-induced aberrations are predominantly concentrated in segment 14 of chromosome III. Chromosome II is the satellite chromosome of the ACB karyotype as a result of the I/III exchange. Owing to this rearrangement, the segment most susceptible to MH-induced aberrations in chromosome I of the standard karyotype is transferred to chromosome III. This transfer involves a corresponding transition of the hot point and a shift in the intra- and interchromosomal distribution of the aberrations. It appears from these findings that the agent-specific concentration of MH-induced aberrations depends more on the nature of the "hot" segment than on its position in the karyotype. However, another possibility must be considered. It is known that MH accumulates in the nucleolus (Gallaghan and Grun, 1961). Consequently, the preservation of the particular susceptibility of the segment to aberrations when transferred from chromosome I to III may be due to the simultaneous transfer of another chromosomal area responsible for the formation of nucleoli — the site of accumulation of the mutagen. The following experiments with E confirm this hypothesis.

Determination of the intrachromosomal distribution of aberrations in the ACB karyotype shows that agent-specific aberrations appear mostly in segments 14 and 19, whereas the majority of aberrations in segment 23 of chromosome V are induced by both E and MH. Thus, the last segment typically shows aberrations which are not agent-specific; moreover, segment 23 is a "hot spot" of aberrations in the term of duplicative deletions (dd) and intercalary deletions (d). For this reason the interchromosomal

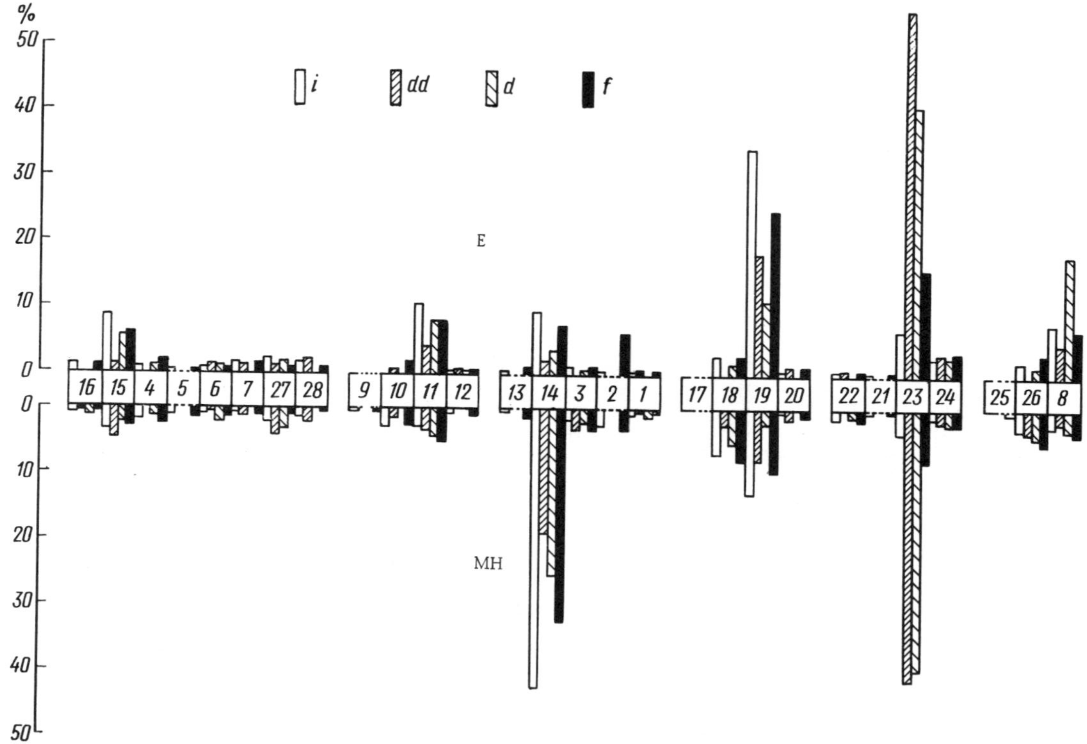

FIGURE 4. Participation of the 28 segments of the ACB karyotype (as %) in chromatid aberrations induced by E and MH. Lettering as in Figure 3.

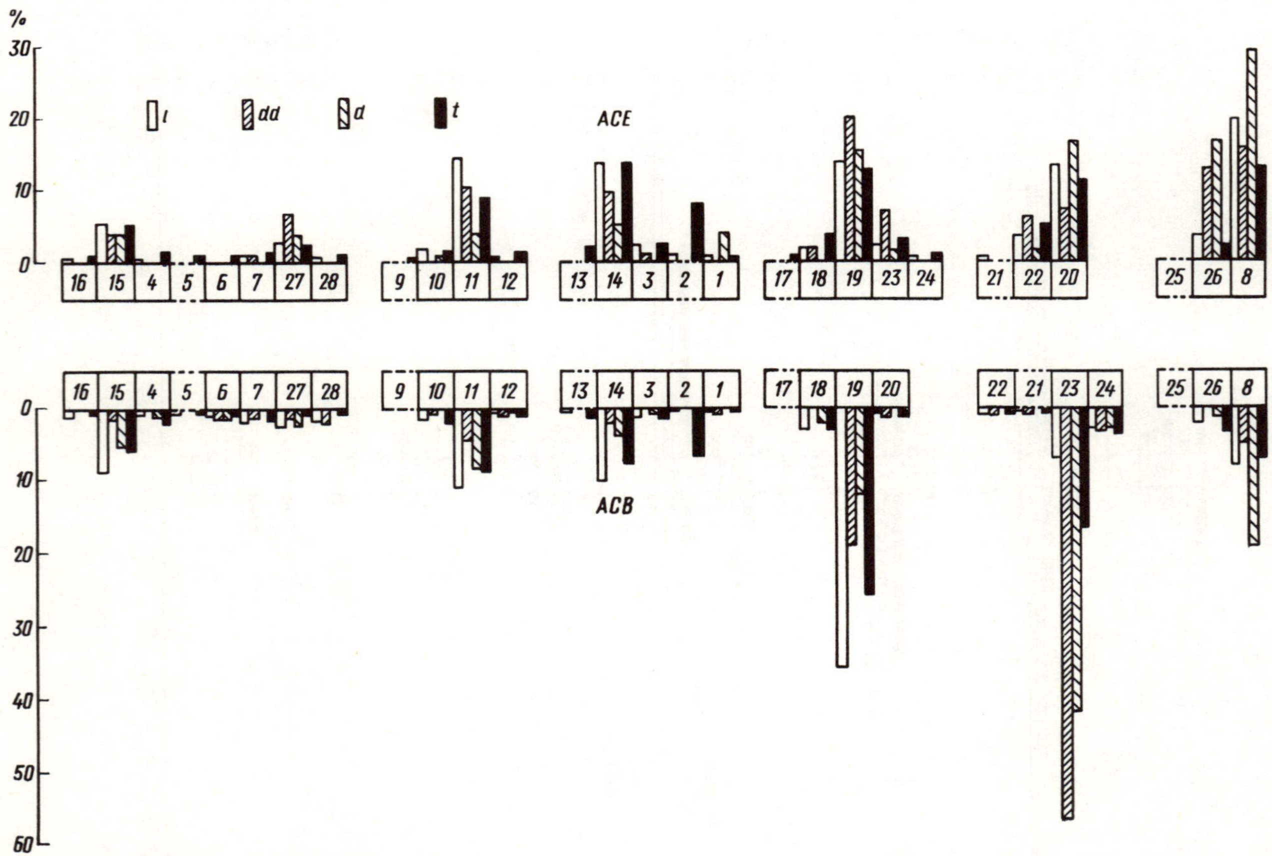

FIGURE 5. Participation of the 28 segments of the ACB and ACE karyotypes (as %) in chromatid aberrations induced by E.

For ACB: i = 1,378, dd = 162, d = 311, t = 649. For ACE: i = 740, dd = 106, d = 80, t = 446. Lettering as in Figure 3.

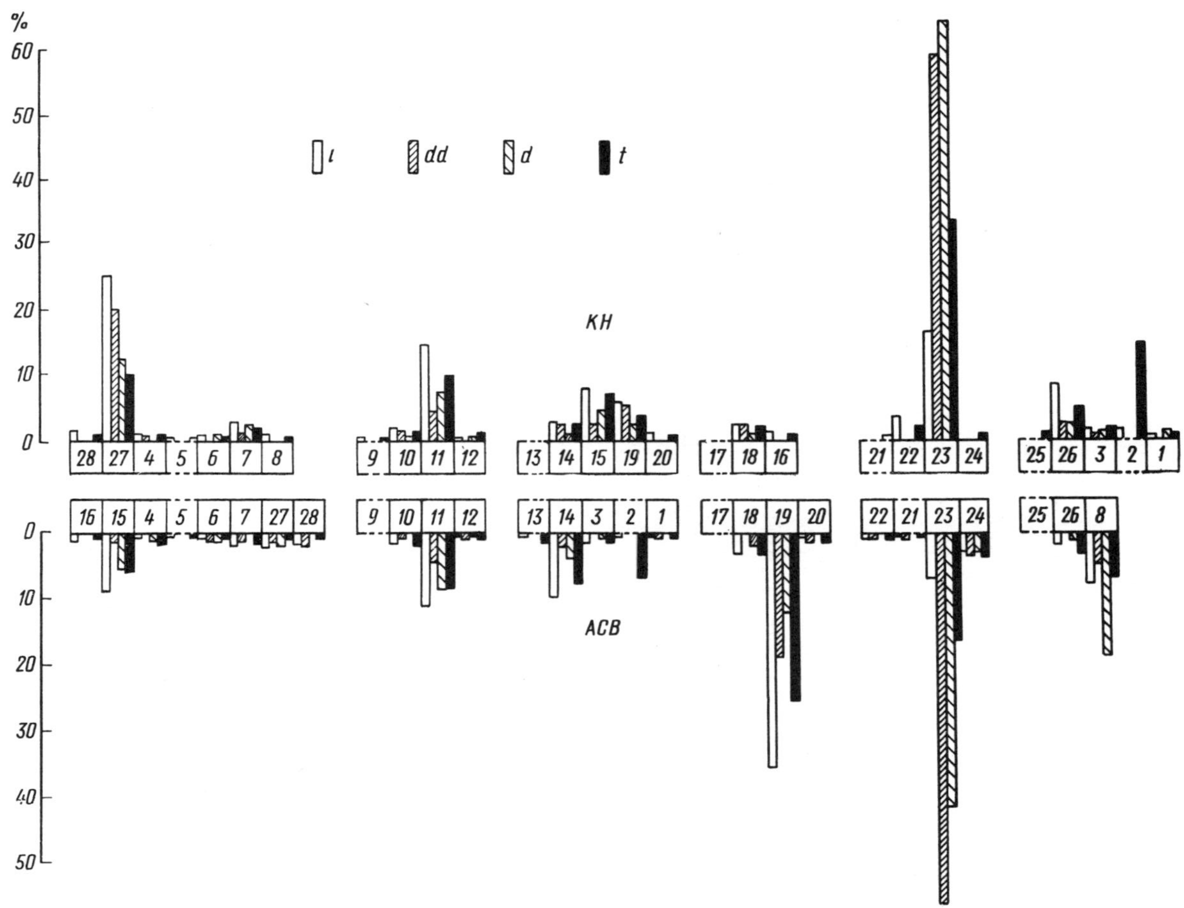

FIGURE 6. Participation of the 28 segments of the ACB and KH karyotypes (as %) in the formation of chromosomal aberrations induced by E.

For ACB: i = 1,378, dd = 162, d = 311, t = 649. For KH: i = 470, dd = 289, d = 213, t = 497.
Lettering as in Figure 3.

distribution of aberrations after induction with E and MH was interpreted in terms of a predominant concentration in chromosome V. It would be premature to discuss the possible causes of such a concentration of a particular type of chromatid aberration; however, the observed effect cannot be attributed to a structural rearrangement caused by an inversion of chromosome V in the ACB karyotype. A similar concentration of aberrations was found in the KH karyotype (Figures 6, 7), which contains a standard chromosome V.

Three other new karyotypes (ACE, BKH, and KH) were examined in order to determine in greater detail the distribution of the aberration "hot spots" showing agent specificity and assess the relationship between the position of a given segment in the karyotype and its capacity to act as a hot spot. These karyotypes were treated with E and the intrachromosomal distribution of aberration hot spots was compared with that of the ACB karyotype.

ACE karyotype

This karyotype (Figure 5) differs from the ACB karyotype with respect to the position of segment 23 (which is situated in chromosome V in ACB and in chromosome IV in ACE). As a result of this transfer, segment 23 ceases to be a hot spot and no longer shows the concentration of dd and d type aberrations characterizing the ACB karyotype. The transfer of segment 20 from chromosome IV (ACB karyotype) to V (ACE karyotype) causes a marked increase in the number of aberrations observed in the latter chromosome after treatment with E.

KH and BKH karyotypes

These two karyotypes differ from one another in only one aspect: KH contains a standard chromosome V, whereas this chromosome is inverted in BKH (as in ACB). Treatment of the KH karyotype with E leads to the following interesting phenomenon. The transfer of segment 27 from one arm of chromosome I to the other arm leads to a sharp increase in the number of aberrations induced by E in this, whereas the transfer of the "hot" segment 19 (ACB and ACE) from chromosome IV to chromosome III causes a marked decrease in the number of aberrations in the latter chromosome.

Comparison of the interchromosomal distribution of aberrations in the KH and BKH karyotypes shows that the distribution of E-induced aberrations is highly reproducible and that the structural changes in chromosome V in the BKH karyotype (as well as in ACB) do not reduce the high frequency of aberrations in segment 23 (Figures 6, 7).

The main conclusion from these preliminary findings concerning the effect of E on the four new karyotypes is that the change in the intrachromosomal distribution of aberrations resulting from the transfer of a segment depends not only on the properties of the segment itself but also on its position in the karyotype. This relationship deserves a thorough study involving a finer subdivision of the chromosome into segments; such work

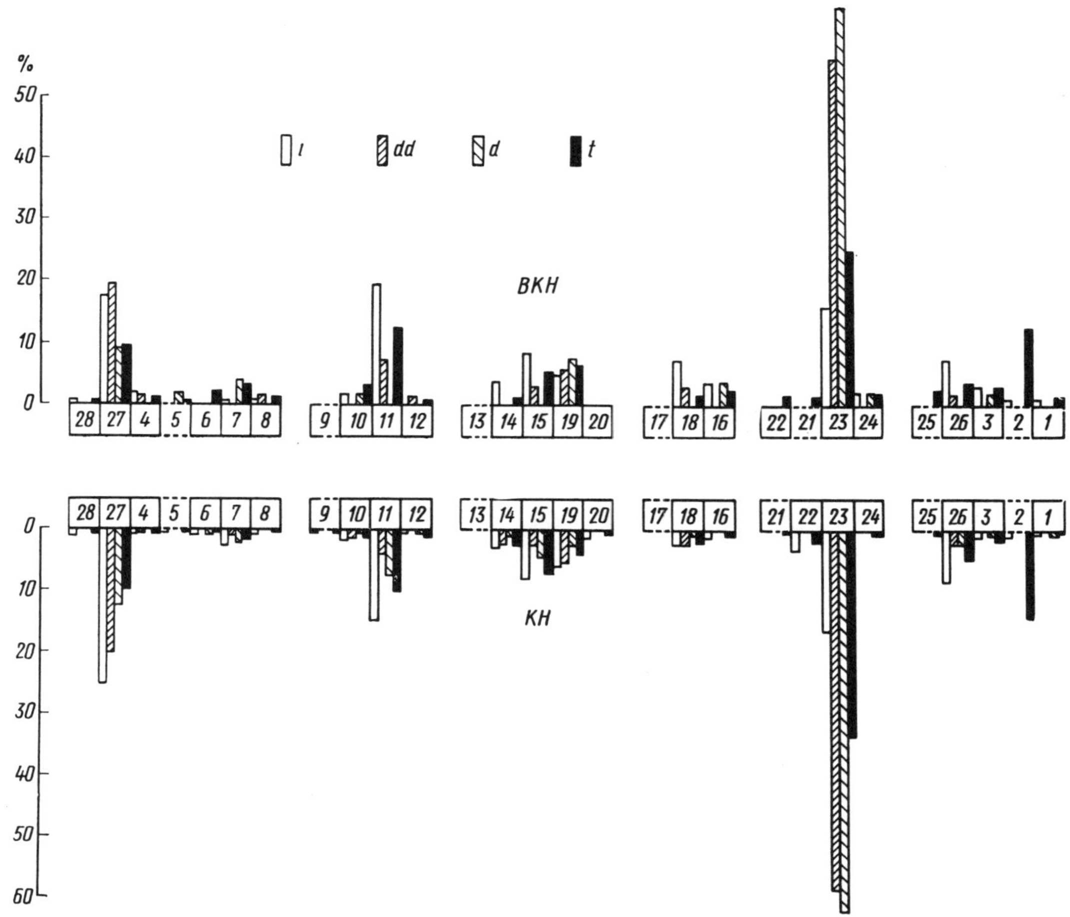

FIGURE 7. Participation of the 28 segments of the KH and BKH karyotypes (as %) in the formation of chromatid aberrations induced by E.

For KH: i = 470, dd = 289, d = 213, t = 487, For BKH: i = 109, dd = 68, d = 55, t = 94.
Lettering as in Figure 3.

is now under way with respect to a number of chromosomal areas. It is assumed that the difference in the number of E-induced aberrations from one chromosomal segment to another stems from the particular behavior of these segments during replication. There is no doubt that the segments most susceptible to chemically induced chromatid aberrations belong to the heterochromatin and replicate at a late stage (Kihlman, 1966; Rieger and Michaelis, 1967).

We are now studying the effect of the transfer of segments on their behavior during replication and the relationship between the frequency of aberrations in a given segment and the time of its replication.

CONCLUSIONS

1. The inter- and intrachromosomal distribution of chromatid aberrations induced by ethanol (E) and maleic hydrazide (MH) is not random and is agent-specific.

2. Experiments with four new karyotypes of V. faba show clearly that the position of E-induced chromatid aberrations depends largely on the properties of the chromosomal segments concerned and on the position of these segments in the karyotype. In all karyotypes treated with MH, transfer of the area of the nucleolar organizer is accompanied by the simultaneous transfer of the "hot" aberration spot. This may be due to the accumulation of MH in the nucleolar organizer. The pattern of the distribution of aberrations is not related to the duration of the restoration period following treatment with the mutagen and remains unchanged in the case of previous treatment with EDTA which sensitizes the cell with respect to both mutagens used.

3. The aberrations are concentrated in the heterochromatin areas of the chromosomes, but the particular chromosomal segments most susceptible to aberrations may vary markedly in accordance with the inducing agent used. Our knowledge on the structure and functions of the heterochromatin is still insufficient for us to draw any definitive conclusions as to the causes of these differences.

BIBLIOGRAPHY

Darlington, C.D. and J. McLeish. — Nature, Vol. 167 (1951), 407.
Evans, H.J. — Int. Rev. Cytol., Vol. 13 (1962), 221.
Evans, H.J. and D. Scott. — Genetics, Vol. 49 (1964), 17.
Gallaghan, I.I. and P. Grun. — J. Biophys. Biochem. Cytol., Vol. 10 (1961), 567.
Kihlman, B.A. — J. Biophys. Biochem. Cytol., Vol. 2 (1956), 543.
Kihlman, B.A. Actions of Chemicals on Dividing Cells. — New Jersey, Prentice-Hall, 1966.
McLeish, J. — Heredity, Vol. 6, Suppl. (1963), 393.
McLeish, J. — Heredity, Vol. 8 (1954), 385.
Michaelis, A., H. Nicoloff, and R. Rieger. — Biochem. Biophys. Res. Commun., Vol. 9 (1962), 280.
Michaelis, A., K. Ramshorn, and R. Rieger. — Naturwissenschaften, Vol. 46 (1959), 381.
Michaelis, A. and R. Rieger. — Züchter, Vol. 29 (1959), 354.
Michaelis, A. and R. Rieger. — Mutation Res., Vol. 6 (1968), 81.

Rieger, R. and A. Michaelis. — In: Chemische Mutagenese, H. Stubbe (Ed.) E. Baur-Gedächtnisvorl, 1, Abh. DAW, Kl. Medizin, **54**, 1960a.

Rieger, R. and A. Michaelis. — Mber. dt. Akad. Wiss. Berl., Vol. 2 (1960b), 290.

Rieger, R. and A. Michaelis. Die Chromosomenmutationen, Genetik, Grundlagen. — Ergebnisse und Probleme in Einzeldarstellungen, Vol.6, p.433. Jena, Fischer, (1967), 433.

Rieger, R., H. Nicoloff, and A. Michaelis. — Biol. Zbl., Vol. 82 (1963), 393.

Rieger, R. and A. Michaelis. — Mutation Res., Vol. 10 (1970), 162.

UDC 575.24

THE MECHANISM OF THE MUTAGENIC EFFECT OF HYDROXYLAMINE: THE EFFECT OF N-METHYLHYDROXYLAMINE

C. JANION and D. SHUGAR
Institute of Biochemistry and Biophysics,
Polish Academy of Sciences, and Department of Biophysics,
Warsaw University, Warsaw

The reaction of N-methylhydroxylamine (MHA) with cytosine yields only one product, incapable of forming the base pairs typical of cytosine or uracil. MHA does not react at all with 5-substituted derivatives of cytosine. The significance of these results with regard to establishing the mechanism of hydroxylamine-induced mutagenesis is discussed.

Mutagenic activity was found not only in hydroxylamine but also in its N- and O-methylated derivatives — N-methylhydroxylamine (CH_3NHOH) and methoxyamine (NH_2OCH_3). Both these hydroxylamine analogs were used in studies of the mechanism of the mutagenic effect of hydroxylamine (Phillips and Brown, 1967; Singer and Fraenkel-Conrat, 1969; Kochetkov and Budowsky, 1969). It was shown that methoxyamine reacts with cytosine (and with cytosine residues in polynucleotide chains) in the same way as hydroxylamine, but the mechanism of action of N-methylhydroxylamine remains obscure.

N-methylhydroxylamine is even more effective than methoxyamine in causing mutations in transforming DNA but is not mutagenic with respect to bacteriophageT4 (Freese and Freese, 1964; Freese et al., 1967). Moreover, N-methylhydroxylamine has no mutagenic effect on tobacco mosaic virus RNA (Singer and Fraenkel-Conrat, 1969). It was shown in vivo at least in one case that the mutagenic activity of N-methylhydroxylamine is much less specific than that of hydroxylamine or methoxyamine, since this agent induced both types of base pair transitions (Malling, 1967). In view of the discrepancy between these findings it was necessary to examine the products formed by the reaction of some cytosine analogs and 4-alkoxyuracil with N- and O-methyl derivatives of hydroxylamine. The results of such an analysis might resolve at least in part some of the above contradictions.

The reaction between NH_2OH and cytosine residues yields the products II and III via the hypothetical intermediary compound I; compound III is unstable and is readily converted into II by the elimination of NH_2OH from the 5,6 bond of cytosine:

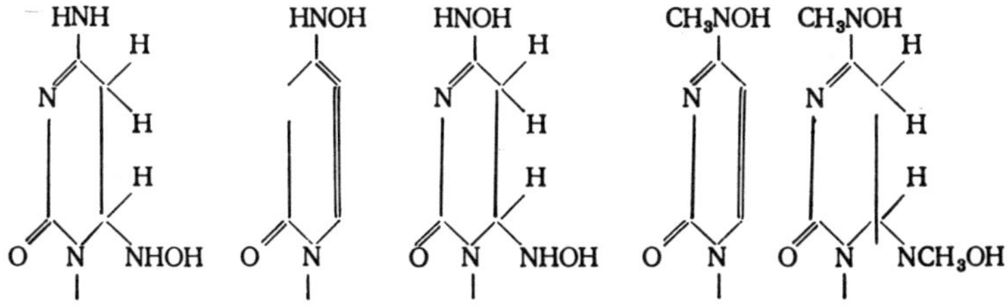

I II III IV V

If the C_5 hydrogen of cytosine is substituted with an alkyl radical (methyl or oxymethyl), the only product of the reaction is the 5-alkyl analog of II (Janion and Shugar, 1965).

It was assumed (Phillips and Brown, 1967) that the mutagenic effect of the product (or products) formed by the action of hydroxylamine on cytosine is due to the formation of a tautomeric imino form (IIa) of this product:

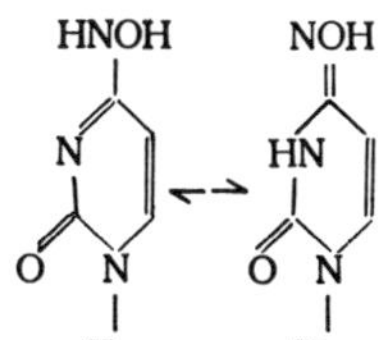

II IIa

Such a form would probably pair like uracil. The fact that N-methylhydroxylamine shows no mutagenic activity with respect to phage T4 was regarded as confirmation of this scheme, since the analogs of types IV and V cannot exist in an imino form (Phillips and Brown, 1967).

The observed mutagenic activity of N-methylhydroxylamine with respect to transforming DNA was attributed to the formation of product VI (similar to I) by reaction of the agent with cytosine; amino-imino tautomerism of the NH_2 group is possible in this case, as shown below.

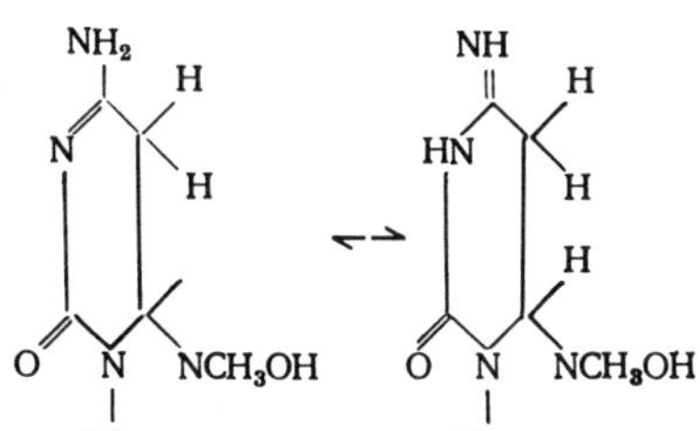

VI VIa

In fact, data on the effect of N-methylhydroxylamine on cytosine analogs are scarce. Phillips and Brown (1967) observed the formation of compound IV (corresponding to II) and probably the highly unstable compound V (corresponding to III).

We were able to obtain several methylated derivatives of compound IV, the properties of which will be described in a subsequent publication. We found that N-methylhydroxylamine does not react with 5-alkylcytosine or with 5-alkyl-4-ethoxy-2-ketopyrimidine, the alkoxy group of which shows an increased tendency to split off in reactions with hydroxylamine (Malling, 1967). This finding fully agrees with the numerous reports on the impossibility of introducing N_4-dialkyl substituents into cytosine or into its glycosides containing substituents at N_3 or C_5 (Kulikowski et al., 1969). The impossibility of obtaining such analogs obviously results from steric hindrances created by the N_3- or C_5 substitutents, which can be clearly visualized in three-dimensional models. It can be assumed by way of analogy that the presence of a 5-methyl or 5-oxymethyl radical in the cytosine ring would greatly hamper the nucleophilic substitution of NH_2 at C_4 with CH_3NOH, which is actually the case. Further support for this viewpoint is given by the observation that the corresponding 6-methyl-4-ethoxy-2-ketopyrimidine reacts readily with N-hydroxylamine, forming a single substitution product which corresponds to product II. Thus, there is nothing surprising about the lack of mutagenic activity of N-hydroxylamine with respect to phage T4, the DNA of which contains 5-oxymethylcytosine only.

As to the reaction between free cytosine or its glycosides and N-methylhydroxylamine, the available data show that it yields the exchange product IV and a relatively unstable compound corresponding to product I. We examined this reaction in greater detail despite the technical difficulties involved, resulting from the instability of CH_3NHOH in an aqueous medium under aerobic conditions owing to the formation of free radicals in the presence of oxygen (Freese et al., 1967). Moreover, the degradation products of CH_3NHOH display some degree of selective absorption at a wavelength of 225 nm (Freese et al., 1966), as is the case with the absorption spectra of compounds of types I and III. Despite these difficulties, it was shown quite clearly that compound IV is the only product of the reaction.

If compound V or an analog of compound I is nevertheless formed, it must be at least as unstable as product I and therefore of no importance as far as mutagenesis is concerned. Thus, compound IV is the only possible product of the reaction. Since product IV can hardly be expected to pair normally with any other base, this means that N-methylhydroxylamine will not exert any mutagenic effect on normal DNA or RNA, as demonstrated in the recent work of Singer and Fraenkel-Conrat (1969).

To sum up, it is evident that N-methylhydroxylamine does not react at all with 5-substituted cytosine derivatives while its reaction with unsubstituted cytosine or its glycoside yields only product IV, which is incapable of normal base pairing. Consequently, this compound cannot be expected to act as a mutagen inducing pairing errors. The mutagenic effect of N-methylhydroxylamine reported in some publications can be attributed to secondary processes resulting from the degradation of this compound in an aerobic medium (Freese et al., 1967).

Another possibility deserves attention in view of some unexpected observations by Ludlum and Wilhelm (1968) and by Singer and Fraenkel-Conrat (1969). The first authors report that an increase in the methylation of N_3-cytosine residues in poly-C enhances the incorporation of U instead of G into the complementary strand in the RNA polymerase reaction which leads to the synthesis of poly-GU. In the second publication, it is shown that the number of tobacco mosaic virus mutants obtained by treatment with nitrosoguanidine corresponds to the number of N_3-methylated cytosine residues. It is known that N_3-methylcytosine is incapable of pairing. Yet a mutagenic effect is reported in both works. These observations must be thoroughly verified before any definitive conclusion is drawn.

We are grateful to K. Myszkowska for her technical assistance.

BIBLIOGRAPHY

Freese, E.B. and E. Freese. — Proc. Natn. Acad. Sci. U.S.A., Vol. 52 (1964), 1289.

Freese, E., E.B. Freese, and S. Graham. — Biochim. biophys. Acta, Vol. 123 (1966), 17.

Freese, E.B., J. Gerson, H. Taber, H.-J. Phaese, and E. Freese. — Mutation Res., Vol. 4 (1967), 517.

Janion, C. and D. Shugar. — Acta biochim. pol., Vol. 12 (1965), 337.

Kochetkov, N.K. and E.I. Budowsky. — In: "Progress in Nucleic Acid Research," Vol. 9 (1969), 403.

Kulikowski, T., B. Zmudzka, and D. Shugar. — Acta biochim. pol., Vol. 16 (1969), 201.

Ludlum, D.B. and R.C. Wilhelm. —J. Biol. Chem., Vol. 243 (1968), 2750.

Malling, H.V. — Biochim. biophys. Acta, Vol. 4 (1967), 559.

Phillips, J.H. and D. M. Brown. — In: "Progress in Nucleic Acid Research," Vol. 7 (1967), 349.

Singer, M. and H. Fraenkel-Conrat. — In: "Progress in Nucleic Acid Research," Vol. 9 (1969), 1.

UDC 575.24

THE EFFECT OF PHENOCOPY-INDUCING COMPOUNDS ON THE FORMATION OF PUFFS IN DROSOPHILA MELANOGASTER

M.N. SAKHAROVA and I.A. RAPOPORT
Institute of Chemical Physics, Academy of Sciences of the USSR, Moscow

The effect of microdoses of $CrCl_3$ and $NaClO_4$ on the formation of puffs in 12-hour old larvae was examined. The results obtained suggest that the effect of phenocopy-inducing compounds on the formation of puffs is associated with feedback-type regulation of an enzymatic function. By blocking an enzyme which is essential for morphogenesis, the inhibitor causes an increase in the concentration of the substrate normally processed by the enzyme. This possibly triggers the synthesis of a new enzyme which transforms the blocked metabolite.

It has long been known that the polytene chromosomes of dipterans form peculiar structures known as puffs and Balbiani rings. That the puffs and Balbiani rings are the morphological expression of gene activity, however, is a comparatively recent finding (Beermann, 1952). A number of scientists have demonstrated the tissue specificity of puffs and the relationship between the sequence of puff activation and particular stages of morphogenesis (Becker, 1959, 1962; Breuer and Pavan, 1955; Kiknadze and Filatova, 1963; Aschburner, 1967; Berendes, 1967). Biochemical studies have revealed an active synthesis of mRNA in the puffs (Pelling, 1964; Ritossa, 1962; Rudkin, 1962; Vogt-Köhne and Karlson, 1963). It was shown in various dipteran species that a number of chemical and physical agents — ecdysone, zinc, magnesium, chloroform, thioacetamide, narcotics, changes in the potassium-sodium balance of the cell, temperature, anaerobiosis, etc. — cause major changes in the formation of puffs (Clever and Karlson, 1966; Kroger, 1964; Ritossa, 1962; Kiknadze and Filatova, 1963; Berendes, 1965).

These experiments have created a new concept of the polytene chromosome as a model for the solution of a number of the major problems of molecular biology up to now studied in microorganisms, such as the storage and realization of information, the regulation of gene activity, differentiation, etc. The puff proved to be a precision system suitable for fine analysis. The possibility of experimentally altering the activity and spectrum of puffs enables visual control of intervention in fundamental processes at the molecular level and at the same time their comparison with the corresponding changes in the cells, organs and the organism as a whole.

The present paper deals with the effect of the enzyme poisons $CrCl_3$ and $NaClO_4$ on the formation of puffs in the chromosomes of the salivary glands of D. melanogaster. These agents were chosen on the basis of earlier works (Rapoport, 1948), according to which chromium chloride and sodium perchlorate induce specific morphological changes in Drosophila—namely, mutation-like morphoses or phenocopies. Chromium chloride induced the formation of melanin tumors in the internal organs of Drosophila larvae and imagoes while sodium perchlorate caused atresia and partial constriction of the genitalia with a rotation by 60—80° in the male and a shortening of the abdomen in the female. It was assumed that the specific morphogenetic changes induced by compounds acting as enzyme poisons result from the inactivation of one or several enzymes involved in morphogenesis. It was therefore interesting to verify whether these compounds induce chromosomal changes associated with the puff reaction.

The work was carried out on the wild type Batumi line D-32 of D. melanogaster. The concentrations of both compounds in the nutrient medium ranged from sublethal — 1.2 M $CrCl_3$ and 3.3 M $NaClO_4$, previously found to induce the highest percentage of individuals showing phenocopies (Rapoport, 1948), to trace amounts — $2.5 \cdot 10^{-5}$ M. While the frequency of phenocopies was directly proportional to the concentration of the agent, this factor exerted no appreciable effect on puff formation. The results presented below were obtained with trace concentrations of the two agents ($2.5 \cdot 10^{-5}$ M), which caused a pronounced effect without inhibiting the activity of the organism.

The preparations were made according to a standard procedure from larvae at the end of the third instar. The puffs were assessed by the commonly accepted criteria of the degree of loosening of the structure and the increase in diameter relative to the neighboring areas.

The puffs were divided into the following four types: 1) loci in which the distinct pattern of discs has disappeared but which do not yet have a larger diameter than the adjacent areas; 2) puffs with a slightly loosened structure and a slightly greater diameter; 3) puffs with a loose structure and a diameter 1.5 times larger; 4) even larger puffs.

Here we shall present only data relating to the autosomes — the right and left arms of chromosomes II and IV. No structural changes whatsoever were observed in chromosome IV.

We found that $CrCl_3$ and $NaClO_4$ exerted no effect on the activity of most of the normally found puffs. However, a considerable proportion of the puffs (25% of the total number of autosomal puffs for both compounds combined) showed changes which can be divided into three categories:

a) appearance of new puffs which are normally absent;

b) enhancement of the synthetic activity of some normal puffs, reflected in their enlargement;

c) depression of the activity of some puffs, which accordingly become smaller than normal.

The set of affected areas and the nature of the induced changes depend largely on the chemical agent used (see table), but some degree of interference nevertheless exists. $CrCl_3$ induced three new puffs: 34D, 54DE, and 83C; $NaClO_4$ — five puffs: 34D, 60B, 60E, 68 B, and 68 C, all lacking in a normally developing stage III larva. However, puff 34D is common, i. e.

induced by both $CrCl_3$ and $NaClO_4$, and is therefore nonspecific. Puffs 60 B and 60 E, both induced by sodium perchlorate, cannot be regarded as specific either since they sometimes occur normally, though only in rare cases; moreover, we found recently that these puffs may be induced by another enzyme poison — sodium thiocyanate. Puffs 68B and 68C are not typical of this developmental stage but are characteristic of younger larvae aged 110 hours; it appears that they remain in an active state during later stages under the influence of sodium perchlorate. Thus, we are left with puffs 54DE and 83C as specifically induced by $CrCl_3$, and puff 64C induced by $NaClO_4$.

General and individual loci responding to $CrCl_3$ and $NaClO_4$

Enzyme poison	New puffs	Puffs in which the activity is experimentally	
		enhanced	decreased
$CrCl_3$ and $NaClO_4$	34D	25AD, 42AE, 43E 62E, 82CF, 83E, 85C, 85D, 85EF	27C
$CrCl_3$	54DE, 83C	21EF, 23E, 32C, 33B, 48B, 56DE, 67F	34A, 91B, 95F, 99B
$NaClO_4$	60B, 60E, 64C, 68B, 68C	47BC, 63E, 66D, 66E, 98F	96E, 97F

Numerous normally occurring puffs show a markedly enhanced synthetic activity under the influence of $CrCl_3$ and $NaClO_4$: 25AD, 42AE, 43E, 62E, 82CE, 83E, 85CD, 85D, 85EF. However, the activation of some complex puffs reveals certain morphological and quantitative differences. 25AD is a complex puff formed out of two normally occurring type 2 puffs, 25B and 25D, which fuse under the influence of $CrCl_3$, the result being a major puffing of the chromosomal area 25AD. In the experiment with $NaClO_4$ these puffs similarly fuse into a single large puff which occupies the smaller area 25BD; 42AD is likewise a complex puff which in the control attains a peak activity in the 42A area and corresponds to type 3; the adjacent puffs 42BCD are small, corresponding to type 1. Both agents cause a marked enlargement of the puff, so that the 42BC area coincides with the puff maximum. Similar transformations occur in the 82CF area, which in the control preparations contains two puffs, 82C and 82F, of types 1 and 3 respectively; in the experimental preparations these fuse into a large complex puff, 82CF, with a maximum at 82F.

Apart from the puffs which are nonselectively activated by either agent, there are groups of puffs in which the activity is enhanced only by one of the agents. In the experiment with $CrCl_3$ these were 21F, 23E, 32C, 33B, 48B, 56DE and 67F; in the experiments with $NaClO_4$ they were 47BC, 63E, 66DE, 66E, and 98F. Puff 27CD is depressed by either agent, contracting to 27C. In larvae treated with $CrCl_3$ the activity of puffs 34A, 91B, 95F, 99B is depressed; in the case of $NaClO_4$ there is a depression of puffs 96E and 97F.

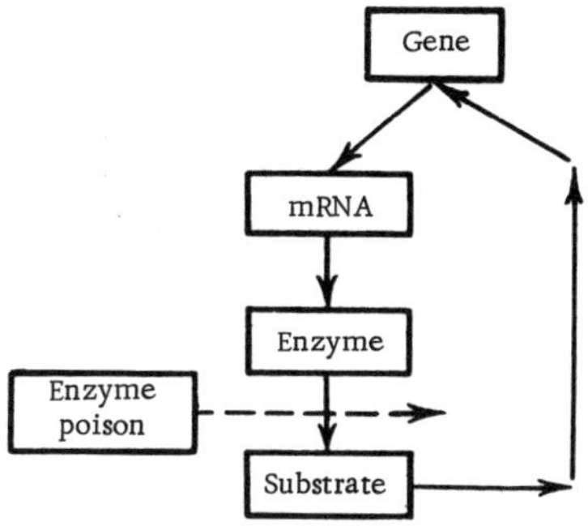

Scheme of the feedback mechanisms causing puffing in the presence of an enzyme poison.

It is evident from these data that trace amounts of each of the enzyme poisons used cause diverse changes in the puff pattern of the given developmental stage, so that analysis of a single agent is not sufficient for the detection of the specific changes among these. Comparisons of the data obtained with the two agents already reveals some degree of interference which is especially important with respect to new puffs since it approximates to a determination of the specific puff associated with a certain phenocopy.

The results of this experiment, showing the direct effect of the agents used at the chromosome level, suggest that the puffing induced by enzyme poisons may be due to feedback regulation of the enzymatic function (see diagram). If the enzyme poison blocks some catalytic activity essential for morphogenesis or the metabolism of the enzyme, the increased concentration of the substrate normally attacked by this enzyme may trigger the synthesis of a new enzyme necessary for the transformation of the blocked metabolite. The total inactivation of the enzyme in our experiment presumably creates suitable conditions for the appearance of a new specific puff, while the partial inhibition of the enzyme, also stimulating processes bypassing the catalytic deficiency, leads to an increase in the synthetic activity in the same puff.

Puffing induced by enzyme poisons as a result of a feedback mechanism (Rapoport et al., 1970) can be regarded as one of the possible explanations of the appearance of new specific puffs, which may be the sites of normally inactive and therefore morphologically invisible syntheses, and also the activation and depression of puffs in comparison with the normal state.

BIBLIOGRAPHY

Aschburner, M. — Chromosoma, Vol. 21 (1967), 398.
Becker, H.J. — Chromosoma, Vol. 10 (1959), 645.
Becker, H.J. — Chromosoma, Vol. 13 (1962), 341.
Beermann, W. — Chromosoma, Vol.5 (1952), 139.

Berendes, H.D. — Chromosoma, Vol. 17 (1965), 35.
Berendes, H.D. — Chromosoma, Vol. 22 (1967), 274.
Breuer, M. and C. Pavan. — Chromosoma, Vol. 7 (1955), 371.
Clever, U. and L. Karlson. — A. Rev. Ent., No. 11 (1966), 1.
Kiknadze, I.I. and I.T. Filatova. — Doklady AN SSSR 152, No. 2 (1963), 450.
Kroger, H. — Chromosoma, Vol. 15 (1964), 36.
Pelling, C. — Chromosoma, Vol. 15 (1964), 71.
Rapoport, I.A. — Trudy Inst. Cytol. Histol. Embryol. Acad. Sci., USSR, Vol. 2, No.1 (1948), 3.
Rapoport, I.A., M.N. Sakharova, Yu. L. Nikiforov, and M.M. Beknazar'yants. — Doklady AN SSSR, Vol. 193 (1970), 441.
Ritossa, F.A. — Experientia, Vol. 18 (1962), 571.
Rudkin, G.T. — Ann. Histochem., Suppl., Vol. 2 (1962), 77.
Vogt-Köhne, and L. Karlson. — Chromosoma, Vol. 14 (1963), 186.

UDC 575.24

THE WAVE KINETICS OF MUTAGENESIS AND THE PROBLEM OF POTENTIAL CHANGES

K.P. GARINA and A.I. KORYTOVA
Institute of General Genetics, Academy of Sciences of the USSR, Moscow;
Dnepropetrovsk State University

The kinetics of the mutation process during the storage of mutagen-treated seeds of barley and Crepis capillaris was studied. A parallelism was found between the change in the frequency of gene mutations and the frequency of chromosomal rearrangements during storage in barley seeds treated with ethylenimine. It is concluded that the same fundamental molecular processes may lead to both gene mutations and chromosomal rearrangements. The combined application of alkylating compounds and radiation during storage causes a transition of the primary lesions into latent potential changes which later develop into actual mutations.

Studies of the mechanism of the appearance and development of mutations occupy an important place in the literature dealing with induced mutagenesis (Sobels, 1963; Kimbal, 1965; Rieger and Michaelis, 1965; Dubinin, 1966, 1969; Dubinin and Dubinina, 1968a, b). In a study of chromosomal rearrangements in seeds of Crepis capillaris, N. P. Dubinin and L. G. Dubinina first demonstrated in 1968 that the mutation process occurring during the storage of seeds treated with alkylating agents shows wave kinetics. This finding suggested the existence of a new category of potential changes which can persist in a metastable state for a long period. Analysis of the ratio between chromosomal and chromatidal mutations indicated the possible role of dark repair in the processes of mutagenesis.

This paper deals with the kinetics of the mutation process during the storage of mutagen-treated seeds with particular reference to chromosomes and gene mutations. Determination of the ratio between gene and chromosome mutations is essential for an understanding of both their nature and of the basic principles of chemical mutagenesis.

The study of the kinetics of the mutation process during the storage of seeds treated with both ethylenimine (EI) and X rays is also of interest.

The ratio between gene and chromosome mutations was determined in the barley variety MOS-121. EI at a concentration of $2.3 \cdot 10^{-3}$ M and $7 \cdot 10^{-3}$ M was used as the mutagen. Air-dried seeds were treated with EI for 3 hours, then washed, dried and stored in a desiccator over granular KOH for 3 to 172 days. For the analysis of chromosome aberrations the seeds were germinated at 3-day intervals during the first two months and every two weeks afterwards. Seeds from a total of 29 storage periods were tested in this manner for each dose. The chromosomal rearrangements were examined by the anaphase method in temporary pressed preparations stained according to Feulgen. On the average, 3,000 anaphases were examined for each point.

For the determination of chlorophyll mutations, the seeds were treated as described above the corresponding number of days before planting and were stored over granular KOH until sown in the field.

The second generation was grown in a greenhouse in racks filled with sand. About 6,000 plants were examined for each period of seed storage.

Seeds of C. capillaris were used for the study of the combined effect of radiation and EI during storage. Air-dried seeds were first exposed to X rays at a dose of 1,500 and 4,500 r and then treated with EI at a concentration of $9.3 \cdot 10^{-3}$ M by the usual method. The treated seeds were stored in a desiccator over granular KOH. At various intervals (from 6 to 126 days) the seeds were germinated in a 0.01% colchicine solution for analysis of the first division metaphases. The general level of mutations and the number of chromosome and chromatid rearrangements were determined. The results showed that the dynamics of the mutation process during the storage of EI-treated seeds follows a distinct wavelike pattern. The curves in Figure 1 show the total number of chromosomal rearrangements after different periods of storage of seeds treated with a low and a high concentration of EI. Statistically significant rises and falls may be observed in the mutation level throughout all the storage periods. The curves might be expected to reach a plateau at longer storage periods. However, even on the 88th and 130th days, there were distinct waves with statistically significant maxima and minima.

After treatment at the low concentration, the largest percentage of mutations (15.12%) was obtained on the 21st day of storage; in seeds treated at the high concentration the percentage of rearrangements reached a peak (63.44%) on the 60th day.

Analysis of the chlorophyll mutations showed that their frequency similarly depends on the duration of seed storage. Thus, statistically significant undulatory changes in the frequency of chlorophyll mutations occurred during the first two months of storage of the seeds treated with the low concentration of EI. These mutations attained a peak (6.52%) on the 18th day (Figure 2).

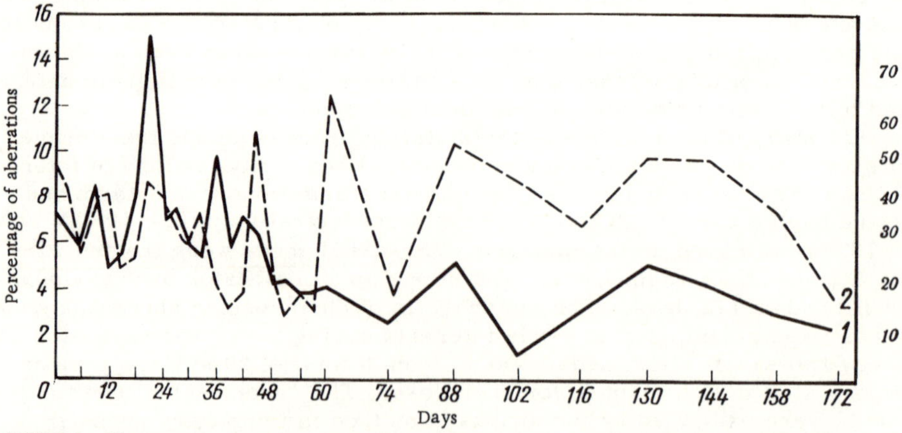

FIGURE 1. Chromosome aberrations during the storage of barley seeds treated with ethylenimine. Concentration of EI ($M \cdot 10^{-3}$): 1) 2.3 (left scale); 2) 7.0 (right scale).

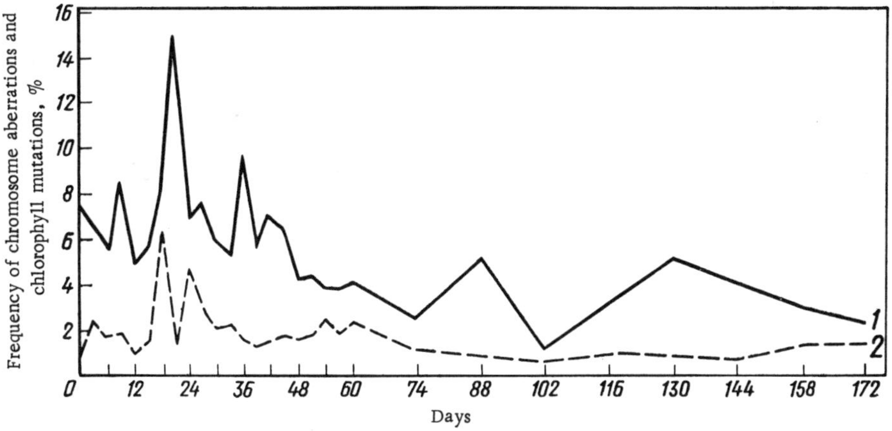

FIGURE 2. Chromosome aberrations (1) and chlorophyll mutations (2) during the storage of barley seeds treated with EI at a concentration of $2.3 \cdot 10^{-3}$ M.

Comparison of the dynamics of the appearance of chlorophyll mutations and the number of chromosome aberrations during different periods of storage of the treated seeds shows a close resemblance between the two phenomena. In 1968 we obtained similar results (Garina and Romanova, 1970).

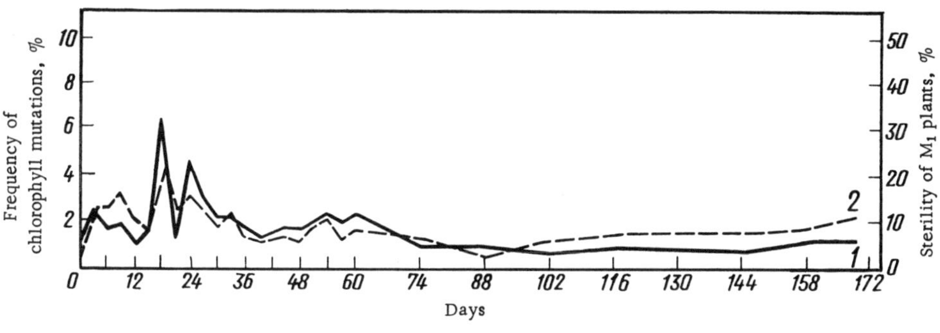

FIGURE 3. Frequency of chlorophyll mutations (1) and sterility of M_1 plants (2) after different periods of storage of barley seeds treated with ethylenimine at a concentration of $2 \cdot 10^{-3}$ M.

In seeds stored for more than two months the frequency of chlorophyll mutations becomes stabilized and the curve levels off to a plateau while the frequency of chromosomal rearrangements still shows statistically significant fluctuations.

Thus, there is a marked similarity in the dynamics of the appearance of gene mutations and chromosome rearrangements during the storage of barley seeds treated with EI. Analysis of this phenomenon suggests that the processes leading to gene mutations and chromosome rearrangements have a common molecular basis. Both types of mutation evidently stem from metastable potential changes differing in their duration. Most of the potential changes induced by the treatment of barley seeds with EI at a concentration of $2.3 \cdot 10^{-3}$ M develop into actual mutations within 18—21 days. The potential changes leading to the appearance of gene mutations are apparently of shorter duration; by the end of the second month after treatment of the seeds these changes have already been completely repaired and the frequency of chlorophyll mutations becomes stabilized at a level practically identical with the initial one.

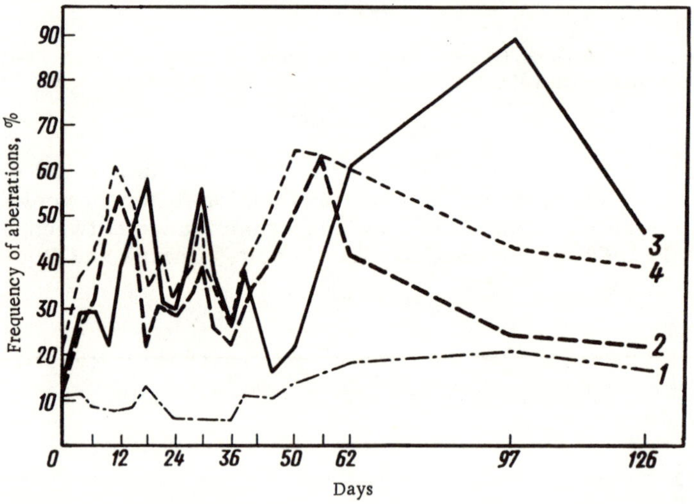

FIGURE 4. Mutability of C. capillaris on storage of seeds treated simultaneously with X rays and ethylenimine:

1) X rays, dose 1,500 r; 2) EI, concentration $9.3 \cdot 10^{-3}$ M; 3) X rays + EI; 4) theoretically expected curve with combined treatment.

A comparison of the sterility rate of first generation flowers and the frequency of chlorophyll mutations is given in Figure 3. The parallel course of the curves suggests that gene mutations are largely responsible for the partial sterility of the first generation plants induced by EI. Similar views have already been expressed by Ehrenberg and Ekberg (Ehrenberg et al., 1961, 1966; Ekberg, 1969).

During the storage of C. capillaris seeds treated with X rays and EI the mutation process similarly shows distinct wave kinetics. Of particular interest are the results of the combined treatment, shown in Figure 4.

The curve obtained for the change in the dynamics of the overall level of mutability during the storage of seeds exposed to a combined treatment differs in its course from the expected curve (closely resembling the EI curve). As shown in Figure 4, the rises and falls in the level of mutability do not coincide (9th, 19th, 45th days). On the 97th day of storage, the seeds exposed to the combined treatment show a marked increase in the total number of chromosomal rearrangements, exceeding not only the peak mutability levels obtained with each mutagen separately but also the additive effect. A similar picture was obtained with respect to treatment with X rays at a dose of 4,500 r.

The results obtained can be interpreted as an indication that with the combined application of X rays and EI a large proportion of the primary lesions may be converted into latent potential changes characterized by great stability. These latent potential changes remain unrepaired for a long period and on the 97th day a mass transition of the latent changes into true mutations is observed.

When X rays at a dose of 4,500 r were used for the combined treatment, the effect was far inferior to that expected. Moreover, the total level of mutability during the storage of seeds exposed to the combined treatment was lower, as a rule, than that induced by X rays alone although above that induced by EI. Here EI apparently weakens the effect of the X rays.

Analysis of the chromosomal rearrangements induced by the combined treatment of C. capillaris showed that these include both chromosome and chromatid aberrations. The rise in the total level of mutability during storage can be attributed mainly to chromosome-type rearrangements. This confirms Dubinin's (1969) hypothesis that the probability of manifestation of potential changes during the G_1 phase increases during the storage of EI-treated seeds and that the lesion can be transmitted under such circumstances to the complementary template of the resting DNA molecule.

All the data obtained in this study undoubtedly confirm the repair hypothesis of mutagenesis.

BIBLIOGRAPHY

Dubinin, N.P. — Genetika, Vol. 7 (1966), 3.
Dubinin, N.P. — Genetika 5, No. 8 (1969), 5.
Dubinin, N.P. and L.G. Dubinina. — Genetika 4, No. 9 (1968a), 5.
Dubinin, N.P. and L.G. Dubinina. — Doklady AN SSSR, Vol. 179 (1968b), 1221.
Ehrenberg, L., A. Gustafsson, and U. Lundqvist. — Hereditas, Vol. 47 (1961), 243.
Ehrenberg, L., U. Radioactivity, S. Osterman, and Sparrman. — Hereditas, Vol. 56 (1966), 277.
Ekberg, J. — Hereditas 63, Nos. 1 — 2 (1969), 257.
Garina, K.P. and N.I. Romanova. — Molec. and Gen. Genet. 106, No.2 (1970), 93.
Kimbal, R.F. — Mutation Res. 2, No. 5 (1965), 413.
Rieger, R. and A. Michaelis. — Nature, Vol. 206 (1965), 741.
Sobels, T.H. — Genetics Today, Vol. 2 (1963), 235.

UDC 575.24

THE EFFECT OF ETHYLENIMINE AND UV AND GAMMA RAYS ON THE APPEARANCE OF COMPLETE AND MOSAIC MUTATIONS AT DIFFERENT STAGES OF THE CELL CYCLE OF CHLORELLA

V.A.SHEVCHENKO and D.R.PYATYSHEV

Institute of General Genetics, Academy of Sciences of the USSR, Moscow

The ratio between sectored and complete mutant colonies at G_2 is the same as at G_1 and increases only with the beginning of mitosis. It is concluded that after the triple synthesis of DNA, the chromosomes of Chlorella respond to the mutagen as a unitary formation. On treatment with ethylenimine $^4/_8$, $^2/_3$ and $^1/_8$ sectors appeared mostly during the S phase, and $^3/_5$, $^5/_8$, $^6/_8$ and $^7/_8$ (aberrant) sectors between consecutive syntheses of DNA. The formation of aberrant sectors is attributed to nonreciprocal recombinations between normal and mutant DNA molecules.

Sectored mutant colonies (mosaics) are widely used as a test in modern molecular genetics. Since the work of Demerec (1946), the appearance of mosaics after a single exposure to various mutagens has been observed in a large number of studies. Many authors (Oster, 1963; Parry and Cox, 1965; Nasim and Sanders, 1968; Auerbach, 1967; Esposito, 1968) have used the sectored mutant colony test in studies of the mutation process in various objects. It was found that mutant sectors often arise not only by the interaction of mutagens with one of the chromatids or by delayed mutagenesis but also as a result of recombination processes.

We shall discuss first some unsolved problems concerning the elaboration of the mosaic test. First of all, a detailed analysis of the ratio between normal and mutant nuclei in the original maternal cell has not yet been carried out in eukaryotes. The existence of such a ratio is evident from the data of Witkin (1951) on the UV-induced lac^- mutation in E. coli at different phases of growth. Another interesting question is whether the induction of mutations involves a transition from complete to mosaic mutations at the time when the synchronous cell cultures enter the S and G_2 phases, just as the transition from chromosome to chromatid aberrations occurs at these stages.

The unicellular alga Chlorella vulgaris, grown in synchronous culture, has proved a convenient model for the study of these questions. Figure 1 shows diagrammatically the cell cycle of this species. The peculiarity of the development cycle of Chlorella is that the maternal cell divides not into two but into 2n cells, usually into 8 cells, n being the number of consecutive replications of DNA in the nucleus of the maternal cell. Since nuclear division takes place only at the end of the cell cycle, we assumed earlier that a temporary polyteny of the chromosomes of Chlorella arises during the period of DNA synthesis (Shevchenko, 1965). This hypothesis agrees well with the electron microscope observations of Muracami and his colleagues (1963), as well as with our own data on the kinetics of the incorporation of ^{14}C-thymidine into the DNA of synchronous cells of Chlorella and the results obtained by Anikeeva (1967) and Kates et al. (1969), indicating that the nuclei and cells divide only after the synthesis of DNA.

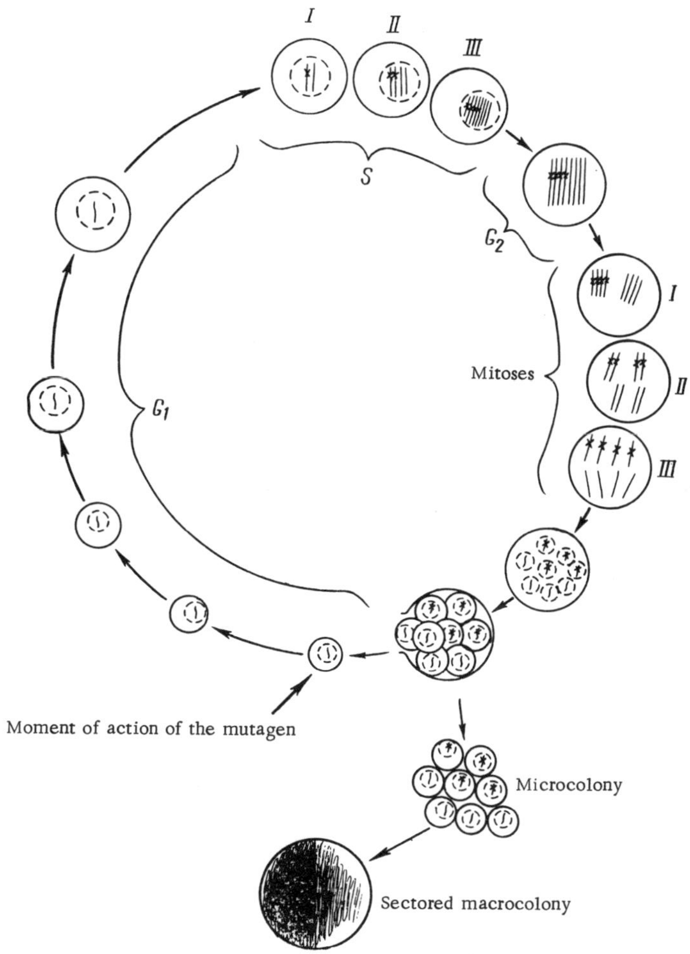

FIGURE 1. Scheme of the life cycle of Chlorella and the appearance of a 1/2 sectored mutation.

The synthesis of DNA in the nucleus is preceded by a corresponding process in the chloroplasts, as can be seen from the incorporation of a radioactive label, which attains a peak in the 4th-5th hour of the cell cycle — long before the synthesis of nuclear DNA, which begins in the 8th-9th hour of the cycle. A peak of DNA synthesis in chloroplasts was earlier observed in synchronous cultures of vegetative cells of Chlamydomonas (Chiang and Sueoka, 1967) and Chlorella (Dalmon, 1969).

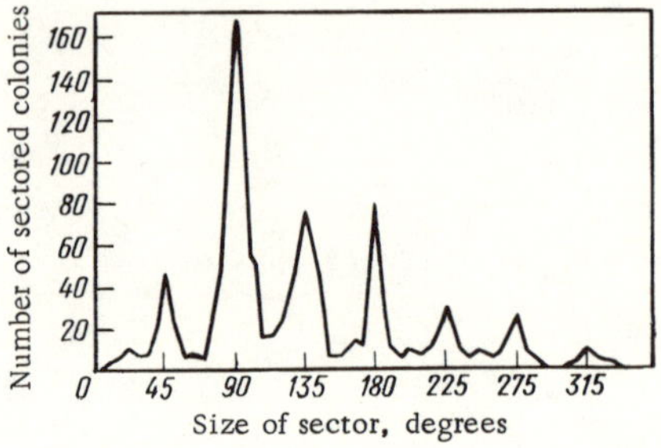

FIGURE 2. Numbers of different types of sectored colony.

In most cases the sectors observed correspond to $1/2$, $1/4$ or $1/8$ of the Chlorella colonies, probably because of the induction of mutations at consecutive DNA syntheses. Apart from this type of sectored colony, however, many colonies are observed with $3/8$, $5/8$, $6/8$ and $7/8$ sectors, the appearance of which cannot be explained by the above scheme. We will refer to such sectored colonies as aberrant. In order to check that the identification of these sectors does not correspond to an error in measurement, we determined the angles of the mutant sectors of Chlorella colonies grown on Tamiiya agar medium after treatment of the young Chlorella cells (G_1 phase) with ethylenimine (EI).

The pigmented mutant sectors arising from the center of the colony and differing only slightly from the norm in growth rate were discounted. The spontaneous frequency of such mutations is $4 \cdot 10^{-5}$. The data amassed in experiments with Chlorella indicates that these are gene mutations.

Measurements of the sector angles for each type of sector gave values closely similar to normal distribution curves (Figure 2). The size of the sector varies widely and is a multiple of 8, i. e. the mutation involves 1, 2, 3 or more nuclei of a total of 8.

Our data show that sectors of the aberrant types ($3/8$, $5/8$, $6/8$, $7/8$) actually exist and are likewise multiples of 8 (the 135°, 225° and 315° peaks in Figure 2).

Let us examine the frequency of total and sectored mutant colonies obtained by exposure to UV rays, X rays and EI (see table). The Chlorella cells were treated with these mutagens at doses corresponding to a 50% survival rate of the cultures.

The number of sectored mutant colonies was lowest after irradiation with X rays, practically corresponding to the number of multinuclear cells in the culture at the time of irradiation. In the case of UV irradiation and spontaneous mutation the proportion of sectored mutant colonies is greater, but only slightly so. Finally, after treatment with EI most of the mutations are sectored.

It is noteworthy that only EI induces a large number of aberrant sectors of the types $^3/_8$, $^5/_8$, etc. (Figure 3). The proportion of such sectors is much lower in the case of spontaneous mutation or exposure to UV rays. The spectrum of sectored colonies induced by X rays is distinguished by the nearly total absence of sectors of the type $^3/_5$, $^5/_8$, etc.

Absolute and relative number of complete and sectored mutant colonies of Chlorella after exposure to various mutagens

Mutagen	Number of colonies examined	Complete			Sectored		
		number	absolute %	relative %	number	absolute %	relative %
Control	91,031	306	0.331	74.5	101	0.11	25.5
X rays	45,588	1,987	4.33	95.5	94	0.21	4.5
UV rays	96,897	5,134	5.33	89.6	603	0.62	10.4
Ethylenimine	49,810	523	1.05	18.2	2,350	4.72	81.8

Let us examine the dynamics of the appearance of complete and sectored mutant colonies of Chlorella at different phases of the cell cycle of a synchronized colony following exposure to gamma rays, UV radiation and EI.

Young Chlorella cells in a liquid medium were placed in the light. Under such conditions the cells pass synchronously through the consecutive phases of the cell cycle and divide every 20—22 hours. Samples taken at 1—2 hour intervals throughout the cycle were irradiated or treated with EI and were seeded on agar for determination of the survival rate of the culture and the proportion of mutant colonies. A parallel determination of the rate of incorporations of ^{14}C-thymidine into DNA was carried out in order to identify the phases of the cell cycle.

The relative number of sectored mutant colonies induced by EI is lowest at the 6th hour of the cycle, which corresponds to the middle of the G_1 phase (Figure 4). The number of sectored colonies subsequently rises during the S phase (9—16 hours) and decreases again at 16—22 hours, during the G_2 phase.

One very important phenomenon must be stressed, namely that the relative numbers of sectored mutant colonies before and after the synthesis of DNA are roughly equal. In other words, the proportion of sectored colonies remains unchanged through three consecutive syntheses of DNA, during which the amount of DNA increases eightfold. This fact may be of crucial significance for the study of the chromosomal structure and the processes of formation of mutations. The gamma irradiation of synchronized Chlorella cultures increases the proportion of sectored mutations only after completion of the mitoses (Figure 4). Ionizing radiation induces a relatively large number of $^1/_8$ sectors during this period.

Exposure of Chlorella to UV rays gives rise to a similar picture. Here the relative number of sectored mutant colonies forms four peaks which can be linked with the consecutive syntheses of DNA (Figure 5). The relative number of sectored colonies remains practically unchanged during the G_1 phase between and after the DNA syntheses (compare with Figure 4).

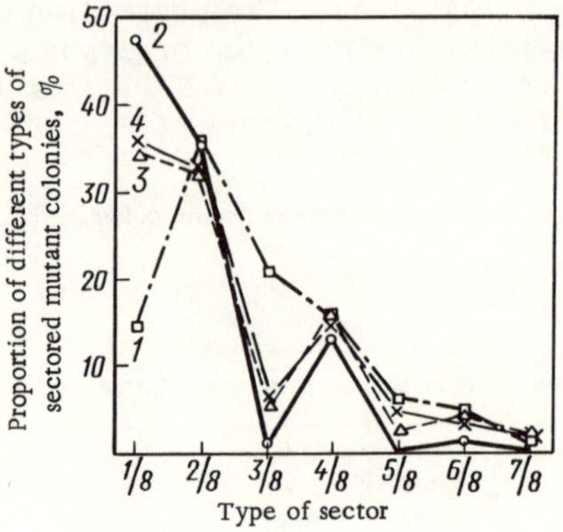

FIGURE 3. Proportion of sectored mutant colonies of different types after exposure to various mutagens:

1) EI; 2) X rays; 3) UV rays; 4) control.

The mutation peak at 4—6 hours, obtained only with UV, coincides with the peak of incorporation of ^{14}C-thymidine into the DNA of the chloroplasts of Chlorella. The mutations appearing during this period (mostly of the dwarf type) probably belong to the category of plastid mutations.

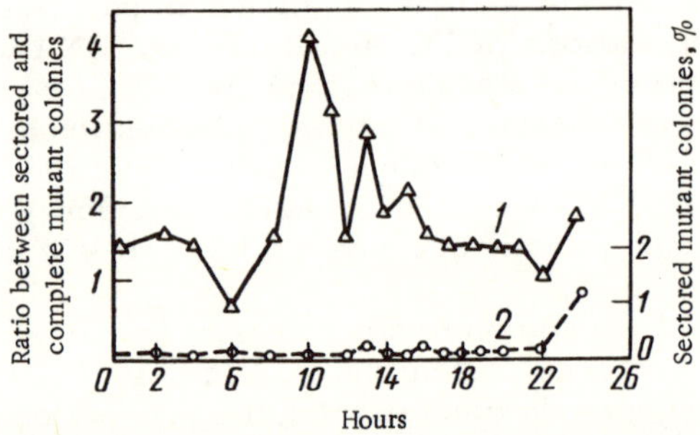

FIGURE 4. Relative number of sectored mutant colonies at different stages of the cell cycle of Chlorella following exposure to EI (1) and gamma radiation (2).

A more detailed analysis of the results obtained showed that the proportions of the different sector sizes vary during the successive phases of the cell cycle (Figure 6). Thus, the proportion of aberrant sectored mutant colonies attains a peak after each DNA synthesis and especially after the third synthesis, whereas the peaks of the sectored mutant colonies of the ½ and ¼ types coincide with the periods of synthesis. Moreover, the

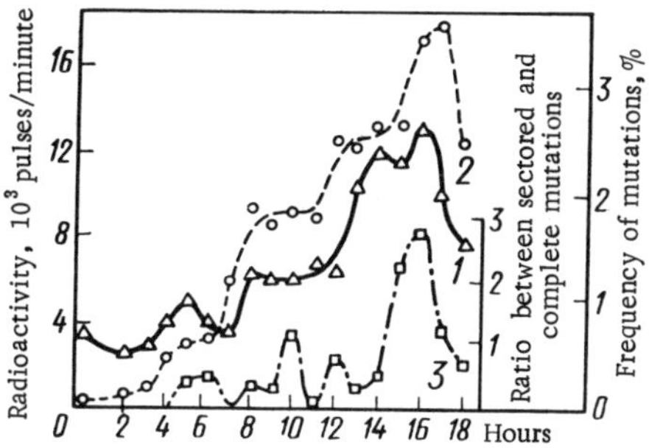

FIGURE 5. The mutagenic effect of UV rays at different phases of the cell cycle of Chlorella:

1) number of mutations; 2) incorporation of ^{14}C-thymidine, added to the nutrient medium before the experiment, into the DNA of Chlorella cells; 3) ratio between sectored and complete mutations.

relative number of complete mutants likewise increases during the periods between consecutive DNA syntheses (Figures 4, 5). This suggests that the aberrant sectors and probably some of the complete mutants arise as a result of recombination processes involving gene conversion (nonreciprocal recombinations). The mechanism of nonreciprocal recombinations can lead to an increase or a decrease in the proportion of mutations among interacting mutant and normal chromatids (Holliday, 1968).

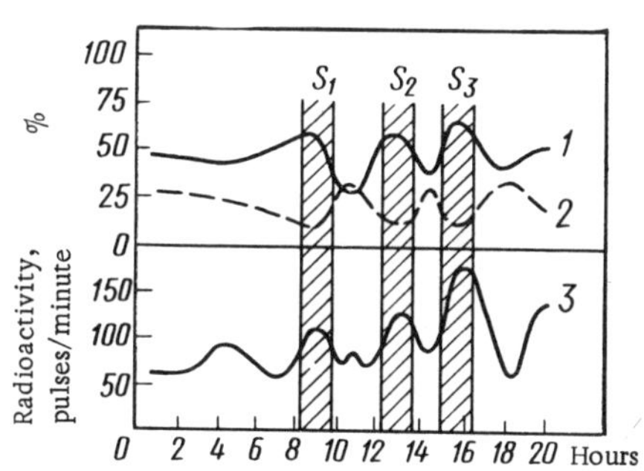

FIGURE 6. Proportion of sectored mutant colonies as a % of the total number of mosaic colonies at different phases of the cell cycle of Chlorella:

1) $1/2 + 1/4$; 2) $3/8 + 5/8$; 3) incorporation of pulse labeled ^{14}C-thymidine.

Precisely these sectored mutations ($3/8$, $5/8$, $6/8$, $7/8$) of Chlorella arise predominantly after treatment with EI — an agent which increases the frequency of recombinations, especially of the nonreciprocal type (Zimmerman and Laer, 1967). It is known that the process of recombination in synchronous cultures takes place after DNA synthesis (Esposito, 1968); this agrees with our own data on the dynamics of the appearance of aberrant sectors. Finally, the kinetics of DNA synthesis and the course of the division of the genetic material of Chlorella suggest the existence of a functioning mechanism of mitotic nonreciprocal recombination.

CONCLUSIONS

The following conclusions can be drawn on the basis of the results obtained. Both as a result of spontaneous mutation and under the influence of various mutagens (especially EI), the unicellular green alga Chlorella (LARG-1 strain) yields not only complete pigmented mutant colonies but also numerous sectored mutant colonies showing various ratios between the normal and mutant parts ($^1/_8$, $^1/_4$, $^3/_8$, $^1/_2$, $^5/_8$, $^3/_4$, $^7/_8$). It was established that the size of the mutant sector reflects the primary ratio between mutant and normal daughter nuclei arising as a result of the three consecutive DNA syntheses in the maternal cell of Chlorella.

The kinetics of the incorporation of ^{14}C-thymidine into the cellular DNA were studied in a synchronous culture of Chlorella. There are four consecutive peaks of incorporation of the radioactive label; the first peak coincides with the synthesis of chloroplast DNA while the other three are associated with the consecutive syntheses of nuclear DNA.

The proportions of sectored mutant colonies induced by EI before and after DNA synthesis are roughly equal. The percentage of sectored mutations induced by ionizing radiation likewise increases only after the mitoses. This means that the chromosomes of Chlorella can respond to treatment with a mutagen as single entities even after three DNA syntheses (until the formation of the daughter nuclei).

In the study of the effect of EI at different phases of the cell cycle, it was found that the $^1/_2$, $^1/_4$, $^1/_8$ sectors appear mainly during periods of DNA synthesis, whereas the aberrant types of sectors — $^3/_8$, $^5/_8$, $^3/_4$, $^7/_8$ — develop after the consecutive syntheses of DNA. It is proposed that the aberrant sectors arise as the result of a mechanism of reciprocal recombinations.

BIBLIOGRAPHY

Anikeeva, I.D. Effects of Ultraviolet Radiation on Chlorella. — Candidate Thesis. Moscow, 1967. (Russian)

Auerbach, C. — Mutation Res. 4, No. 6 (1967), 875.

Chiang, K.S. and N. Sueoka. — J. Cell. Comp. Physiol., Suppl. 70, No. 1 (1967), 89.

Dalmon, I. — C. r. hebd. Séanc. Acad. Sci., Paris, D 248, No. 4 (1969), 689.

Demerec, M. — Proc. Natn. Acad. Sci. U.S.A., Vol. 32 (1946), 36.

Esposito, R.E. — Genetics, Vol. 59 (1968), 191.

Holliday, R. — In: Replication and Recombination Genetic Material. Canberra, Austral. Acad. Sci., p.157. 1968.

Kates, J. R., K.S. Chiang, and R.F. Jones. — Expl. Cell. Res. 49, No. 1 (1968), 121.
Muracami, S., V. Morimura, and A. Takamija. — Japan. Soc. Plant Physiol., Vol. 65, 1963.
Nasim, A. and A.S. Sanders. — Mutation Res. 6, No. 3 (1968), 475.
Oster, I.O. — In: "Repair from Genetic Radiation Damage," p. 51, 1963.
Parry, I.M. and B. S. Cox. — J. Gen. Microbiol., Vol. 40 (1965), 235.
Shevchenko, V.A. — Radiobiologiya, Vol. 5 (1965), 25
Witkin, E.M. — Cold Spring Harbor Sympos. Quant. Biol., Vol. 16 (1951), 357.
Zimmerman, F.K. and U. Laer. — Mutation Res., Vol. 4 (1967), 377.

UDC 575.24

THE MECHANISM OF THE MUTAGENIC EFFECT OF N-NITROSO-N-ALKYLUREA

A.M. SEREBRYANYI
Institute of Chemical Physics, Academy of Sciences of the USSR, Moscow

Nitrosoalkylureas (NAU) exert a complex mutagenic effect and cause two types of changes in genetic material resulting from alkylation or carbamoylation reactions.

N-Nitroso-N-alkylureas (NAU) are potent mutagens and carcinogens which undoubtedly act as alkylating agents. N-Nitroso-N-methylurea (NMU) can methylate nucleic acids both in vivo (P. Brookes, 1966; Swann and Magee, 1968) and in vitro (Kononova et al., 1968; A. M. Serebryanyi et al., 1969; Loveless and Hampton, 1969). On the other hand, there are indications that the biological effect of NAU cannot be attributed entirely to alkylation reactions.

Studies of the mutagenic activity as a function of the mutagen structure have shown that an increase in the length of the radical from C_1 to C_6 does not lead to the loss of mutagenic activity in any of the NAUs tested whereas in other alkylating agents only the first members of the series are active (Rapoport, 1966).

The mechanism of the cytogenetic effect of NMU differs from that of other alkylating agents since the effect of NMU is inhibited by NaN_3 or by treatment under anaerobic conditions (Mueller, 1965; Michaelis et al., 1965).

E. coli strains resistant to methylmethanesulfonate have been shown to be sensitive to NMU. The effect of NMU in *E. coli* is inhibited by NaN_3 (Rosenkranz et al., 1968).

There is a radical difference between the mutagenic and inactivating effects of NMU, nitrosoethylurea (NEU) and the corresponding methylsulfonates on phage T2. In the opinion of the authors the effects observed cannot be attributed to alkylation (Loveless and Hampton, 1969).

In contrast to the effect of alkylating compounds, the treatment of seedlings of *Crepis capillaris* and some other objects with NMU causes not only chromatid but also chromosome aberrations (Grigoreva and Zoz, 1970).

Together, these data present strong evidence that the changes induced by NAU in DNA differ from those caused by alkylating agents.

Our search for a possible mechanism led us to the carbamoylation reaction. All urea derivatives with acceptor substituents of the form $RNHCONH_2$, where $R= NO_2$ or $R= CH_3CO$, can carbamoylate a reactive substrate:

$$R'NH_2 + H_2NCONHNO_2 \rightarrow RNHCONH_2 + N_2O + H_2O.$$

In order to establish the probability of the occurrence of such a reaction between NMU and DNA we synthesized NMU containing radioactive carbon in the carbonyl moiety of the molecule, $CH_3N(NO)C^*ONH_2$, and examined the transfer of the label of DNA. The reaction was performed in NaCl-citrate buffer, pH 7, at 37°. The molecular ratio between NMU and R_{DNA} was 100:1. A transition of the label to DNA was observed:

$$DNA + CH_3N\,(NO)\,C^*ONH_2 \rightarrow DNA\,C^*ONH_2\cdot$$

The potential sites of carbamoylation in DNA are the phosphate groups and the base amino groups. We examined the reaction between cytidine and NMU. Carbamoylcytidine (1) and 1-methylcarbamoylcytidine (2) were isolated from this reaction:

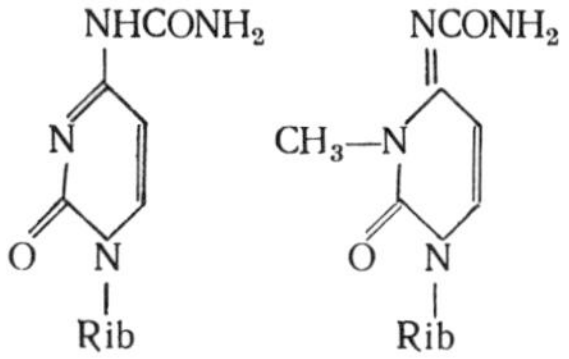

(1) (2)

These compounds hydrolyze into uridine and 1-methyluridine respectively. 1-Methylcarbamoylcytidine hydrolyzes with particular ease.

Altogether the data obtained present convincing evidence that NMU reacts not only as an alkylating agent but also as a carbamolyating agent.

Is the carbamoylation reaction significant as regards the biological properties of NMU? In our studies of the mutagenic action of NMU we have not yet been able to distinguish between the methylating and the carbamoylating effects of this compound. To answer this question we tested the mutagenic activity of some compounds acting strictly as carbamoylating agents, namely isocyanic acid and its potassium salt, KNCO. Seeds of *Pisum sativum* were either treated with the vapor of a solution of isocyanic acid in ether or immersed in this solution or in a solution of KNCO in acetate buffer at pH 4.2. The results obtained are shown in the table. Data on the mutagenic activity of NMU and NEU (N-nitroso-N-ethylurea) are included for comparison. Both the compounds tested clearly possess a mutagenic activity comparable to that of NMU. The results of the mutagenic activity of isocyanic acid in the gaseous phase are especially interesting. Here the induction of mutations was observed in conjunction with a higher germination rate than in the controls.

Effect of NMU, NEU, HNCO and KNCO on Pisum sativum

Treatment	M_1			Percentage of mutant seeds in M_2, %
	germination rate, %	survival rate, % of seedlings	sterility, % of harvested plants	
HNCO, solution in ether				
0.14%	78	100	0	18
0.07%	75	100	3.0	18
0.035%	71	100	1.5	17
0.014%	65	100	0	18
vapor, 1 hour	81	100	1.0	5
vapor, 2 hours	81	100	7.0	5
vapor, 64 hours	76	100	0	16
Control (ether)	63	100	0	3
NMU, solution				
0.005%	73	92	20.6	15
0.01%	83	89	24.5	23
NEU, solution				
0.025%	63	71	50.5	23
0.05%	7.5	58.8	100	—
KNCO, solution				
0.1%	46	78	46	11
0.2%	76	72	76	23
Control (water)	79	88	11	1

Similar results were obtained in a study of the mutagenic effect of KNCO on Arabidopsis thaliana Heynh. At pH 3 the mutagenic activity of KNCO was 30%, close to that of ethylenimine.

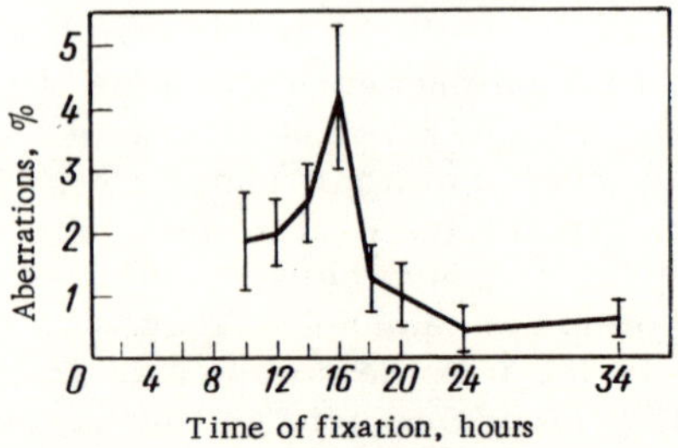

Cytogenetic effect of KNCO on cells of C. capillaris

The cytogenetic effect of KNCO on seedlings of Crepis capillaris was also examined according to the standard procedure. The seedlings were treated with a $5 \cdot 10^{-3}$ M solution of KNCO in phosphate-citrate buffer at pH 5.0. The results are shown in the figure. KNCO clearly belongs to the category of delayed action mutagens. It is distinguished by its ability to induce chromosome as well as chromatid aberrations.

As noted above, NMU similarly induces both chromatid and chromosome aberrations in contrast to alkylating compounds. The data obtained indicate that the carbamoylation reaction is responsible for the appearance of chromosome aberrations in C. capillaris seedlings treated with NMU. Preliminary experiments on the mutagenic activity of KCNO with respect to Actinomyces rimosus indicate that this compound also has mutagenic properties.

BIBLIOGRAPHY

Brookes, P. — Cancer Res., Vol. 26 (1966), 1994.

Grigorova, N. V. and N. N. Zoz. — In: "Khimicheskii mutagenez i selektsiya." Moscow, "Nauka," 1970.

Kononova, S.D., S.T. Sizova, V.V. Surkov, and L. L. Gumanov. — In: "Spetsifichnost' khimicheskogo mutageneza," p. 76. Moscow, "Nauka," 1968.

Loveless, A. and Ch. L. Hampton. — Mutation Res., Vol. 7 (1969), 1.

Michaelis, A., J. Schoneich, and R. Rieger. — Chromosoma, Vol. 16 (1965), 101.

Mueller, A.J. — Naturiwssenschaften, Vol. 52 (1965), 213.

Rapoport, I.A. — In: "Supermutageny," p. 18. Moscow, "Nauka," 1966.

Rosenkranz, H.S., M. Bitoon, and R.M. Schmidt. — J. Natn. Cancer Inst., Vol. 41 (1968), 1099.

Serebryanyi, A.M., M.A. Smotryaeva, and K.E. Kruglyakova. — Izv. AN SSSR, biol. ser., No. 4 (1969), 607.

Swann, P.F. and P.N. Magee. — Biochem. J., Vol. 110 (1968), 39.

UDC 575.24:576.851.5

INDUCED MUTAGENESIS IN A rec⁻ STRAIN OF BACILLUS SUBTILIS

A.A. PROZOROV, L.S. NAUMOV and G.V. SAVCHENKO
Institute of Genetics and Selection of Industrial Microorganisms
Central Administration of the Microbiological Industry,
Council of Ministers of the USSR, Moscow

The mutagenic effect of diethylsulfate and novembichin on a rec^- strain of Bacillus subtilis was shown to be greater than on the original strain with an unimpaired capacity for recombination.

In earlier publications we described B. subtilis mutants with an impaired capacity for recombination and an increased sensitivity to UV radiation, mitomycin C and methylmethanesulfonate (Prozorov, 1966; Prozorov et al., 1968). The rec^- mutations were localized in the same part of the chromosome in this group of mutants (Prozorov et al., 1968). Later we found that this group of strains shows a much greater frequency of spontaneous mutations (Prozorov and Barabanshchikov, 1967). This, however, did not apply to all mutations but only to some of them, distinguished by a specific localization (Prozorov and Barabanshchikov, 1968a). The transforming capacity of the respective loci varied according to the origin of the DNA — namely, from the original strain or from a rec^- strain (Prozorov and Barabanshchikov, 1968b). We assumed that this complex of properties in the rec^- strains examined resulted from a disturbance of the reparative synthesis of DNA — a process involved in recombination (Prozorov and Barabanshchikov, 1968b). If this assumption is correct, the repair of some of the lesions caused by mutagens in the rec^- strains should induce mutations in excess of those obtained in the original strain.

This work examines the frequency and characteristics of the mutations induced in the rec^-149 strain and in the original strain by treatment with diethylsulfate (DES) and novembichin, a mustard gas analog.

For a comparison of the frequency of induced mutations we determined the total number of auxotrophic mutations per viable cell of the population. This approach necessitates the examination of each colony obtained and is therefore much more laborious than the assessment of induced mutagenesis by the frequency of one or two mutations distinguished by a selective advantage (for example, resistance to an antibiotic in a medium containing this compound or mutations to prototrophy). However, it gives a better idea of the mutability of the entire bacterial genome since the loci involved in auxotrophic mutations are very numerous and are scattered more or less evenly throughout the chromosome. The results of one of the experiments are given in the following figure. The frequency of mutations obviously increases with the DES dose, reaching a peak at a survival rate of 0.1 — 0.5 %. The mutation frequency curve for the rec⁻ strain is steeper; at each dose, the mutation frequency is 2 — 4 times greater than in the original strain. Mortality at a given dose of the mutagen is greater in the rec⁻ strain than in the original strain. Even at equivalent doses, however, mutagenesis in the rec⁻ strain exceeds that in the original strain. At a survival rate of 2 %, for example, auxotrophic mutants constitute about 16 % of the total number of cells in the rec⁻ strain, compared with 8 % in the original strain (see figure).

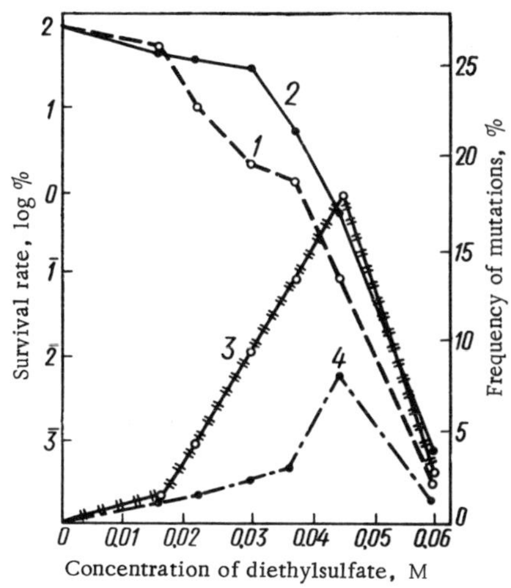

Induced mutagenesis and survival rate of cells of the rec⁻ 149 strain and the original strain on treatment with diethylsulfate.

Survival rate: 1) rec⁻149 strain; 2) original strain.
Percentage of auxotrophs: 3) rec⁻149; 4) original strain.

Thus, DES induced a greater number of mutations in the rec⁻ strain than in the original strain. However, the mechanism of action of DES is not quite clear, and it is not yet known whether the reparative synthesis of DNA participates in the repair of lesions caused by this agent. We therefore decided to test the response of both strains to the mutagenic action of compounds known to trigger repair synthesis. Hanawalt (1965) showed that repair synthesis takes place in cells treated with mustard gas analogs. Accordingly, we tested the action of novembichin, which belongs to the latter category of compounds. As in the case of DES, cells of the rec⁻ strain proved more sensitive to this mutagen than cells of the original strain. Since novembichin induced only a few mutations, we have summed up the results of all the experiments in the figure. The mutation frequency in the rec⁻ strain increases with the dose of novembichin to a level several times higher than in the original strain, towards which this compound is almost devoid of mutagenicity. In other words, the experiments with novembichin confirm the assumption that certain disturbances of the repair synthesis of DNA are characteristic of the rec^-149 strain.

Our results show that the mutagenic effect of DES and novembichin is greater in the rec^- strain than in the original strain with an unimpaired capacity for recombination. This contradicts the data of Miura and Tomizawa (1968) and of Likhoded and his colleagues (1969), who found that UV irradiation of rec^- strains of *E. coli* causes fewer mutations than in strains with a normal capacity for recombination. The discrepancy may be due to the difference in the bacterial objects and mutagens tested and in the procedures used for calculating the mutations.

BIBLIOGRAPHY

Hanawalt, Ph. — Biochem. Biophys. Res. Commun. 19, No. 3 (1965), 462.

Likhoded, L. Ya., A.A. Kiryushkina, and A.G. Skavronskaya. — Genetika 5, No. 5 (1969), 137.

Miura, A. and J. Tomizawa. — Molec. and Gen. Genet. 103, No. 1 (1968), 1.

Prozorov, A.A. — Genetika 2, No. 8 (1966), 99.

Prozorov, A.A., R.R. Azizbekyan, B.I. Barabanshchikov, and N.N. Belyaeva. — Genetika 4, No. 2 (1968), 97.

Prozorov, A.A. and B.I. Barabanshchikov. — Doklady AN SSSR 176, No. 6 (1967), 1422.

Prozorov, A.A. and B.I. Barabanshchikov. — Genetika 4, No. 5 (1968a), 80.

Prozorov, A.A. and B.I. Barabanshchikov. — Genetika 4, No. 12 (1968b), 5.

UDC 575.591

VIRUSES AND THE ORGANISM. THE FORMATION OF HYBRID GENOMES AND THE CONSEQUENCES TO BE EXPECTED AT THE CELL, ORGANISM AND POPULATION LEVELS (A HYPOTHESIS)

L.B. MEKLER
Institute of Experimental and Clinical Oncology
Academy of Medical Sciences of the USSR, Moscow

The possible types of interaction between the cellular and viral genomes and their integration are discussed. It is suggested that there may be mechanisms exerting a control over foreign genetic information and the union of viral and cellular genomes. The hypothesis formulated provides an integral explanation for a number of important phenomena in virology.

Among the multitude of phenomena involved in the interaction of viruses and cells it is possible to distinguish two groups, based respectively on 1) processes in which the cell does not control the replication of the viral nucleic acid; 2) processes in which the cell controls to a various extent the replication of the viral nucleic acid (lysogeny, cell transformation induced by DNA-containing and some RNA-containing viruses, formation of incomplete viruses, induction of an interferon).

Why does the encounter between identical virions, say type 7 adenovirus, and the same kind of cell sometimes trigger one type of process and sometimes another? What is the mechanism by which the cell achieves control over the replication of the foreign nucleic acid? Is this control mechanism the same in all the cases mentioned above? These key problems of virology remain unsolved (Luria and Darnell, 1967). The present communication is an attempt to find a solution on the basis of the existing data.

Basic hypothesis

The crucial consequences of the penetration of the virus into the cell are as follows: 1) an increase in the amount of genetic information in the cell; 2) the appearance in the cell of two "control centers" for the translation of genetic information.

However, the increase in the amount of genetic information in the cell and the appearance of two "control centers" for the translation of this information do not necessarily connote an independent operation of the genomes. In principle, such a "double" system can function in three ways: 1) the genome of the first system (virus) controls the function of the genome of the second system (cell); 2) the genome of the second system controls the function of the genome of the first system; 3) there is mutual control of the two genomes (Mekler, 1965). Such an interaction of the genomes probably takes place in reality.

It is well known that the amount of cellular DNA varies between organisms at different evolutionary stages. Consequently, a change in the amount of genetic information in the cell is not a process confined to virus-cell interactions. The increase in the amount of genetic information per cell in the course of evolution did not impair the function of the cells or kill them. Obviously, the increase in genetic information in the course of evolution was achieved in such a way that the new genetic locus, once established in the cell, came under the control of the pre-existing genome. This means that at least the second type of genome interaction does exist in reality.

What change in the new genome brings it under the control of the old one inside the cell? This question may be answered by comparing the structural organization of the genomes of organisms on different evolutionary levels with the structural organization of the "double genome" of the virus-cell system. Obviously, the principal difference between the two cases is that the genome growth occurring during evolution led to a new, structurally integral genome, whereas in the virus-cell system the viral and cellular genomes are structurally segregated.

As regards the possible consequences of this difference, all that can be said is strictly hypothetical; in our view, the lengthening of the DNA molecule which took place in the course of evolution automatically brought the new locus under the control of the general repression system of the pre-existing genome — a process which requires a structural unity between the new genome locus and the old DNA molecule.

It follows logically from this hypothesis that the actual trend of the virus-cell interaction, i. e. according to the first or the second type, depends on whether fusion of the viral genome and the cellular DNA into a single molecule is at all possible and indeed has actually occurred. The process involved here is essentially similar to the "integration" of the viral genome with that of the cell, as demonstrated in lysogeny and postulated in the case of virus-induced malignant transformations of cells (Dulbecco, 1960; Zil'ber and Abelev, 1962).

What is the molecular mechanism of genome integration? Is this mechanism the same in, for example, lysogeny and malignant transformation as in the protective action of the interferon? We would rather expect the existence of different mechanisms in view of the marked differences between these phenomena. Lysogeny and cell transformation involve a specific immunity to recurrent infection and a hereditary transfer of this property to the daughter cells. In contrast, the protection conferred by the interferon against foreign genetic information is neither specific nor transmitted to the progeny (Luria and Darnell, 1967).

Thus, there are at least two distinct mechanisms by which discrete molecules of nucleic acid fuse into a single molecule. These mechanisms must meet the following requirements. Firstly, they must be able to unite nucleic acids not only of the same type but also of different types, i. e. DNA and RNA, into a single structural system. Secondly, they must provide an explanation of the observed differences in the specificity and inheritability of the two above-mentioned types of immunity toward nucleic acids foreign to the cell.

As will be shown below, these requirements can be met by assuming that one of the mechanisms involves a process which can most simply be termed temporary hybridization with double-strand completion (Mekler, 1968) enabling the information contained in different nucleic acids, i.e. RNA and DNA, to be encoded into a single nucleic acid molecule (unlike the mechanism proposed by Campbell, 1962), while the second mechanism merely joins two nucleic acid molecules into a structure which is morphologically and functionally integral but does not constitute a continuous polynucleotide strand. For a rough assessment of the probability of the above hypothesis we shall consider the phenomena which can be expected on this basis at the virus, cell, organism, and population levels.

Phenomena to be expected at the viral and subcellular levels

Let us consider the events which may be triggered by the infection of the cell with a virus of which the nucleic acid contains areas complementary to certain sections of some of the cellular DNA molecules. One consequence of such a homology may be the formation of a hybrid molecule. On the basis of the law of mass action the probability of the formation of hybrids and their lifetime will be proportional to the concentration of viral nucleic acid in the cell. Since the normal cell lacks DNA molecules identical to those of the virus, it is obvious that only part of the viral nucleic acid strand can form a double-stranded structure with the cellular DNA (Figure 1). This, however, creates the possibility of completion of the double-stranded structure of the hybrid molecule. The composition and structure of the completed section of the cellular DNA will obviously be determined by the viral nucleic acid, while the composition and structure of the corresponding section of the viral nucleic acid will be determined by the cellular DNA. This series of events may give rise to two new nucleic acid molecules: 1) a molecule of cellular DNA which now includes a nucleotide sequence capable of reproducing all the genetic information contained in the original viral nucleic acid; 2) a molecule of viral nucleic acid now with an additional section carrying the genetic information for one of the proteins present in the organism to which the given cell belongs. The original strand of viral nucleic acid may subsequently break up, showing that there is no real integration between the viral and cellular genomes. This mechanism operates equally well with both viral DNA and viral RNA (see Figure 1), causing in each case a hereditary variability of both partners.

What changes in the properties of cells and viruses can be expected as a result of such a transcription of viral and cellular genetic information into the genome of the respective partner? As regards the cell, one of the consequences, as postulated above, is that the viral genome is subjected to the control of the induction repression system governing the function of the cellular genome. In other words, when the nucleic acid of the transforming virus appears in the transformed cell, it behaves like the normal cellular nucleic acids. This means that all the genetic information present in the cell, including that of viral origin, is transcribed strictly according to the laws governing the biosynthesis of cellular mRNA.

It is well known that only one of the DNA strands (the plus-strand) is transcribed in the cell, while the section complementary to this strand is replicative and completely repressed (Medvedev, 1968; Mahler and Cordes,

1968). On the other hand, it has been shown in many viruses that the genetic information coding for the synthesis of early proteins is stored in the plus-strand of the viral nucleic acid (in the case of double-stranded nucleic acid), whereas the minus-strand contains the genetic information for the synthesis of structural proteins (Cohen and Hurwitz, 1967; Guha et al., 1967, 1968; Mekler et al., 1970). It may be assumed, therefore, that when the genetic code for the structural proteins of the virus is located in the replicative strand of the host cell DNA, its genetic information will be fully repressed and only a certain amount of viral RNA (+ strands) and early proteins will be synthesized in the cell according to the viral genome (Figure 2). This state of affairs obviously corresponds to a virus-induced cell transformation, including viral oncogenesis. On the other hand, when the viral genetic information determining the synthesis of early proteins is located in the replicative strand of the host cell DNA, it is evident that the cell will synthesize the structural proteins of the virion but no viral nucleic acid. At the level of the virus, this leads to the formation of incomplete virions. Conversely, the normal mechanisms of the cell will induce and carry out the synthesis of those viral proteins which have become an integral part of the structure of the transformed cell.

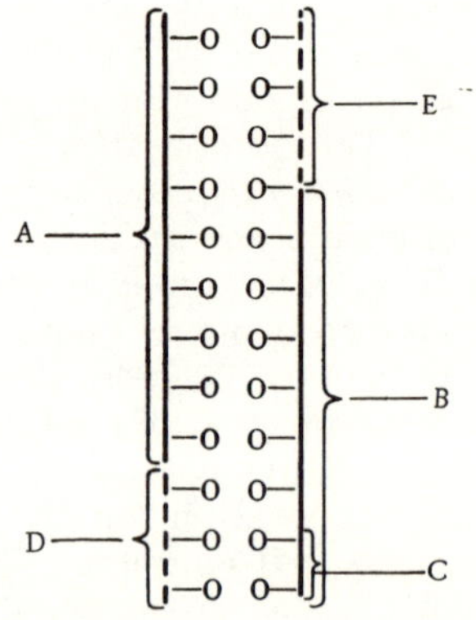

FIGURE 1. Scheme of the process of temporary hybridization with double-strand completion:

A) original strand of cellular DNA; B) original strand of viral DNA or RNA; C) viral nucleic acid locus responsible for the antigenic mimicry of the virus-host system; D) completed locus of the cellular DNA effecting the subsequent transformation of the cell; E) completed locus of the viral nucleic acid causing subsequent changes in the properties of the virus.

As for the virus, its acquisition of a locus of the cellular genome may result in the appearance of antigenic determinants analogous to those of the cell on the viral envelope; this creates a population of virions manifesting antigenic mimicry toward the organisms to which the given cell belongs (Mekler, 1965, 1967).

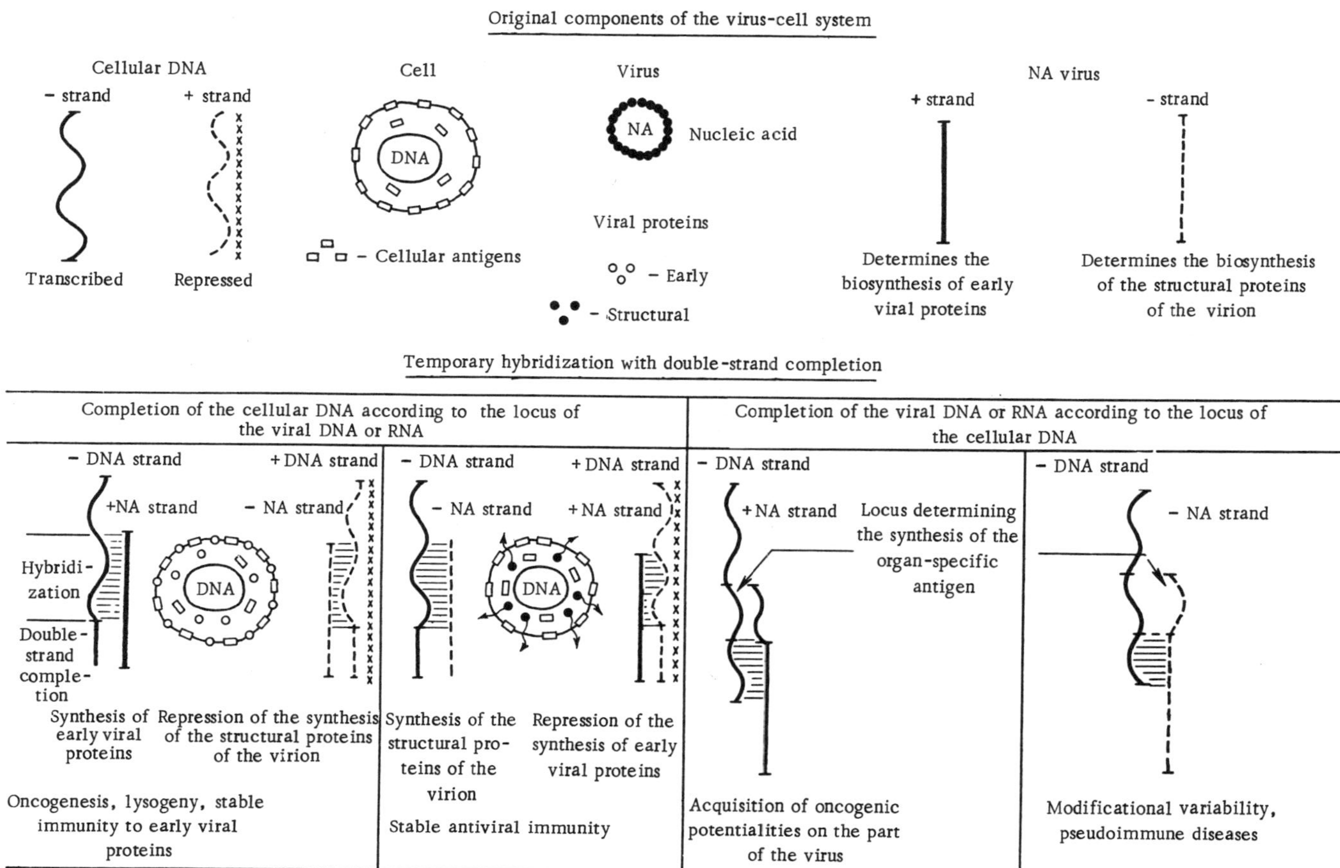

FIGURE 2. Hypothetical variations of the interaction of viral and cellular genomes and the respective biological effects.

Phenomena to be expected at the cell and organism levels

Structures which are antigenically viral but correspond morphologically to the cell may appear in the nuclear and cytoplasmic membranes of the cell under the influence of viruses belonging to certain taxons (Mekler, 1965, 1967, 1968). In the conditions outlined here, this phenomenon yields a population of viable cells the cytoplasmic membrane of which differs from that of the original cell (Figure 3). This difference may in principle be of two types, depending on whether the virus inducing the cell transformation displays antigenic mimicry with respect to the host organism (Mekler, 1968). If the virus is not characterized by antigenic mimicry, it can only induce the formation of antigenic determinants foreign to the given organism on the cytoplasmic membrane. At least two processes can be expected in this case.

1. The original homologous cells of the tissue or organ to which the original cell belongs lose the capacity to control the growth of the cell population descended from the transformed cell.

2. The organism induces an immune reaction causing elimination of the cells bearing the heterologous antigens.

Depending on the nature of the predominant process, the cells containing heterologous antigens will either lose or form a morphological structure analogous to a benign tumor (Mekler, 1968). On the other hand, more ominous events may be expected if the cell transformation is induced by a virus displaying antigenic mimicry toward the host organism. The antigenic mimicry itself may be caused by the presence in the viral genome of a locus identical with respect to its genetic information to any one of the active loci of the DNA of some cell of the organism. Since the virus inducing the transformation can in principle penetrate into the cells of a large variety of organs or tissues, the result may be something akin to transduction, i. e. a transfer by the virus of genetic information from one category of cells to another within the organism.

The organism, however, represents a community of differentiated cells which differ from one another in their organ or tissue specificity. In view of this the virus can supply the cell with the genetic information responsible for the organ or tissue specificity of a different kind of cell belonging to another tissue or organ of the given organism. In this case, antigenic determinants of the organ or tissue specificity of other tissues or organs of the given organism as well as antigenic determinants alien to the organism may appear on the cytoplasmic membrane of the transformed cell under the influence of the virus (see Figure 3). As a result, the cell becomes capable of specific contact with adjacent cells or with heterologous organs or tissues situated elsewhere in the organism. This process is manifested as an invasive growth of the transformed cells, accompanied by metastases (Mekler, 1968).

The above outline does not cover all the phenomena which can be expected at the level of the organism if the process of temporary hybridization and double-strand completion takes place in the cells. At least two other phenomena are possible: 1) the creation of a stable, lasting immunity toward the virus whose genome is "imprinted" in the cellular DNA according to the above mechanism; 2) the development of a pseudoimmune disease, i. e. a condition pathogenetically analogous to autoimmune disease but possessing a viral etiology (Mekler, 1965, 1967).

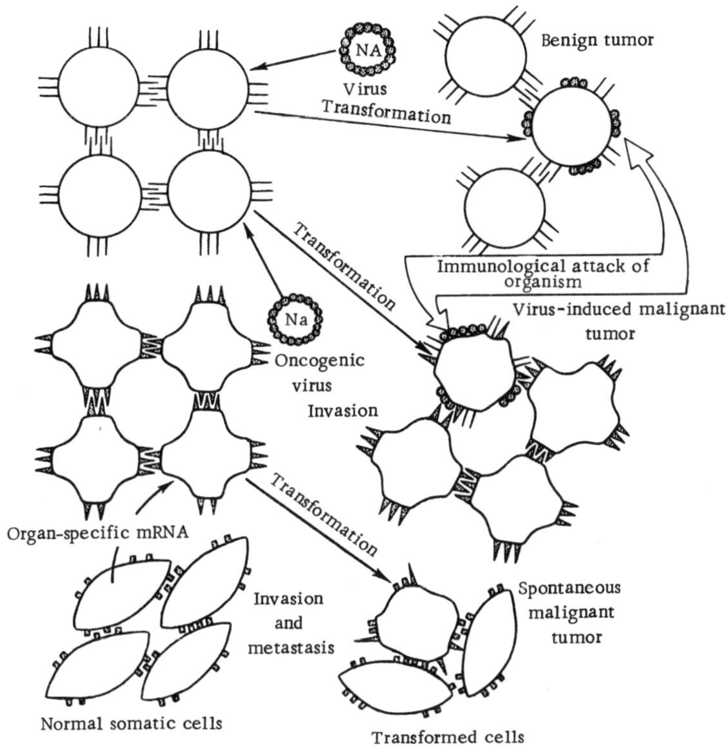

FIGURE 3. Changes in the antigenic topography of the cell surface leading to the formation of tumor cells according to the theory of oncogenesis (Mekler et al., 1968).

As noted above, one of the possible consequences of temporary hybridization with double-strand completion is the formation of viable cells producing the structural proteins of the virion but not its infectious nucleic acid (+ strand) or early proteins (see Figure 2). In principle it is conceivable that the structural proteins of the virion synthesized by such a cell might pass into the blood and there maintain a constant high level of immunity against the virus inducing the synthesis of these proteins. This may explain why some viruses, such as parotitis, measles and smallpox, cause a stable immunity which often lasts a lifetime, whereas others, for example, the influenza virus, do not induce such an immunity.

Moreover, the formation in the organism of populations of virions enclosed by a capsule containing antigenic determinants analogous to the organ or tissue-specific antigenic determinants of the transbarrier tissues necessarily triggers an immune reaction in the organism against the corresponding tissue, the result being an autoimmune type of illness. Bearing this consequence in mind, it is not difficult to conceive that brain diseases such as subacute sclerotic panencephalitis, multiple sclerosis and schizophrenia, found to be associated with autoimmune processes

(Pette et al., 1965; Heath et al., 1967) and suspected to be of a viral origin (Bychkova, 1966; Mekler et al., 1960), do indeed possess such an etiology and pathogenesis (for a more detailed account see Mekler, 1965, 1969).

Phenomena to be expected at the level of populations of organisms

Without entering into a discussion of other aspects of this question, treated in detail elsewhere (Mekler, 1967), we shall examine the case in which the surface of the modified virions bears antigenic determinants which will be tolerated by the organism subsequently infected with this particular virus (see Figure 2). Naturally, the appearance of such virions in a population of sensitive organisms will cause an epidemic because of the absence of immunological protection against the infection. At the same time the presence of a common locus in the nucleic acid of the given virus and cell — an obligatory condition if our interpretation of antigenic mimicry is correct — will very probably cause a second cycle of temporary hybridization with double-strand completion, but this time in the cells of an organism belonging to another species or to a different genetic group of the same species, another blood group, for example.

The result of a bilateral exchange of genetic information may well be an acute blastomatous process. On the other hand, if the temporary hybridization with double-strand completion modifies only the virus, the cell will die. At the same time a new population of virions with new superficial antigenic determinants necessarily arises. If the new antigenic determinants are identical with the antigenic determinants of the normal tissues, the state of tolerance toward these virions will persist and the process will increase. If the new antigenic determinants are identical with the antigenic determinants of the transbarrier tissues, the infection process must cease owing to the induction of immunological protection in the organism. Two or three weeks later, however, such patients develop autoimmune complications because of the simultaneous lesion of the corresponding transbarrier tissue. The former case obviously resembles the diseases caused by the leukosis and Rous sarcoma viruses, whereas the latter case is reminiscent of the diseases induced by the influenza virus. These viruses indeed display antigenic mimicry with respect to their hosts (Luria and Darnell, 1967).

The interferon

We examined above the mechanism by which the cell exercises control over the functioning of the genetic information introduced by foreign nucleic acids containing a locus complementary to the cellular DNA. However, the original hypothesis postulates the existence of a second mechanism uniting any foreign nucleic acid with the cellular DNA to form a morphologically integral structure. This is obviously the mechanism associated with the interferon, forming the basis of immunity. This mechanism must consequently explain not only the species specificity of the interferon but also the fact that the immunity conferred by the interferon is effected by certain macromolecules, the biosynthesis of which is induced in the interferon-containing cell when this is infected with the virus (Marcus and Salb, 1966).

Comparison of the original hypothesis with these data leads to the conception that the interferon is a particular kind of protein possessing two receptors: 1) a specific receptor permitting the interferon to bind only the DNA of cells of the given species; 2) a nonspecific receptor enabling the interferon to unite with any nucleic acid. On this basis the interferon would probably act as a bridge, uniting the foreign nucleic acid and the cellular DNA into a single morphological structure which would break down on cell division since it does not constitute a continuous polynucleotide strand. Further development of this idea leads to the conclusion that the macromolecules synthesized in interferon-containing cells under the stimulus of foreign nucleic acids are actually the repressors which, according to the basic hypothesis outlined here, are synthesized in the cell in response to the formation of an integral morphological structure uniting the cellular DNA and the foreign nucleic acid.

The hypothesis formulated here provides a universal explanation for a number of key phenomena in virology. Although this constitutes no more than indirect support for our hypothesis, it fully justifies its experimental verification.

BIBLIOGRAPHY

Bychkova, E.N. — In: "Rasseyannyi skleroz i drugie demieliniziruyushchie zabolevaniya," p.55. Minsk, "Belarus'", 1966.

Campbell. — Adv. Genet., Vol. 11 (1962), 101.

Cohen, S.N. and J. Hurwitz. — Proc. Natn. Acad. Sci. U.S.A., Vol. 57 (1967), 175.

Dulbecco, R. — Cancer Res., Vol. 20 (1960), 751.

Guha, A., W. Salser, and W. Szybalski. — Fedn. Proc. Fedn. Am. Soc. Exp. Biol., Vol. 27 (1967), 646.

Guha, A., M. Tabaczynski, and W. Szybalski. — J. Molec. Biol., Vol. 35 (1968), 207.

Heath, R.G. and I.M. Krupp. — Arch. ges. Psychol., Vol. 16 (1967), 1.

Luria, S.E. General Virology. 2nd edition. — New York, Wiley, 1967.

Mahler, H.R. and E.H.Cordes. Basic Biological Chemistry. — Harp. Har-Row,. 1968.

Marcus, P. and J. Salb. — Virology, Vol. 30 (1966), 502.

Medvedev, Zh. A. Molecular-Genetic Mechanisms of Development, pp.10, 81. Moscow, "Meditsina," 1968. (Russian)

Mekler, L.B. — Voprosy Virusologii, No. 4 (1965), 499.

Mekler, L.B. — Nature, Vol. 206 (1965), 343.

Mekler, L.B. — In: Materialy simpoziumov kluba immunologov, No. 2 (1967), 79.

Mekler, L.B. — Zhurn. Vses. Khim. Obshch. im. D.I. Mendeleeva, Vol. 13 (1968), 431.

Mekler, L.B. Surface Structures of the Virus and Cell. — Author's Summary of Thesis for the Degree of Candidate of Biological Sciences. Pushchino, 1969. (Russian)

Mekler, L.B., N.N. Lapteva, D.V. Lozovskii, and T.I. Bolezina. — Doklady AN SSSR, Vol. 130 (1960), 1148.

Mekler, L.B., M.A. Shlankevich, and V. Ya. Shevlyaghin. — Arch. ges. Virusforsch., Vol. 30 (1970), 309.

Pette, E., K. Mannweiler, O. Palacios, and B. Mütze. — Ann. N.Y. Acad. Sci. **122**, art.1 (1965), 417.

UDC 575.24

CYTOGENETIC ANALYSIS OF THE ACTION OF INHIBITORS OF DNA SYNTHESIS IN RELATION TO THE FUNCTIONAL ACTIVITY OF THE GENOME

E.D. AINGORN
Institute of General Genetics
Academy of Sciences of the USSR, Moscow

The effect of inhibitors of DNA synthesis on lymphocytes of human peripheral blood not in the S phase was examined in vitro. Since chromosomal breaks occur only when inhibitors are introduced into stimulated cells, it is concluded that activation of the genome is accompanied by reorganization of the genetic material, one of the stages of which is the synthesis of DNA.

The cytogenetic analysis of intrachromosomal processes is a valuable means of obtaining information on the structural-functional organization of the genonema of higher organisms. Almost all known genetic processes may involve a transient disintegration of the DNA molecule without, however, causing fragmentation of the chromosome. It appears that any defects formed in the DNA molecule are normally eliminated by an additional synthesis. This suggests that the detection of DNA synthesis outside the S phase of the mitotic cycle, by whatever method, can be interpreted as an indication of the occurrence of a genetic process which is not identical with replication.

In view of these considerations we examined the effect of inhibitors of DNA synthesis outside the S phase.

In vitro experiments with lymphocytes of human peripheral blood have demonstrated the efficiency of 5-aminouracil (AU) as regards the induction of chromosome breaks when applied during the G_1 and G_2 phases. On the other hand, the frequency of chromosome aberrations was not above the control level in cells treated with AU before their stimulation with phytohemagglutinin.

Essentially the same results were obtained in experiments with 5-fluorodeoxyuridine (FUDR) in which the lymphocytes were treated one hour before and one hour after stimulation.

There were no chromosome breaks in stimulated cells treated with FUDR in the presence of an equimolar amount of thymidine — an indication of the specificity of the effect of FUDR. Since chromosome breaks resulting from treatment with both inhibitors appeared only when these were introduced into stimulated cells, we conclude that the activation of the genome is coupled with a reorganization of the genetic material, DNA synthesis being one of the stages of this reorganization, as is the case with other genetic processes.

This hypothesis was confirmed in experiments with actinomycin D at doses not affecting DNA synthesis. The direct inhibitory action of the antibiotic on the genome activity of the stimulated lymphocytes reduced the cytogenetic effect of the inhibitor of DNA synthesis (FUDR) to the control level.

UDC 575.591

THE MOLECULAR GENETIC ANALYSIS OF HEMOGLOBIN MUTANTS IN MAN

A.A. STREKALOV
Institute of Medical Radiology
Academy of Medical Sciences of the USSR, Obninsk

The substitutions of amino acids in abnormal hemoglobins can be attributed to the replacement of one nucleotide in the codon determining the particular amino acid. Substitutions of the C → T type are more frequent than any others; this conclusion applies to transitions and transversions alike.

Most of the abnormal human hemoglobins known today result from gene mutations in which the change in the codons has led to single substitutions of amino acids in various polypeptides. More than 150 such hemoglobin mutants have been deciphered and described in recent publications; comparative analysis of this material sheds light on the peculiarities of these amino acid substitutions and the corresponding nucleotide changes in the DNA codons.

Comparison of the ratio between the occurrence of different amino acids in the substitutions (per 100 substitutions) and their percentage concentration in the hemoglobin — a measure of the "relative mutability" of the amino acids — shows that the substitutions involve mainly polar amino acids. The substitutions are more frequent in the non-helical sections of the polypeptide chains than in the helical sections. This indicates the relative stability of the helical sections and their importance in the three-dimensional structure of the molecule. Substitutions are more frequent in areas of the polypeptide chains close to the amino terminal than in the vicinity of the carboxyl terminal. This suggests that the amino acids situated near the C-terminal are more stable and therefore play a greater role in the formation of the tertiary and quaternary structures of hemoglobin and in the functioning of the molecule than do the amino acids in the section close to the N-terminal. Fewer mutants have been found in the α chain than in the β chain; this agrees with the concept of the different evolutionary age of the respective structural genes.

All amino acid substitutions can be attributed to point mutations, i. e. to substitutions of a single nucleotide in the codon determining the particular amino acid. Almost all triplets found in bacteria and phages could participate in substitutions which may be interpreted as proof of the universality of the code and its degeneracy in the human genome. Some amino acids determined by four or six codons are coded by different triplets in various parts of the structural gene, the difference consisting in an alternate coding by purine or by pyrimidine at the third nucleotide, as can be seen from the nature of the observed substitutions. It is evident from the frequency of occurrence of the different nucleotides in the DNA codons that the C → T substitution is much more common than all the others which occur with apparently equal frequency; this conclusion applies to transitions and transversions alike since their frequencies do not differ significantly from the average. Study of the distribution of the mutants along the gene reveals the existence of two nonmutating zones analogous to those found in the genes determining the proteins in tobacco mosaic virus and the DNA polymerase of phage T4B. This phenomenon possibly reflects inequalities in the mutability of the different parts of the structural gene regardless of the function of the protein controlled by the gene.

UDC 575.24

SOME ASPECTS OF THE MECHANISM OF CHROMOSOMAL REARRANGEMENTS

Yu.A. MITROFANOV, V.V. OTRADNOVA and A.Z. VOSKONYAN
Institute of General Genetics
Academy of Sciences of the USSR, Moscow

Treatment of cells of Crepis capillaris in phase G_1 with deoxyadenosine (DA) caused no mutagenic effect. In the S and G_2 phases, DA induced fragmentation and pulverization of chromosomes. DA modified the effect of radiation and nitrogen mustard, causing a decrease in the frequency of rearrangements and unions and an enhancement of fragmentation.

The mutagenic and modifying effect of deoxyadenosine (DA) was studied on cells of Crepis capillaris. The mutagenic effect of DA was determined at several concentrations (in M): $10^{-1}, 5 \cdot 10^{-3}, 8 \cdot 10^{-3}, 2 \cdot 10^{-3}$. None of these concentrations produced a mutagenic effect when applied to phase G_1 cells. Treatment with DA during the S and G_2 phases caused fragmentation of the chromosomes, especially in the S phase. The cytogenetic effect of DA (10^{-1} M, $5 \cdot 10^{-2}$ M, $8 \cdot 10^{-3}$ M) during phase S is characterized by the appearance of cells in which the chromosomes are massively fragmented and pulverized. When applied during the G_2 phase, DA causes chromatid breaks and especially isolocus breaks of chromosomes without reunions. The modification of the mutagenic effect of radiation or nitrogen mustard by DA consists in an increase in the number of fragmentations and a decrease in the frequency of rearrangements involving reunions. DA exerted a modifying effect on the formation of both chromatid and chromosome rearrangements, the effect on the latter being less pronounced. The modifying effect of DA on the mutation process was especially pronounced during the S phase, though modification in the G_2 phase was always evident. A brief application of DA to G_1 cells irradiated or treated with nitrogen mustard during the same phase did not enhance the mutagenic effect, apparently because of the ineffectiveness of the short inhibition of DNA repair synthesis during the G_1 phase.

UDC 575.24

THE LOCALIZATION OF *rIIts* MUTANTS INDUCED BY DIFFERENT MUTAGENS

G.F. NESTEROVA and A.A. ZAPADNAYA
Institute of General Genetics
Academy of Sciences of the USSR, Moscow

Two "hot points" were discovered in the rII gene. Three of the mutations examined lie to the left of the rIIA gene at a distance approximately equal to the size of this gene.

One hundred and eighty-two rIIts mutants were isolated on the basis of phenotypic criteria: in E. coli B at 27° — phenotype r^+, at 42° — phenotype r; in E. coli K (λ) at 27° — positive growth, at 42° — no growth. Ninety-seven of these mutants proved slightly diffuse and were mapped together with the two rIIts mutants isolated by V. N. Krylov. Of the 97 mutants examined, 31 were induced by hydroxylamine (HA), 38 by nitrosoguanidine (NG), and 28 by 2-aminopurine (AP). The localization of the mutants in the relatively small segments of the rII region was determined by the drop method of Benzer, using the standard set of deletions. The additional mapping was carried out by the standard cross method. Sixty-six mutations were mapped in gene A, eight mutations in gene B. It appears that the protein controlled by gene B is much less susceptible to amino acid substitutions leading to a change in its thermal stability. Crosses of the mutants within each segment revealed two "hot spots" — at the beginning and at the end of the A gene: in the Alb2 segment (8 mutations induced by NG and 1 mutation induced by AP) and in the A6c segment (5 mutations induced by AP, 5 by NG, and 3 by HA). The existence of these "hot spots" suggests that the A polypeptide chain contains two areas in which amino acid substitutions lead to a change in the conformation of the protein multimer controlled by the gene A. The three mutagens used were equally effective in inducing rIIts mutations. The percentage of ts mutants among the rII mutants was 10, 16, and 20 in the case of HA, NG and AP respectively. There were no marked differences in the spectrum of the mutations induced by the three mutagens used as regards position.

Twenty-seven mutants which were indistinguishable from the rIIts mutants in phenotype (plaque morphology, harvest and growth cycle in restrictive conditions) yielded recombinations on crossing with all the deletions used for mapping. These mutants were similarly crossed with all the deletions located outside the B gene. Three mutants of this group were mapped by crosses with the deletions 1272 and 187. Because of the limited number of crosses the distances indicated on the map are not additive, and their magnitude is somewhat underestimated. It is obvious, nevertheless, that the mutations studied lie to the left of the rIIA gene at a distance roughly equal to the size of the gene. The existence of "external" mutants may reveal a chromosomal area in the neighborhood of the rII region, adjacent to the A gene and functionally related to the rII region.

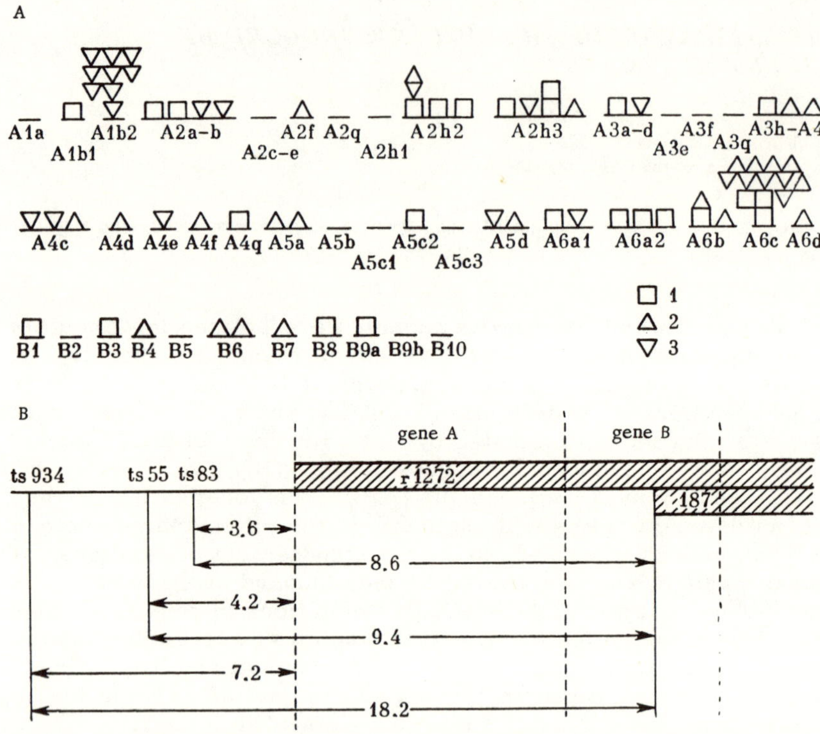

Localization of rIIts mutants:

A: 1) mutants induced by hydroxylation; 2) mutants induced by nitrosoguanidine; 3) mutants induced by 2-aminopurine; each symbol designates an independently isolated mutant.
B: mutants possessing the rIIts phenotype and located to the left of the rII A gene; the results of three experiments are given for the rII ts 55 mutant.
In segment Alb2, 8 of the mutations were induced by nitrosoguanidine, and 1 mutation by 2-aminopurine.

TRANSCRIPTION AND TRANSLATION OF GENETIC INFORMATION

UDC 575

PROPERTIES OF NEWLY FORMED D-RNA* AND THE STRUCTURE OF THE OPERON IN EUKARYOTES

Ch. KUTELL', A.P. RYSKOV, V.L. MANT'EVA, E.R. AVAKYAN
and G.P. GEORGIEV
Institute of Molecular Biology
Academy of Sciences of the USSR, Moscow

The authors present experimental support for their hypothesis on the structure of the operon (namely, the existence of informative and noninformative loci within a given operon and the predominant localization of the informative zone at the 3' end of the D-RNA precursor molecule). The existence of additional short recurrent sequences of unknown function in the distal parts of the operon is also shown.

HYPOTHESIS ON THE STRUCTURE OF THE OPERON IN EUKARYOTES

According to the hypothesis proposed by one of us in an earlier publication (Georgiev, 1969), the operon of eukaryotes consists of two kinds of sequences: acceptor cistrons mainly adjoining the promotor (acceptor or noninformative zone), and structural cistrons, located distally from the promotor (informative or structural zone) (Figure 1). Many of the acceptor sites correspond to operators, reacting with regulatory proteins.

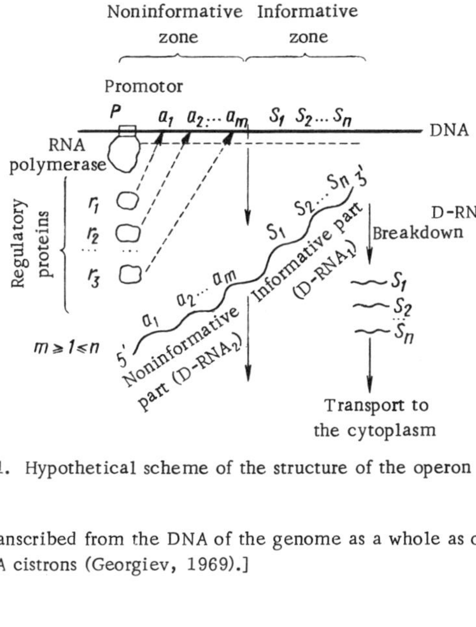

FIGURE 1. Hypothetical scheme of the structure of the operon in eukaryotes.

* [D-RNA denotes RNA transcribed from the DNA of the genome as a whole as opposed to R-RNA transcribed from the ribosomal DNA cistrons (Georgiev, 1969).]

The noninformative loci can also be represented by other, "service" polynucleotide sequences (terminators, areas recognized by specific nucleases, etc.). In this case they may also be located among the structural genes. If the RNA polymerase meets no obstacles, the noninformative part of the operon is transcribed first and the structural genes only afterwards. This process yields a giant D-RNA precursor which contains replicas of the informative zone (mostly at the 3' end) as well as of the noninformative zone (mostly at the 5' end). The noninformative part of the molecule (D-RNA_2) breaks down, while the informative part (D-RNA_1, or true mRNA) migrates to the cytoplasm (see Figure 1). This hypothesis explains many phenomena related to transcription in eukaryotes and lends itself to experimental verification.

RESULTS

Isolation of the D-RNA fraction

The object of this study was newly formed D-RNA. This was isolated by a slightly modified hot phenol fractionation procedure (Georgiev and Mant'eva, 1962; Arion et al., 1967). A suspension of Ehrlich cancer cells or a rat liver homogenate in 0.14 M NaCl was treated with phenol at 4°. The interphase obtained ("phenol nuclei") was treated with phenol two or three times at 4° and once at 40° and 55°. The interphase was then suspended in a solution containing 1% dodecylsulfate and 0.14 M NaCl and was shaken with an equal volume of phenol at 65° or 75° for 15 minutes. Part of the D-RNA passed into the aqueous phase under these conditions. The "phenol nuclei" interphase was once again treated with phenol, but this time at 85°. The yield of nuclear D-RNA was 95—100%.

The D-RNA obtained at 65° (or 75°) and 85° was further deproteinated with phenol and chloroform (to the maximum elimination of protein) and after two reprecipitations with alcohol and treatment with DNase it was centrifuged in a sucrose gradient with dodecylsulfate. The RNA was collected in fractions and aliquots were taken for determination of the optical density and radioactivity. Since a number of additional purification steps after treatment with DNAase (Georgiev and Mant'eva, 1962; Arion et al., 1967) were eliminated, the time from the start of D-RNA isolation to gradient ultracentrifugation did not exceed one day. The RNA preparations were treated with bentonite if they were to be stored for a longer time. Figure 2 shows sedimentograms of rat liver D-RNA after prolonged labeling with $Na_2H^{32}PO_4$ (three intraperitoneal injections during the course of 40 hours). The procedure described enables the D-RNA to be preserved in a nondegraded state — the peak of the 85° fraction lies at 30—35 S, and much of the material sediments within the 40—70 S zone. Even at 85°, heat treatment in a solution containing 1% dodecylsulfate, 0.14% M NaCl and phenol causes only a negligible degradation of RNA (see below). In the 65° (or 75°) fractions the peak appears in the 18 S zone, though some of the material sediments in the 40—70 S zone.

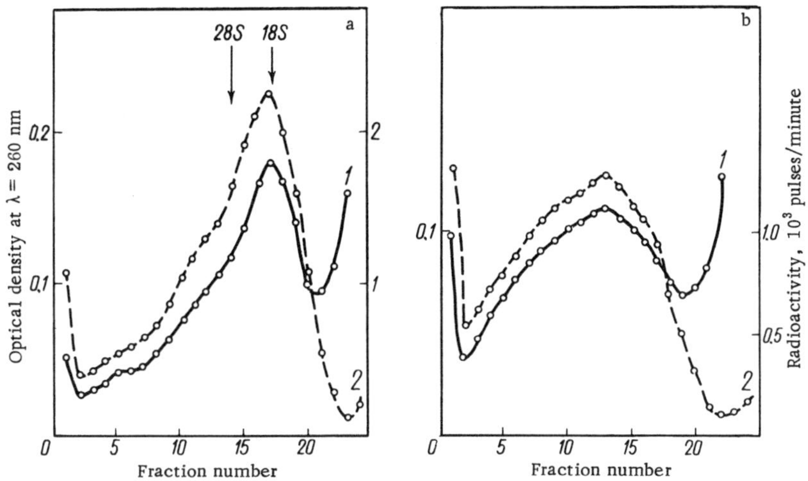

FIGURE 2. Sedimentogram of rat liver D-RNA (^{32}P-labeled, 36 hours) isolated by hot phenol fractionation: a) at 70°C; b) at 85°C; 1) optical density (D_{260}); 2) radioactivity.

The RNAs obtained resemble DNA in nucleotide composition (G+C/A+U= =0.8—0.95). This method yields a giant nuclear D-RNA fraction (M. W. = $= 2 \cdot 10^6$) as well as a lower molecular weight fraction (M. W. $=4-8 \cdot 10^5$).

Evidence that only giant D-RNA is a primary product of transcription

It is known from experiments with cell-free systems and viral RNAs that the primary transcription product bears a triphosphate residue at the beginning of the strand (at the 5' end) (Hatlen et al., 1969; Takanami, 1966). On alkaline hydrolysis this residue yields tetraphosphates (nucleoside tetraphosphates — pppXp).

We found that the alkaline hydrolysate of newly formed D-RNA from animal cells (labeled for 2—3 hours with $Na_2H^{32}PO_4$) contains pppXp and pXp residues. Only the hydrolysates of heavy D-RNA contain pppXp residues. Conversely, hydrolysates of the light fraction are comparatively rich in pXp residues and lack pppXp residues (see below).

	Heavy D-RNA	Light D-RNA
Xp, 10^6 pulses/minute	1.44	0.6
pXp		
pulses/minute	187	600
% of the total activity	0.013	0.1
pppX +ppXp		
pulses/minute	590	25
% of the total activity	0.041	0.004

This means that the true beginnings of the strands (pppXp) are found only in the heavy RNA, i. e. only the heavy RNAs of molecular weight $2 \cdot 10^6$ are primary products of transcription. As a result of transformation the heavy RNAs are split into shorter strands which correspond to cytoplasmic RNA.

Evidence for the presence of D-RNA$_1$ and D-RNA$_2$ within the same giant precursor RNAs

Some information may be obtained from an analysis of the results of an actinomycin chase. If actinomycin is added after brief radioactive labeling, there is no further synthesis of RNA and a considerable part of the preformed D-RNA is observed to break down (Harris, 1963; Gvozdev and Tikhonov, 1964; Lerman et al., 1965). In many cases the transport of D-RNA to the cytoplasm is also disturbed as a result of still unknown factors. If the transport of D-RNA is disturbed, the study of the changes of D-RNA with time will shed light on the mechanism and significance of the process of D-RNA transformation. As shown in Figure 3 and the table, about $^2/_3$ of the newly formed D-RNA of the 85° fraction breaks down after the addition of actinomycin. This breakdown is accompanied by a roughly threefold decrease in the average molecular weight (the sedimentation coefficients are reduced by a factor of 1.7). Control experiments involving the extraction of the total cellular RNA (Scherrer et al., 1963) showed that the difference in molecular weights is genuine and unrelated to the degree of possible degradation of D-RNA during treatment with phenol at 85°. The specific activity of the nuclear D-RNA in the chase-type experiments is the same as in the controls. As shown in the preceding section, only the giant D-RNAs are primary transcription products.

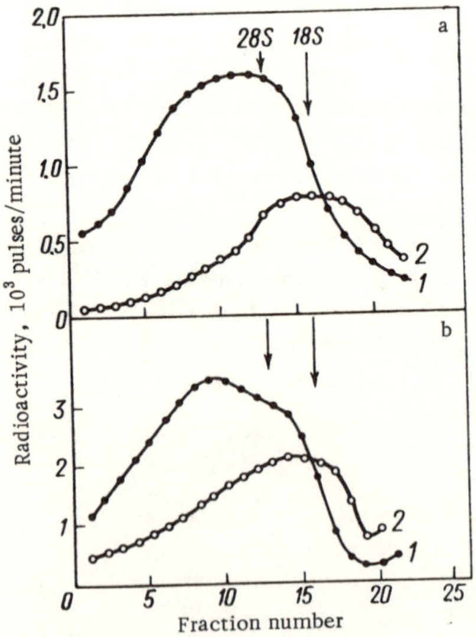

FIGURE 3. Sedimentograms of D-RNA of rat liver (a) and Ehrlich mouse ascites carcinoma (b) before (1) and after (2) the chase. Labeling with ^{14}C-orotic acid for 45 minutes.

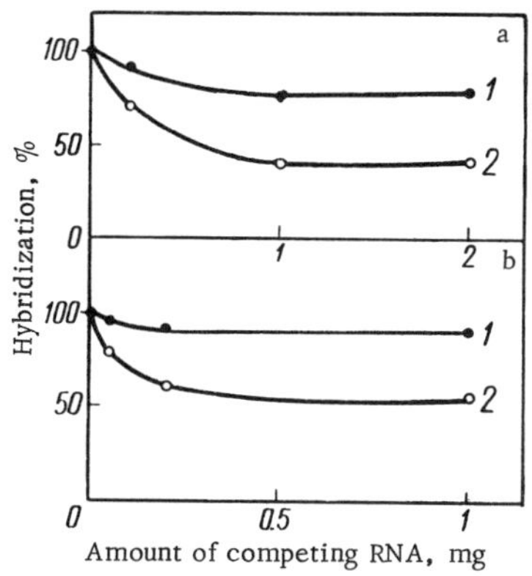

FIGURE 4. Competitive hybridization of ^{14}C D-RNA from rat liver (a) and Ehrlich mouse ascites carcinoma (b) in the presence of unlabeled polysomal homologous RNA and homologous DNA.

DNA:RNA = 50:1; 1) ^{14}C D-RNA control; 2) ^{14}C D-RNA chase.

Therefore, the threefold decrease in the average molecular weight accompanying the breakdown of $^2/_3$ of the total D-RNA while the specific activity remains unchanged can only be explained on the basis that almost all the molecules of the newly formed D-RNA undergo partial breakdown in the nucleus.

Changes in the total radioactivity, molecular weight (MW) and competitive hybridization in chase-type experiments

Variant	Total activity, % of control	MW (in the region of the peak) $\cdot 10^4$	Specific radioactivity, pulses/minute · mg RNA $\cdot 10^6$	Competition with excess polysomal RNA, %
Rat liver				
Control	100	1.9	1.47	−218
Chase	36	0.67	1.31	−46
Ehrlich mouse ascites cells				
Control	100	3.6	0.24	−23
Chase	40	1.2	0.22	−60

Experiments on the competitive hybridization of DNA and D-RNA in the presence of unlabeled polysomal RNA (using D-RNA from before and after the chase) showed that the competition between cytoplasmic mRNA and nuclear D-RNA increases markedly after the chase (Figure 4). In other words, the breakdown of part of the giant D-RNA is accompanied by its enrichment in the sectors corresponding to true mRNA. This finding shows that both D-RNA_1 and D-RNA_2 (cytoplasmic mRNA and disintegrating nuclear D-RNA) are parts of the same precursor molecules, as postulated by our hypothesis.

Evidence that true mRNA is located predominantly at the 3′ end of the giant precursor molecule

According to the above hypothesis, the segment situated at the 3' end of the precursor D-RNA corresponds to mRNA.

In order to test this assumption, different fractions of D-RNA evenly labeled with ^{32}P or ^{14}C (isolated from rat liver cells or Ehrlich mouse ascites cells) were oxidized with periodate and then reduced with ^{3}H-sodium borhydride. This led to the incorporation of tritium into the terminal nucleoside at the 3' end of the D-RNA. Next the D-RNA was subjected to alkaline hydrolysis and nucleoside fraction containing 3'-terminal nucleosides was separated from the nucleotides. Nucleoside derivatives can be separated from other radioactive compounds by paper chromatography in various solvent systems (Leppla et al., 1968).

It was found that 70—75% of the 3' ends consist of adenosine, about 15–20% of uridine and about 5–10% of guanosine and cytidine. The composition of the 3' ends is the same in the heavy and light D-RNAs — an indication of their possible conservation during the transformation of the D-RNA in the nuclei.

The final experiment consisted in the hybridization of these D-RNAs, labeled totally with ^{14}C or ^{32}P and terminally with ^{3}H, with DNA and competitive hybridization in the presence of polysomal RNA, in order to determine the position of true mRNA in the precursor molecule (see below).

	Fractions of rat liver D-RNA	
	light, 18 S	heavy, >30 S
Hybridization, %		
total (determined by ^{14}C)	13.2	13.4
3' end determined by adenosine (^{3}H)	25.6	26.6
Competition, %*		
total (determined by ^{14}C)	25.8	0
3' end determined by adenosine (^{3}H)	54.6	32.6

* Competitor — RNA from rat liver polysomes.

The percentage of hybridization determined by ^{3}H was higher in all fractions than that determined by ^{32}P, which indicates that different operons bear similar or identical terminal RNA sequences. In the case of competition between polysomal RNA and heavy D-RNA, the percentage of hybridization determined by ^{32}P is only slightly reduced whereas the hybridization determined by ^{3}H drops considerably. Hence it may be concluded that the mRNA transferred to the cytoplasm and incorporated into the polysomes is located at the 3' end of the heavy RNA precursor molecule.

The data presented demonstrate that true mRNA lies at the 3' end of the precursor molecule, as postulated by our hypothesis. They also prove that D-RNA_1 (the precursor of true mRNA) and D-RNA_2 (strictly nuclear D-RNA) form parts of the same molecules of newly formed D-RNA.

DISCUSSION AND CONCLUSION

We have presented here experimental evidence in support of our hypothesis on the structure of the operon — namely, the presence of informative and noninformative loci within the same operons and the predominant location of the informative zone at the 3' end of D-RNA.

Our hypothesis also postulated the predominant localization of recurrent DNA sequences in the area adjacent to the promoter.

It was suggested that these recurrent sequences correspond to multiple acceptor loci; by reacting with these, regulatory agents can induce mass effects at once in a large number of operons.

Data obtained in experiments with UV irradiation show that the unfinished D-RNA molecules which do not contain a 3' end are rich in replicas of the recurrent sequences of DNA (Mant'eva and Arion, 1969). We were able to demonstrate in the present work that replicas of the recurrent sequences are also present at the 3' end of the operon on the basis that the percentage of hybridization determined by ^{3}H is higher than that determined by ^{32}P.

This sector may represent a "service sequence" responsible for termination in the synthesis of D-RNA or for the protection of D-RNA during its transformation. This topic requires a separate study. The existence of a comparatively short recurrent sequence located at the end of the operon, replicas of which are transferred to the cytoplasm, explains the capacity for hybridization of cytoplasmic D-RNA — a property which could not be understood on the basis of the original model of the hypothesis.

To sum up, our findings on the structure of newly formed D-RNA confirm the basic postulates of our previously published hypothesis on the structure of the operon of eukaryotes; they also prove that the distal parts of the operon include additional short recurrent sequences, the function of which remains unknown.

BIBLIOGRAPHY

Arion, V.Ya., V.D. Mant'eva, and G.P. Georgiev. — Molekulyarnaya Biologiya, Vol. 1 (1967), 689.

Georgiev, G.P. and V.L. Mant'eva. — Biokhimiya, Vol. 27 (1962), 949.

Georgiev, G.P. — J. Theoret. Biol., Vol. 25 (1969), 473.

Harris, H. — In: "Progress in Nucleic Acid Research," Vol. 2 (1963), 20.

Hatlen, G.E., F. Amaldi, and G. Attardi. — Biochemistry, Vol. 8 (1969), 4989.

Leppla, S.H., B. Bjoraker, and R.M. Bock. — In: "Methods in Enzymology," Vol.XII, B.Grossman and K.Moldave (Eds.), p.230. New York—London, Acad. Press, 1968.

Lerman, M.I., E.A. Vladimirtseva, V. V. Terskikh, and G. P. Georgiev. — Biokhimiya, Vol.30 (1965), 375.

Mant'eva, V.L. and V. Ya. Arion. — Molekulyarnaya Biologiya, Vol. 3 (1969), 294.

Scherrer, K., H. Latham, and J.E. Darnell. — Proc. Natn. Acad. Sci. U.S.A., Vol. 49 (1963), 240.

Takanami, M. — Cold Spring Harbor Sympos. Quant. Biol., Vol. 31 (1966), 611.

UDC 575:577.1

RNA POLYMERASE AND THE CONTROL OF RNA SYNTHESIS

R. B. KHESIN
I.V.Kurchatov Atomic Energy Institute, Moscow

Original data are presented showing that apart from the Jacob-Monod repressor mechanism for the regulation of the synthesis of RNA and proteins there exists another mechanism the operation of which is based on the ability of RNA polymerase to recognize the promotors of particular genes.

The formation of mRNA is the first step of the multistage process of regulation of protein synthesis. The regulation of RNA synthesis is best known in bacteriophage in which the gene groups involved and the types of RNA produced vary from one developmental stage to another. Three groups of mRNA are formed successively during the development of T-even phages: 1) "pre-early" RNAs, formed immediately after infection even if the synthesis of proteins and DNA is totally inhibited; 2) "early" RNAs, formed only in conjunction with normal protein synthesis; the replication of viral DNA is not necessary for the synthesis of "early" RNAs; 3) "late" RNAs, the formation of which requires normal protein synthesis as well as one or two replications of viral DNA.

What are the mechanisms responsible for switching RNA synthesis from one group of genes to another? Many laboratories are engaged in the solution of this problem. Yet no conclusive answer has been found so far. The repressor theory of Jacob and Monod explained the switching off of individual genes or operons on the basis of the binding of specific repressor proteins with DNA; removal of the repressors from the DNA would then activate the genes. This theory has been confirmed and reinforced by the isolation of the repressor of the lactose operon and the repressor of the lambda phage.

As long ago as 1962 it became clear, however, that this repressor mechanism is not the only means for the control of the synthesis of mRNA, at least in the T-even phages. We found that even if the viral DNA used in experiments on the RNA polymerase of *Escherichia coli* is thoroughly purified and freed from proteins which might include repressors, the enzyme transcribes only certain genes — the genes functional in the cell during the early stage of infection (Khesin et al., 1962). Late genes are not transcribed in a system containing only DNA and RNA polymerase. This means that the control of mRNA synthesis and the selection of genes for transcription can be effected in some cases without the participation of repressors. Consequently, RNA synthesis may be controlled not only by repressor mechanisms but also by other mechanisms involving the specific interaction of only two components — RNA polymerase and the DNA template — and based on the ability of the enzyme to "recognize" particular points on the DNA.

We shall discuss here some of the mechanisms exerting a qualitative and quantitative control over RNA synthesis based on the structural peculiarities of RNA polymerase and on factors reacting directly with the enzyme, rather than with the template as is the case with repressors. Results in this field are still fragmentary and only a few topics will be reviewed here.

Various procedures including electron microscopy and biochemical experiments with labeled RNA polymerase have shown that even in the presence of excess enzyme the enzyme molecules binding to DNA are separated by distances of about 1,000 nucleotide pairs (Crawford et al., 1965; Richardson, 1966; Kamzolova et al., 1969).

The binding of the enzyme with DNA is not affected by the presence in the medium of nucleosidetriphosphates (NTP), or rifampicin-type antibiotics, inhibiting RNA polymerase, although the enzyme-DNA bond is strengthened with the initiation of RNA synthesis.

All these data show that the position of RNA polymerase on DNA is not accidental and that the enzyme cannot begin the transcription from any point on the DNA. Thus, the specificity of RNA synthesis is predetermined to some extent by the binding of enzyme and template, which can thus be regarded as the first stage in the regulation of the process. However, the number of points of attachment of RNA polymer on phage DNAs far exceeds that of the early genes to be transcribed. This means that only some of the RNA polymerase molecules attached to DNA can carry out an effective initiation and synthesis of RNA strands. The exact proportion is determined both by the properties of the DNA and the conditions of synthesis. It is known that the number of RNA strands initiated by RNA polymerase varies greatly according to the concentration of NTP and the temperature, although these factors do not influence the number of enzyme molecules attached to the DNA (Nikiforov, 1969, 1970; Bremer and Mueller, 1969; Mueller and Bremer, 1969).

A decrease in the concentration of NTP reduces the number of RNA molecules initiated. It thus appears that the points of attachment of RNA polymerase on DNA are not equivalent; effective synthesis of RNA begins only on some of them, the exact number depending on the conditions.

Thus, the assortment of newly formed RNAs is determined not only at the stage of binding of the enzyme to DNA but also during the initiation of the RNA strands. The influence of the concentration of NTP on this process may be biologically significant. It is known, for example, that amino acid starvation in bacteria inhibits the synthesis of ribosomal RNAs but only slightly reduces the synthesis of mRNAs (Edlin and Broda, 1968). Moreover, amino acid starvation of cells depletes the NTP pool (Cashel and Gallant, 1968). It may be justifiably assumed that this does not affect the synthesis of RNA on all gene groups equally.

Why does RNA polymerase initiate RNA synthesis at some points of attachment to the template but not at others?

It may be postulated that the attached RNA polymerase undergoes activation by a structural rearrangement of its molecules only at some specific points of the DNA, distinguished by particular properties. This hypothesis is suggested by the results of our experiments performed in collaboration with O. B. Astaurova (Khesin et al., 1967), according to which the enzyme undergoes a change on binding with native DNA even in the absence of NTP. This modified enzyme shows a greater stability to various salts, urea, guanidine, phenethyl alcohol, high temperature, and even trypsin. This modification of the enzyme depends on the structure of the DNA; it does not take place on single-stranded denatured DNA.

It is not yet clear, however, whether the RNA polymerase becomes activated after attachment to the template or is already potentially active in the free state. Together with M. F. Shemyakin and S. L. Stvolinskii we therefore examined the ability of E. coli RNA polymerase to synthesize polyribonucleotides (in particular poly-A) without a template. It emerged that pure RNA polymerase cannot synthesize poly-A in the absence of a template, or does so at a negligible rate. This means that the free enzyme is inactive. It appears, however, that the limiting stage is the initiation of polynucleotide strands, the stage at which the first nucleotides are bound, since the addition of oligonucleotides (oligo-A) with a free 3'-OH group stimulates the polymerization of ATP, although this process is very slow even under these conditions. Can RNA polymerase be activated by a change in its molecular structure enabling it to initiate RNA synthesis even in the absence of a template? As shown in Figure 1, enzyme preparations containing guanidine at concentrations which inhibit the synthesis of RNA and poly-A on DNA have a markedly greater capacity for the template-free synthesis of poly-A. It became clear by ultracentrifugation in a glycerol concentration gradient, however, that the guanidine-stimulated synthesis of poly-A in the absence of a template is due mainly to enzyme precursors or breakdown products with a sedimentation coefficient of 8 — 10 S, rather than intact enzyme molecules (Figure 2).

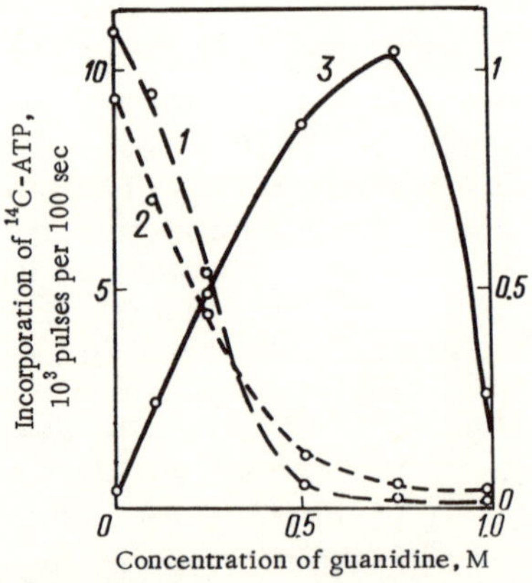

FIGURE 1. Effect of guanidine on the synthesis of RNA and poly-A by E. coli RNA polymerase:

1) RNA synthesis on phage T2 DNA. The incubation medium contained phage T2 DNA, unlabeled GTP, CTP and UTP, and ^{14}C-ATP (left scale). 2) poly-A synthesis on phage T2 DNA. The incubation medium contained T2 DNA and ^{14}C-ATP but no unlabeled NTP (left scale). 3) poly-A synthesis in the absence of a template. The incubation medium contained ^{14}C-ATP but no DNA or unlabeled NTP (right scale).

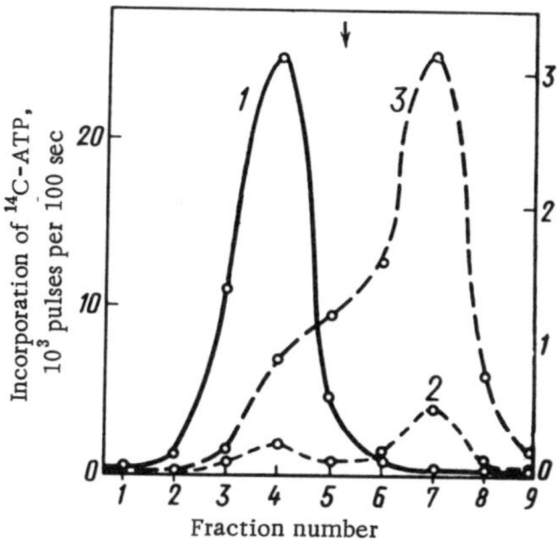

FIGURE 2. Synthesis of poly-A on T2 DNA and without a template on incubation of fractions obtained by ultracentrifugation of E. coli RNA polymerase in a glycerol concentration gradient:

1) synthesis of poly-A in the presence of DNA; T2 DNA and ^{14}C-ATP but no other NTP added (left scale); 2) synthesis of poly-A without DNA; only ^{14}C-ATP added (right scale); 3) synthesis of poly-A without DNA in the presence of guanidine chloride; ^{14}C-ATP and guanidine chloride at a final concentration of 0.5 M in the sample added (right scale). The arrow indicates the position of the catalase peak (11.3 S).

These data are not yet conclusive and require a further check for the absence of traces of other enzymes in the RNA polymerase preparations. It is obvious, however, that entire molecules of RNA polymerase cannot actively synthesize poly-A in the absence of a template. Activation of the enzyme molecules involves structural changes induced inside the cell by the action of binding with DNA. We suggest, therefore, that the role of DNA is not restricted to the complementary fixation of NTP during RNA synthesis, when the template participates directly in the reaction. The second function of the template is the activation of RNA polymerase molecules, which undergo structural changes on binding with the DNA. In this respect DNA can be regarded as a "conformational cofactor."

It is evident from these data that the selectivity of RNA synthesis on DNA, confined to the transcription of only certain genes, is determined by two factors: 1) the specific way in which the enzyme selects its binding site on the DNA, and 2) the capacity of the enzyme to become activated only at some of the binding sites on DNA, these being the only sites where efficient synthesis of RNA strands begins. The first factor has been proven, whereas the second remains hypothetical.

The selection of binding sites on DNA and the capacity for activation at these sites obviously depend on the structure of the enzyme molecules. For this reason the mechanism regulating RNA synthesis by the specific interaction between enzyme and template cannot be understood without a study of the molecular structure of RNA polymerase, the functions of its components and their possible changes in different physiological states — for example, during the different stages of development of bacteriophage.

We examined these problems by means of three mutations affecting the properties of E. coli RNA polymerase (the mutations were obtained by T. S. Il'ina, E. V. Sokolova, S. Z. Mindlin and E. S. Kalyaeva): 1) a temperature-sensitive mutation, sX, which according to S. L. Stvolinskii, partly impairs the ability of the polymerase to bind with DNA (Khesin et al., 1968); 2) mutations conferring resistance to rifampicin, which acts at the stage of RNA strand initiation (Toccini-Valentini et al., 1968, etc.); 3) mutations leading to the resistance of RNA polymerase to streptolydigin which inhibits the elongation of RNA strands (Schleif, 1969).

Experiments involving the mutant and normal enzymes and the complementation of their components in vitro (Khesin, Shemyakin, et al., 1968) showed that the RNA polymerase consists of two basic components with a molecular weight of about 300,000 and less than 100,000 respectively. These two components readily dissociate and reunite with one another. Burgess and Zillig studied the polypeptide structure of RNA polymerase (Burgess, 1969; Zillig et al., 1969). The large component (core or minimal enzyme in their terminology) consists of four polypeptides: β, β', and two α; the existence of another polypeptide, ω, is doubtful. Burgess refers to the smaller subunit as the δ factor.

All these types of mutation concern the properties of the larger component and do not affect the smaller subunit, as shown by the recombination of the larger and smaller components of the RNA polymerase of different mutants, both in the USSR, by Zh. M. Gorlenko and M. F. Shemyakin (Khesin et al., 1969) and elsewhere (Mauro et al., 1969). In the absence of the smaller subunit, the larger component binds with phage DNA but synthesizes little or no RNA on it under such conditions; RNA synthesis requires the addition of the smaller subunit, which does not bind directly with the DNA. If the complex composed of the larger component and DNA is separated by centrifugation, the addition of the smaller subunit will activate it. According to O. B. Astaurova, the association of the smaller subunit with the larger component markedly strengthens the bond of the latter with DNA. It follows from these findings that the larger component performs all the major functions of RNA polymerase i. e. binding with DNA and the initiation and growth of RNA strands. In other words, the larger component bears both the catalytic center and the binding site with respect to DNA, i. e. the "recognition" center. The smaller subunit or δ factor performs a regulatory function by activating the larger component. Travers and Burgess (1969) showed that the δ factor functions during the initiation stage only and then splits away from the enzyme molecule. Free small subunits of RNA polymerase were indeed found in the soluble phase of bacterial lysates (E. S. Bogdanova, Yu. N. Zograf, I. A. Bass and M. F. Shemyakin, 1970).

In view of the fact that RNA polymerase consists of two components, what is the significance of this structure for the control of RNA synthesis, above all quantitatively?

It has long been known that phage infection lowers the rate of RNA synthesis in cells. We found some time ago that there is a simultaneous decrease in the activity of RNA polymerase. E. S. Bogdanova, Yu. N. Zograf et al. (1970) studied the properties of the soluble phase of infected cells.

They found that a new protein, exerting a characteristic effect on RNA polymerase, accumulates in the cells about 10 minutes after infection with T2 or T4 phage. The presence of this protein exerts little influence on the activity of the complete enzyme in which the smaller subunit has already combined with the larger component. This inhibitor, however, totally prevents the association of the smaller subunit with the free large component, thus making activation of the latter impossible. Yu. N. Zograf was able to show that the inhibitor acts stoichiometrically rather than catalytically. The inhibitor binds with the large component, lacking the δ factor. In the presence of chloramphenicol, inhibiting protein synthesis, the infected cells contain free small subunits but produce no inhibitor. This suggests that the inhibitor is coded by the viral genome. The inhibitor is also synthesized in a restrictive host infected with early amber mutants of phage T4 (amN 122, for example). Thus, the inhibitor may be regarded as an early protein. It prevents the association of the δ factor with the minimal enzyme and is therefore referred to as the anti-δ factor. Thus, we are dealing here with a negative regulation of RNA synthesis, acting directly on the structure of the RNA polymerase and in particular on the association of the enzyme components, while not blocking the DNA template, as do repressors or the termination factor (ρ-factor) discovered by Roberts (1969).

So far we have spoken about the quantitative control of the rate of RNA synthesis by the interaction of polymerase components. It is becoming increasingly evident that the association of the smaller subunit with the larger component does not merely affect the specific capacity of the latter to select a particular template i. e. particular sections of the DNA for transcription. Many workers have found that in the absence of the smaller subunit the larger component synthesizes RNA on thymus DNA much more rapidly than on the DNA of T-even phages, whereas the opposite is true of the reconstituted enzyme (Burgess, 1969). Thus, the association of the smaller subunit makes the enzyme much more active towards phage DNA than towards the foreign thymus DNA. These findings suggest that substitution of the smaller subunits of the polymerase will alter the specificity of RNA synthesis if these subunits differ from one another. Hence the question arises whether a substitution of the smaller subunits (δ factor) of RNA polymerase takes place during the development of bacteriophage.

Some evidence in support of this assumption was obtained with T-even phages. It appears that the bacterial enzyme with a bacterial δ factor recognizes and transcribes only the "pre-early" genes but under normal conditions does not pass on to the next group, i.e. to the "early" genes. Travers (1969) has shown that a new δ-like factor appears in cells soon after their infection with T4 phage. The binding of this factor to the larger component of bacterial RNA polymerase apparently enables the latter to transcribe the second group of genes — namely, the "early" genes, although the "late" genes as before are not transcribed. The synthesis of RNA on this third group of "late" genes begins only after the formation of early RNAs and proteins. This process requires protein synthesis, one or several replications of DNA and the functioning of genes 33 and 55. The mechanism of this switch remains obscure.

It would be interesting to determine whether the switching on of the "late" genes is accompanied by changes in the structure of RNA polymerase.

The first question that arises in this connection is whether the RNA polymerase components found at this stage of phage development differ from the enzyme components in noninfected cells.

A. D. Gol'dfarb in our laboratory investigated this on E. coli ts19, a temperature-sensitive mutant with a large modified component of RNA polymerase (Bogdanova et al., 1969). Transfer of noninfected mutant cells to a high temperature caused a rapid drop in the rate of RNA synthesis (Figure 3). Similar results were obtained with the transfer of cells soon after their infection with phage T4. However, the transfer of infected cells, halfway through the infection process, caused only a slight decrease in the rate of RNA synthesis. This means that the synthesis of RNA during the late stage of phage development is independent of the nature of at least one of the bacterial genes coding for RNA polymerase.

Conversely, several workers (Geiduschek and Sklar, 1969; Haselkorn et al., 1969; Mizuno and Nitta, 1969) found that the addition of rifampicin to infected cells of the normal strain halfway through the infection period inhibits phage development. However, phage reproduction was not inhibited in rifampicin-resistant mutant cells. Similar effects are observed in mutants resistant to another antibiotic – streptolydigin, also specifically inhibiting RNA polymerase. This is evident from our studies (Sokolova et al., 1970) on the effect of streptolydigin and rifampicin on the development of phage T2 in cells of the wild strain and in those of the double mutant rif-r stl-r, bearing simultaneously mutations conferring resistance to both these antibiotics (Table 1). Both agents inhibit the development of the phage in the wild bacterial strain but exert only a slight effect on its reproduction in the mutant cells. It follows that at least one polypeptide of the large component of the RNA polymerase participating in the synthesis of phage RNA at the late stage of phage development is coded by the bacterial, rather than the viral, genome.

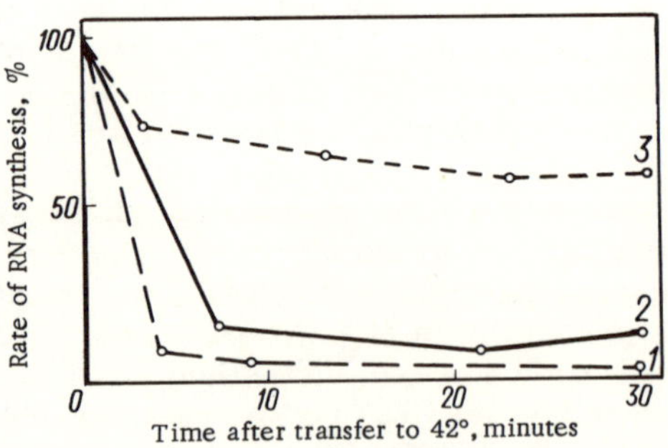

FIGURE 3. RNA synthesis after the transfer of E. coli ts19 from 30° to 42°:

1) noninfected cells; 2) cells transferred to 42° 2 minutes after infection with phage T4; 3) cells transferred to 42° 15 minutes after infection with phage T4.

Direct electrophoretic analysis of the polypeptides of the large component of the RNA polymerase of infected cells by Zillig and his colleagues (Seifert, et al., 1969) led to the conclusion that one polypeptide is modified. Other workers (Hall et al., 1969), however, did not confirm this finding. Thus, the question of whether the RNA polymerase of the cell undergoes a qualitative change halfway through the development period of T-even phages remained open. E. S. Bogdanova and A. D. Gol'dfarb tried to solve this problem by recombining preparations containing large components and small subunits isolated from noninfected cells and from cells at the late stage of infection (Table 2).

TABLE 1. Effect of antibiotics on the intracellular development of phage T2 and on the activity of RNA polymerase in extracts of noninfected E. coli rif-s stl-s and rif-r stl-r

Antibiotic (100 μg/ml)	Addition of the antibiotic, minutes after phage adsorption	rif-s stl-s	rif-r stl-r
		Yield of phage per infective center, %	
None	—	100	100
Rifampicin	0	1	72
	10	6	—
Streptolydigin	0	1	79
	10	5	—
		Incorporation of ^{14}C-UTP into RNA, %	
None		100	100
Rifampicin		1	98
Streptolydigin		6	85

It was found that the large component of the RNA polymerase of noninfected cells is more or less equally activated by small subunits originating from noninfected and infected bacteria. In contrast, the large component of the RNA polymerase of infected cells is much more strongly activated by the homologous subunits isolated from infected bacteria than by preparations from noninfected cells. Clearly, the development of phage T2 involves changes in both constituents of the enzyme — the large component and the small subunit. The latter appears to be replaced by another subunit coded by the viral genome. The nature of the change in the large component remains obscure. One possibility is that the modification takes place at the level of the presynthesized protein molecules.

The alteration of RNA polymerase with infection affects the specificity of the RNA synthesis performed by this enzyme in vitro on phage T2 DNA. This was discovered in experiments involving the hybridization of DNA with labeled RNAs synthesized by different enzymatic preparations: according to E. S. Bogdanova and A. D. Gol'dfarb, RNA polymerase preparations isolated from noninfected cells and from cells at different stages of infection produce different assortments of RNA when incubated with phage DNA.

TABLE 2. Complementation between RNA polymerase components isolated from noninfected and T2-infected cells of E. coli B

Source of small subunits	Source of large components — E. coli B cells	
	noninfected	15 minutes after infection
None	1,500*	114*
Ultracentrifugate from noninfected E. coli B	6,850	1,018
Enzyme from noninfected E. coli ts19	4,020	568
Enzyme from E. coli ts19 extracted 15 minutes after infection with phage T2	3,330	2,780

* Incorporation of ^{14}C-UTP, pulses per 100 seconds per sample.

Evidently, the control of RNA synthesis during the development of bacteriophage is based at least in part on changes in the structure of RNA polymerase and in the ability of this enzyme to recognize the different promotors on the viral DNA. This mechanism of positive control differs radically from the negative control exercised by the repressors and termination factors. Such mechanisms, based on the structural peculiarities of RNA polymerase, probably operate not only in phage-infected cells but also in normal ones. For example, it was found that the δ factor of B. subtilis disappears during sporulation (Losick and Sonenshein, 1969). Animal cells may possess simultaneously at least three distinct RNA polymerases — two in the nucleus and one in the mitochondria.

Thus, the control of RNA synthesis in the cell may depend on the regulation of the synthesis and assembly of the components of RNA polymerase. Hence the need for a study of the mechanism of the genetic control of the synthesis of RNA polymerase components and their interaction during the assembly and functioning of the enzyme molecules. Data in this field are unfortunately still scant; we shall discuss below the results obtained in our laboratory in order to show the development of this trend of research.

The first task is to determine the structure of the operon which controls the synthesis of RNA polymerase. For this purpose we used the above-mentioned mutations affecting RNA polymerase. Mapping showed that the markers are located in the following order: thi rif-s tsX arg H (Khesin et al., 1969); the stl-r mutations lie between rif-r and arg H (Sokolova et al., 1970). However, T. A. Balagurova found later that the rif-r mutations may be located quite far apart: numerous rif-s are obtained on transduction between a recipient and a donor bearing different rif-r mutations. Further studies are expected to show whether the mutations conferring rifampicin resistance on the RNA polymerase are confined to the thi-stl area or lie also between stl and arg H. An interesting discovery by M. I. Ovadis is that mutations to the rif-r phenotype, in which the cells form colonies in the presence of the antibiotic, do not increase the resistance of RNA polymerase to rifampicin. One example of this is the mutation rif-r 3. These mutations, however, do not cross at all with the typical rif-r 5 mutation, making the RNA polymerase resistant to rifampicin. Consequently, both mutations are located in the same gene and there are no grounds for assuming that rif-r 3 affects the permeability of the bacteria to the antibiotic. The conditions under which RNA polymerase operates in living cells and in vitro are probably so different that the properties of the enzyme are not always clearly revealed.

The data obtained on the mapping of the different rif-r mutations provide a basis for determining whether they always affect the same polypeptide — the β subunit (Rabussay and Zillig, 1969). Zh. M. Gorlenko has begun to investigate this problem by studying the recombination of the RNA polymerase subunits obtained after decomposition of the enzyme with urea, according to the method of Lill and Hartmann (1970). Other experiments under way involve merodiploids heterozygous with respect to different RNA polymerase mutations (see below). The first results showed clearly that at least the rif-r 18 and stl-r 18 mutations affect one polypetide. This is evident from the fact that the heterozygous cells rif-s stl-sF' rif-r 18 stl-r 18 contain RNA polymerase molecules of only two types — one of them resistant to both rifampicin and streptolydigin, the other sensitive to both antibiotics. Enzyme molecules resistant to one inhibitor and sensitive to the other are totally absent, although such molecules could well arise by a free recombination of subunits during the formation of the enzyme in the cell. It appears that the stl-r mutations affect the β polypeptide in the same way as the rif-r mutations. Similar experiments are now being carried out with other rif-r, stl-r and tsX mutations.

Thus, the question of the location of the mutations affecting RNA polymerase proved to be highly complicated. The reason for this is not that phenotypically analogous mutations can be located in different parts of the comparatively large chromosomal region examined, but also that the various types of RNA polymerase mutations interact in a very peculiar way. For example, the mutations rif-r 1 or rif-r 5 yield a lethal combination with tsX. On the other hand, this same tsX mutation partly corrects another mutation to rifampicin resistance; M. I. Ovadis was able to show by equilibrium dialysis that the protein of the double mutants tsX rif-r 76 binds rifampicin to a much greater extent than does the enzyme of cells carrying only the rif-r 76 mutation. The tsX mutation proved to be a suppressor of the rif-r 76 mutation. All these relationships greatly distort the results obtained by crossing strains bearing RNA polymerase mutations. At the same time they indicate that RNA polymerase has a very labile structure and that a change in one of its parts induces structural changes in other parts of the molecule.

Besides the reciprocal action of the different types of mutation, we examined the principles of dominance of the different alleles causing monotypic changes, such as rif-r and rif-s or stl-r and stl-s. Using the method of Low (1968), T. S. Il'ina created for this purpose a set of episomes bearing different alleles of RNA polymerase structural genes; this work was done with the Reeves strain, in which chromosomal transfer to the recA recipient begins from the ilv-met B region. The following evidence shows that, as a rule, sensitivity to the antibiotics dominates phenotypically over resistance. This agrees with the findings of Austin and Scaife (1970) on the domination of rif-s over rif-r and in part with those of Yura and Igarashi (1968) on mutations conferring resistance to streptovaricin — a rifampicin analog. The alleles of sensitivity to the antibiotics dominate regardless of whether they lie in the episome or in the chromosome.

Chromosomal markers	Episomal markers	Phenotype*
rif-s stl-s	rif-s stl-s	rif-s stl-s
rif-s	rif-r 1	rif-s
rif-r 1	rif-s	rif-s
rif-s	rif-r5	rif-s
rif-r 5	rif-s	rif-s
rif-r 76	rif-s	rif-s
stl-r 18	stl-s	stl-s
rif-s stl-s	rif-r 18 stl-r 18	rif-s
rif-r 18 stl-r 18	rif-s stl-s	rif-s stl-s

* In a number of cases the sensitivity to streptolydigin could not be determined because of the impermeability of the cells to this agent.

Zh. M. Gorlenko and M. I. Ovadis found that the cells of merodiploids heterozygous for the rif or stl alleles contain roughly equal numbers of enzyme molecules resistant and sensitive to the corresponding antibiotic (Table 3). This means that the RNA polymerase genes situated in the chromosome and in the episome function at an equal rate. These results were perplexing — why was it that even in the presence of a large amount of enzyme resistant to the antibiotic the cells were still unable to multiply and form colonies? To answer this question we examined the synthesis of RNA in mixtures of enzymes obtained from rif-r stl-r and rif-s stl-s. strains. In the presence of excess DNA template both enzymes operate simultaneously, and addition of the antibiotic lowers the rate of RNA synthesis to a level corresponding to the maximum activity of the resistant mutant RNA polymerase (Table 4). The presence of the sensitive enzyme from the wild strain has practically no effect on RNA synthesis after addition of the antibiotic: the different molecules of RNA polymerase function independently of one another in the presence of excess DNA.

The results are quite different if the incubation mixture contains an insufficient amount of DNA template. Under such conditions the enzyme molecules compete for template even if the presence of the antibiotic inactivates the sensitive RNA polymerase. As shown in Table 4, the enzyme resistant to the antibiotic displays only a fraction of its peak activity when mixed with the sensitive RNA polymerase. The reason for this is that the sensitive enzyme binds with DNA even though the antibiotic prevents it from synthesizing RNA. At the same time, the sensitive enzyme blocks some of the promoters and prevents them from uniting with RNA polymerase molecules resistant to the antibiotic. Under in vivo conditions, the presence of the antibiotic therefore causes the temporary or prolonged switching off of some genes and consequently the death of the cells. Indeed, the incorporation of ^{14}C-uracil into the RNA of rif-r rif-s cells is inhibited as early as 5—10 minutes after the addition of rifampicin (incorporation stops even sooner in the case of haploid rif-s cells). A mechanism securing the dominance of sensitivity to antibiotics over resistance to them as a result of competition between enzyme molecules for template probably exists also in other systems apart from RNA polymerase. Such a competition for

TABLE 3. Effect of rifampicin and streptolydigin on the synthesis of RNA in extracts of rif-s stl-s/F' rif-r 18 stl-r 18 heterogenotes

Antibiotic	Chromosome					
	rif-s stl-s		rif-r 18 stl-r 18		rif-s stl-s +rif-r 18 stl-r 18 episome	
First experiment						
Not added	2,870	100**	2,301	100**	2,695*	100**
Rifampicin	51	1.8	2,536	110	1,349	50.4
Streptolydigin	60	2.1	1,688	73.3	1,005	37.3
Rifampicin and streptolydigin	10	0.3	1,888	81.7	891	33.2
Second experiment						
Not added	2,270	100	2,628	100	1,735	100
Rifampicin	26	1.1	2,324	88.8	875	50.3
Streptolydigin	51	2.2	2,030	78.5	921	53.2
Rifampicin and streptolydigin	22	1.0	1,782	67.9	846	48.8

* Pulses per 100 seconds.
** As % of the value in the absence of antibiotics.

TABLE 4. Effect of rifampicin and streptolydigin on the synthesis of RNA during the incubation of mixtures of enzymes isolated from E.coli rif-s stl-s and rif-r stl-r

Experiment	T2 DNA, μg/ml	Enzymes*		Incorporation of ^{14}C-UTP, pulses per 100 seconds per sample		
		rif-s stl-s	rif-r stl-r	without antibiotics	with rifampicin	with streptolydigin
1	100	+	—	20,837	58	530
		—	+	26,667	28,502	22,245
		+	+	37,839	24,749	25,548
	2	+	—	1,221	1	25
		—	+	5,108	4,681	3,617
		+	+	3,753	1,552	1,465
2	2	+	—	4,988	22	331
		—	+	3,362	2,887	3,129
		+	+	6,705	1,794	1,649
3	6	+	—	25,618	31**	
		—	+	18,710	15,545	
		+	+	34,622	7,874	

* The symbol "+" indicates the presence of enzyme, " — " its absence.
** Rifampicin and streptolydigin were added together in experiment 3.

template must exist also in other enzyme systems associated with a template, ribosomes for example, distinguishing them from nontemplate enzymes.

Thus, the study of RNA polymerase mutations reveals the interaction of the different molecules of the enzyme. In some cases certain RNA polymerase molecules can act as repressors of other molecules.

Our main aim was to determine the factors controlling the synthesis of RNA polymerase. It might be expected that mutations affecting the quantity of RNA polymerase in the cell could be obtained by the use of merodiploids heterozygous with respect to the alleles of resistance to the antibiotics. If the expected mutation lowers the number of molecules of the sensitive enzyme, the resistance allele in the trans position will begin to dominate.

On the basis of these considerations S. Z. Mindlin embarked on a search for mutations leading to the domination of rif-r over rif-s. She seeded met B argH$^+$ stl-s rif-s thi/F' met B argH stl-r rif-r thi heterogenotes on a medium containing rifampicin. Some of the selected variants appeared to be of particular interest to us. Their properties can be outlined as follows. The stl-s rif-s/F' rif-r mutant cells had a stl-r rif-r phenotype. On removal of the episome with acridine, the bare chromosome gave rise to the met B argH$^+$ stl-s rif-s thi$^+$ phenotype; these cells proved to be also temperature-sensitive because of a new mutation, designated tsM. The RNA polymerase isolated from haploid cells bearing the tsM mutation did not differ from the enzyme of the wild strain rif-r tsM$^+$ in any of the properties examined (resistance to rifampicin, sensitivity to high temperature, ability to bind DNA, optimum concentration of NTP, etc.). As shown in Table 5, rifampicin completely inhibits the synthesis of RNA during the incubation of enzymes isolated from cells bearing tsM mutations. On the other hand, if the stl-s rif-s tsM recipient acquired any of the episomes bearing stl-r and rif-r, the result was again dominance of stl-r rif-r.

TABLE 5. Effect of rifampicin and streptolydigin on the synthesis of RNA in extracts of heterogenotes bearing the mutations rif-r and stl-r in the episome and rif-s, stl-s and tsM in the chromosome, inducing the dominance of rif-r and stl-r

Chromosome	Episome	Phenotype	Incorporation of ^{14}C-UTP in the presence of rifampicin and streptolydigin, % of the value in the absence of antibiotics	
			first experiment	second experiment
rif-r 18 stl-r 18	None	rif-r stl-r	83.7	75.7
rif-s stl-s ts$^+$	None	rif-s stl-s	0.6	0.9
rif-s stl-s ts$^+$	rif-r 18 stl-r 18	rif-s stl-s	27.3	35.0
rif-s stl-s tsM22	rif-r 18 stl-r 18	rif-r stl-r	77.5	73.6
rif-s stl-s tsM65	rif-r 18 stl-r 18	rif-r stl-r	91.4	—
rif-s stl-s tsM71	rif-r 18 stl-r 18	rif-r stl-r	101.4	77.9
rif-s stl-s tsM143	rif-r 18 stl-r 18	rif-r stl-r	—	92.3

Extracts of such stl-s rif-s tsM/F' stl-r rif-r cells contain predominantly a streptolydigin- and rifampicin-resistant enzyme which accounts for nearly all the RNA polymerase (Table 5). This qualitatively differentiates the stl-s rif-s tsM/F' stl-r rif-r merodiploids from the original stl-s rif-s tsM/F' stl-r rif-r merodiploids, containing roughly equal amounts of stl-s rif-s and stl-r rif-r enzymes. Mapping of the four tsM mutations by transduction showed that all of them are located near argH.

The nature of the tsM mutations remains obscure. Experiments comparing the effect of the tsM mutations on the same rif- alleles occupying cis- and trans- positions with respect to tsM are in progress.

How can we explain the fact that haploid tsM cells are not significantly poorer in RNA polymerase than tsM^+ cells? An answer to this question can be found on the basis of the following data. It is known from the literature that the quantity of various enzymes depends on the number of their structural genes and is doubled in merodiploids bearing an episome with these genes (Pittard and Ramakrishnan, 1964). Yet we were unable to detect a doubling of the activity of RNA polymerase in merodiploids bearing episomes with RNA polymerase genes. Even the addition of isolated large components (core enzyme) and small subunits (δ factor) of RNA polymerase to the extracts did not create any appreciable differences. Moreover, I. A. Bass found that the activity of RNA polymerase is not affected by various physiological conditions, such as amino acid starvation, etc.

This suggests that the amount of RNA polymerase in the cell is regulated to a constant level which remains unchanged even in the presence of a double set of the pertinent genes in merodiploids. However, only further research can show to what extent this hypothesis is valid.

Much remains to be discovered about the control of the synthesis of RNA polymerase. However, even now it is clear that research in this field is essential for an understanding of the complex mechanism of control of the synthesis of different types of RNA in the cell. These mechanisms are numerous indeed. They involve not only repressors and termination factors but also the switching off of RNA polymerase by agents inhibiting the union of its components, conditions affecting the function of the enzyme and in particular the concentration of substrates, the specificity with which the enzyme recognizes the promoters of DNA, depending on the properties and structure of the enzyme and finally, the regulation of the synthesis of RNA polymerase components. Only a comprehensive approach will reveal fully the laws of transcription — the fundamental stage in the manifestation of the genome.

BIBLIOGRAPHY

Austin, S. and J. Scaife. — J. Molec. Biol. 49, No. 1 (1970), 263.

Bogdanova, E.S., Yu. N. Zograf, I.A. Bass, and A.D. Gol'dfarb. — In: "Struktura i geneticheskie funktsii biopolimerov," p.234. Moscow, Inst. Atom. Energy, 1969.

Bogdanova, E.S., Yu. N. Zograf, I.A. Bass, and M.F. Shemyakin. — Molekulyarnaya Biologiya 4, No. 3 (1970), 435.

Bremer, H. and K. Mueller. — J. Molec. Biol., 43, No. 1 (1969), 109.

Burgess, R.R. — J. Biol. Chem. 244, No. 22 (1969), 6168.

Burgess, R.R., A.A. Travers, J.J. Dunn, and E.K.F. Bautz. — Nature 221, No. 5175 (1969), 43.

Cashel, M. and J. Gallant. — J. Molec. Biol. 34, No.2 (1968), 317.

Crawford, L.V., E.M. Crawford, J.P. Richardson, and H.S. Slayter. — J. Molec. Biol. 14, No. 2 (1965), 593.

Edlin, G. and P. Broda. — Bact. Rev. 32, No. 3 (1968), 206.

Geiduschek, E.P. and J. Sklar. — Nature 221, No. 5183 (1969), 833.

Hall, B.D., G. Hager, K.L. Fields, and F. Beguin. — Bull. Soc. Chim. biol. 51. Nos. 10 — 11 (1969), 1397.

Haselkorn, R., M. Vogel, and R.D. Brown. — Nature 221, No. 5183 (1969), 836.

Kamzolova, S.G., V.F. Manyakov, M.F. Shemyakin, O.B. Astaurova, N.A. Kiselev, and R.B. Khesin. — Molekulyarnaya Biologiya 3, No. 1 (1969), 74.

Khesin, R.B., O.B. Astaurova, M.F. Shemyakin, S.G. Kamzolova, and V.F. Manyakov. — Molekulyarnaya Biologiya 1, No. 5 (1967), 736.

Khesin, R.B., Zh. M. Gorlenko, M.F. Shemyakin, S.L. Stvolinskii, S.Z. Mindlin, and T.S. Il'ina. — Molekulyarnaya Biologiya 3, No. 6 (1969), 920.

Khesin, R.B., S. Z. Mindlin, Zh. M. Gorlenko, S.L. Bogdanova, and T.P. Afanas'eva. — Biokhimiya 27, No. 6 (1962), 1092.

Khesin, R.B., M.F. Shemyakin, Zh. M. Gorlenko, S.Z. Mindlin, and T.S. Il'ina. — Molekulyarnaya Biologiya 2, No. 6 (1968), 898.

Lill, U.Y. and G.R. Hartmann. — Biochem. Biophys. Res. Commun. 39, No. 5 (1970), 930.

Losick, R. and A.L. Sonenshein. — Nature 224, No. 5214 (1969), 35.

Low, B. — Proc. Natn. Acad. Sci. U.S.A. 60, No. 1 (1968), 160.

Mauro, E., L. Snyder, P. Marino, A. Lamberti, A. Coppo, and G.P. Tocchini-Valentini. — Nature 222, No. 5193 (1969), 533.

Mizuno, S. and K. Nitta. — Biochem. Biophys. Res. Commun. 35, No. 1 (1969), 127.

Mueller, K. and H. Bremer. — J. Molec. Biol. 43, No. 1 (1969), 89.

Nikiforov, V.G. — In: "Struktura i geneticheskie funktsii biopolimerov," p.255. Moscow, Inst. Atom. Energy, 1969.

Nikiforov, V.G. — FEBS Letters 9, No. 3 (1970), 186.

Pittard, J. and T. Ramakrishnan. — J. Bact. 88, No. 2 (1964), 367.

Rabussay, D. and W. Zillig. — FEBS Letters, Vol. 5 (1969), 104.

Richardson, J.P. — J. Molec. Biol. 21, No. 1 (1966), 83.

Roberts, J.W. — Nature 224, No. 5225 (1969), 1168.

Schleif, R. — Nature 223, No. 5210 (1969), 1068.

Seifert, W., P. Qasba, G. Walter, P. Palm, M. Schachner, and W. Zillig. — Europ. J. Biochem. 9, No.3 (1969), 319.

Sokolova, E.V., M.I. Ovadis, S.M. Gorlenko, and R.B. Khesin. — Biochem. Biophys. Res. Commun. 41, No. 4 (1970), 870.

Toccini-Valentini, G.P., P.Marino, and A.J. Colvill. — Nature 220, No.5164 (1968), 275.

Travers, A.A. — Nature 223, No. 5211 (1969), 1107.

Travers, A.A. and R.R. Burgess. — Nature 222, No. 5193 (1969), 537.

Yura, T. and K. Igarashi. — Proc. Natn. Acad. Sci. U.S.A. 61, No. 4 (1968), 1313.

Zillig, W., E. Fuchs, P. Palm, D. Radussay, and K. Zechel. — Sympos. "RNA Polymerase and Transcription," p.151, 1969.

UDC 575:577.1

NUCLEAR-CYTOPLASMIC RELATIONSHIPS IN A CELL-FREE SYSTEM CONTAINING ISOLATED LIVER NUCLEI

G.A.DVORKIN, V.F. EFIMOVA and L.I. KOGANITSKAYA
Institute of General Genetics
Academy of Sciences of the USSR, Moscow

It was shown by competitive RNA-DNA hybridization that when liver cell nuclei are incubated in the cytoplasm of hepatoma cells the synthesis of RNA in the nuclei is specifically inhibited. The RNA synthesized does not differ from that isolated from the tumor cells. Conversely, when nuclei of hepatoma cells are incubated in the cytoplasm of liver cells the spectrum of RNAs synthesized is somewhat expanded. It is suggested that the cytoplasm of the tumor cells contains a high molecular weight factor responsible for the specific restriction of RNA synthesis in liver cells.

The control of transcription in animal cells remains obscure even though a number of hypotheses have been proposed in relation to this problem (Georgiev, 1970; Britten and Davidson, 1969). The cytoplasm obviously performs a crucial role in the manifestation of the genetic information in the cell. This is evident above all from the works of Gurdon and Brown (1965) and Harris (1968). However, biochemical analysis of the nuclear-cytoplasmic relationships described in these works is hampered by serious methodological difficulties arising mainly from the scarcity of homogeneous material, especially as regards the acquisition of heterokaryons.

Thompson and McCarthy (1968) developed a new methodological approach, examining the effect of the cytoplasm on isolated nuclei of animal cells. They determined the rate of RNA synthesis in isolated nuclei and the changes in this occurring on incubation of the nuclei in homologous and heterologous cytoplasm. The nature of this phenomenon, however, was not examined.

The aim of the present work was to study the specific features of transcription in isolated nuclei and the role of the cytoplasm in this process. Hybridization and competitive hybridization procedures were used to determine the characteristics of the DNA-like RNA (D-RNA) synthesized. It was shown that the D-RNA synthesized in isolated nuclei incubated in the cytoplasm of homologous cells (rat liver and Seidel ascites hepatoma cells) is the same as that formed in intact cells (on the basis of competitive hybridization with DNA).

Incubation of liver cell nuclei in the cytoplasm of hepatoma cells specifically inhibits the intranuclear synthesis of RNA, the result being a narrower spectrum of RNA molecules in comparison with the RNA synthesized in nuclei incubated in homologous cytoplasm.

Competitive hybridization experiments showed that the RNA synthesized under such conditions (in vitro) does not differ from that isolated from the tumor cells. Conversely, the incubation of hepatoma nuclei in the cytoplasm of liver cells widens somewhat the spectrum of RNA molecules synthesized. The cytoplasm of tumor cells apparently contains a high molecular weight factor which specifically restricts RNA synthesis in liver nuclei. This restriction of RNA synthesis takes place on the frequently recurring DNA cistrons.

Finally, it was shown that the cytoplasm of liver cells decreases the permeability of the membranes of isolated nuclei to RNA, whereas the cytoplasm of hepatoma cells enhances it.

METHODS

The experimental material used was rat liver and Seidel ascites hepatoma cells. All manipulations concerned with the isolation of nuclei and cytoplasm were carried out in the cold.

Liver cell nuclei were isolated according to Pogo et al. (1966), with a minor modification. The liver was extracted from rats starved for 24 hours, placed in a chilled solution containing 0. 32 M sucrose and 0.003 M $MgCl_2$, and homogenized in 10 volumes of this solution in a PT-2 tissue homogenizer for 2 minutes. The homogenate was filtered through three layers of gauze and centrifuged for 7 minutes at 800 g. The sediment was resuspended in 30 volumes of a solution containing 2.2 M sucrose and 0.001 M $MgCl_2$ at a density of 1.270 g/cm^3, after which it was homogenized in the same blender for 1 minute. The suspension was layered onto the solution used for the preparation of the tissue suspension and was centrifuged at 20,000 g for one hour. The sediment, containing the nuclei, was resuspended in a solution containing 0.01 M Tris buffer at pH 8.3, 0.004 M $MgCl_2$ and 0.25 M sucrose, and was centrifuged for 6 minutes at 1,000 g. The nuclei were frozen and stored at —30°. The morphology of the nuclei was established by microscopical examination.

Liver cell cytoplasm was obtained as follows. Livers of starved rats were placed in a solution containing 0.2 M Tris buffer at pH 8.0, 0.05 M $MgCl_2$, and 0.03 M β-mercaptoethanol, and were homogenized in a Potter glass homogenizer with a Teflon pestle at a tissue : solution ratio of 1 :1. The homogenate was centrifuged at 12,000 g for 15 minutes. The supernatant was again centrifuged at 105,000 g for one hour and the freshly prepared supernatant fluid obtained was used as the cytoplasm in the incubation mixture.

Seidel hepatoma cells were obtained from the ascitic fluid of crossbreed rats on the 5th day after implantation of the tumor (the tumor was implanted by an intraperitoneal injection of 0.5 ml of ascitic fluid into rats weighing 100 — 120 g). The cells were washed in 0.14 M NaCl solution containing 0.01% heparin and were centrifuged at 200 g for 3 minutes. Washing was repeated 3 — 5 times until total removal of the blood.

Hepatoma nuclei were isolated as described for liver cell nuclei.

Seidel cell cytoplasm was obtained as follows. After the centrifugation of the washed cells the sediment was suspended in the same solution as that used in the preparation of liver cytoplasm in a 1:1 ratio, and the cells were disrupted in a Potter homogenizer. The further isolation of the cytoplasm was carried out as described for the liver cytoplasm.

The in vitro incubation of nuclei for the determination of the intranuclear synthesis of RNA was carried out at 37° for 5 minutes. The complete incubation mixture consisted of liver or hepatoma nuclei (0.5 — 0.7 mg nuclear DNA), the cytoplasm of these cells (7 mg protein), the four nucleoside triphosphates (ATP, GTP, UTP, and CTP) each at a concentration of 0.27 mM, one of them (ATP or UTP) ^{14}C-labeled (0.25 μcurie) and buffer solution (0.2 M Tris buffer pH 8.0, 0.05 M $MgCl_2$, 0.03 M β-mercaptoethanol). The total volume of the sample was 0.25 ml. The protein was determined according to Lowry et al. (1961) and the DNA according to Dische (Georgiev, 1968).

The nuclei were isolated from the sample and the D-RNA was extracted from them by thermal fractionation (Georgiev and Mant'eva, 1962). The radioactivity of the preparations was determined in a MARK-1 scintillation counter. In many experiments the D-RNA was extracted from the complete incubation mixture including the nuclei. D-RNA from liver and hepatoma cells was obtained by the thermal fractionation method and liver DNA by the phenol method (Georgiev, 1968). For the isolation of ^{14}C-RNA from the liver and hepatoma each rat was injected intraperitoneally with 300 — 400 μcuries of orotic acid together with 50 μg of actinomycin D (to inhibit the synthesis of ribosomal RNA) 4 hours before being killed.

RNA-DNA hybridization was carried out in DNA gels cross-linked by UV irradiation (Arion and Georgiev, 1967) at a DNA:RNA ratio equal to 50 in double-strength SSC (0.3 M NaCl and 0.03 M sodium citrate) at 65° for 36 hours. Various amounts of unlabeled competitor RNA, apart from the labeled RNA, were added to the DNA in the competitive hybridization experiments. The total amount of RNA in all samples was equalized by an appropriate addition of RNA from E. coli.

RESULTS AND DISCUSSION

D-RNA synthesis in isolated nuclei

As shown in the figure, RNA synthesis proceeds on an appreciable scale in isolated nuclei of liver and hepatoma cells. The addition of liver cytoplasm to the incubation medium stimulates RNA synthesis not only in liver nuclei but also in hepatoma nuclei. Moreover, the enhancement of RNA synthesis is greater in hepatoma nuclei than in liver nuclei. In contrast to liver cytoplasm, the cytoplasm of the tumor cells caused no increase in RNA synthesis in the hepatoma nuclei and even inhibited the process in liver nuclei. This effect of the tumor cytoplasm differs from that described by Thompson and McCarthy (1968), who found that the influence of tumor cytoplasm (Ehrlich ascites cells and L cells) on the nuclei exceeded that of the cytoplasm of normal cells on the nuclei of rat liver cells and hen erythrocytes. A special test in the presence of a hepatic ribonuclease inhibitor (Gribnau et al., 1970) showed that the lack of enhancement (or the slight enhancement) of RNA synthesis on addition of hepatoma cytoplasm is not due to any ribonuclease activity of the latter. Liver polysome preparations* were used in many experiments as a test for the lack of cytoplasmic RNase activity; the polysomes did not break down during incubation of the nuclei with the cytoplasm.

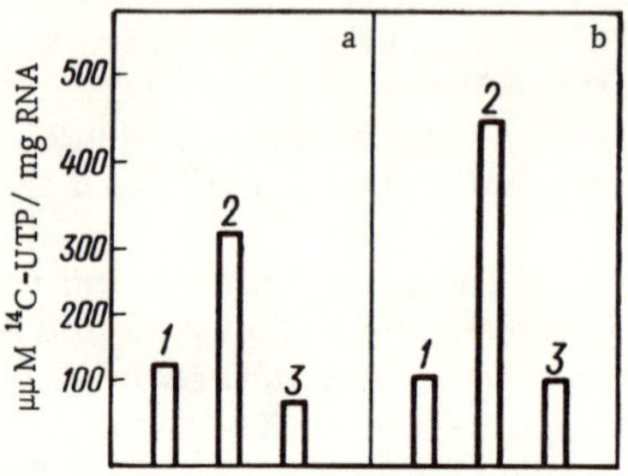

Incorporation of ^{14}C-ATP into the D-RNA of liver cell nuclei (a) and Seidel hepatoma nuclei (b) under different conditions of incubation.

Incubation of the nuclei in the presence of: 1) Tris buffer; 2) liver cytoplasm; 3) hepatoma cytoplasm.

Subsequently we examined the RNA of isolated nuclei incubated under different conditions. However, we were particularly interested in the comparison of the RNAs synthesized in liver and tumor nuclei in vivo and in vitro respectively. This study was based on competitive RNA-DNA hybridization.

Comparative analysis of the RNAs synthesized in isolated nuclei and in cells in vitro

The hybridization experiments were performed with ^{14}C-labeled and unlabeled RNAs isolated from liver and from hepatoma, as well as from previously isolated liver nuclei incubated in vitro in liver cytoplasm (N_L^L), and from Seidel cell nuclei incubated in homologous cytoplasm (N_S^S).

The results of the hybridization of D-RNA (^{14}C-labeled) with DNA in the presence of different competing RNAs are given in the following table.

* Kindly supplied by M.I. Lerman.

Source of ^{14}C-RNA	Source of competitor RNA	Hybridization, %	Source of ^{14}C-RNA	Source of competitor RNA	Hybridization, %
N_L	—	11.0	N_S	—	9.1
N_L	N_L^L	2.0	N_S	N_S^S	2.3
N_L	N_L	1.2	N_S	N_S	2.0
N_L^L	N_L	1.4	N_L	N_S	8.1
N_L^L	N_L^L	2.2	N_S	N_L	2.1

Note. N_L and N_S designate nuclei isolated from liver cells and from Seidel hepatoma cells respectively; N_L^L and N_S^L — liver cell nuclei incubated in liver cytoplasm and hepatoma cytoplasm respectively; N_S^L and N_S^S — Seidel cell nuclei incubated in liver cytoplasm and hepatoma cytoplasm respectively.

The D-RNA isolated from N_L^L clearly competes well with the D-RNA from liver cells, and vice versa. Similar results were obtained with the RNA from N_S^S and from Seidel cells. Special experiments were performed in order to determine the effect of the annealing temperature on the specificity of competition in the hybridization of rat liver DNA with D-RNA from different sources.

Source of unlabeled competitor RNA	Hybridization, %	Competition, %
At a temperature of 60°		
None	14	0
Rat liver	1.2	91
Seidel hepatoma	9	36
E.coli	11	21
At a temperature of 65°		
None	11	0
Rat liver	1.2	89
Seidel hepatoma	7	36
E.coli	10.8	2

Note. DNA:RNA = 50, duration of annealing 36 hours; competitor RNA:D-RNA (^{14}C-labeled) = 50.

The results obtained indicate that the RNAs synthesized in isolated nuclei of liver and hepatoma are the same as those formed in the respective intact cells. Naturally, this conclusion applies only to the RNAs isolated in the conditions of competitive hybridization (concentration of nucleic acids, duration of annealing) described above, i. e. under conditions in which hybridization takes place only at frequently recurring sections of the DNA. These areas are apparently not structural but perform an auxiliary function, regulating the transcription of the structural genes (Anan'eva et al.,

1968; Church and McCarthy, 1968). Our conclusion on the similarity between in vitro and in vivo transcription applies in particular to RNA formed on these sections.

It would be interesting to compare the spectra of the RNAs synthesized in liver and hepatoma cells, and more specifically their competitive relationships as revealed in hybridization tests. As shown above, liver RNA competes efficiently with Seidel hepatoma RNA, reducing the hybridization of the latter from 9.1 to 2.1%. Hepatoma RNA, however, reduces the hybridization of liver RNA to a much lesser extent, from 11 to 8.1%.

Hepatoma RNA clearly competes with only a fraction of the liver RNA, suggesting that the set of RNA molecules synthesized in hepatoma cells is "poorer" than that found in lever cells. Since the liver and hepatoma presumably possess identical genomes, it appears that hepatoma cells transcribe less of the genome than do liver cells. In other words, transcription is restricted in the tumor cells. This restriction apparently concerns the regulatory (recurrent) parts of the genome. We assumed that this restriction of transcription is due to the influence of the tumor cytoplasm. This assumption was tested in vitro by incubating isolated nuclei of liver cells in the cytoplasm of hepatoma cells and performing competitive hybridization experiments with the RNA synthesized under such conditions. The results of these experiments are given below.

Source of ^{14}C-RNA	Source of competitor RNA	Hybridization, %	Source of ^{14}C-RNA	Source of competitor RNA	Hybridization, %
N_L	N_L	1.2	N_S	N_S	2.0
N_L	N_S	8.1	N_S	N_L	2.1
N_L	N_L^L	2.0	N_S	N_S^S	2.3
N_L	N_L^S	4.6	N_S	N_L^S	2.8
N_L	N_S^L	6.4	N_S	N_S^L	2.3
N_L	N_S^S	7.3			

The RNA from liver nuclei incubated in hepatoma cytoplasm competes with labeled liver RNA to a lesser extent than does RNA from N_S^L (see lines 3 and 4 of the above table). At the same time N_S^S RNA competes with liver RNA less efficiently than does N_S^L RNA (lines 5, 6). All the RNAs examined compete well with hepatoma RNA (lines 7—11).

It follows that the spectrum of RNAs synthesized in isolated nuclei is determined by the cytoplasm in which these nuclei are incubated. Moreover, the cytoplasm of hepatoma cells apparently exerts a specific restrictive effect on transcription in liver nuclei and narrows the spectrum of newly formed RNAs. These findings agree with the data obtained in vivo on liver and tumor cells, as do the above data on the intranuclear synthesis of RNA in vivo and in vitro.

The cytoplasmic factor in Seidel hepatoma cells responsible for the restriction of transcription in liver cell nuclei

As stated above, the ribonuclease activity of the hepatoma cytoplasm was tested by incubating liver nuclei in the cytoplasm of hepatoma cells in the presence of liver cytoplasm, which contains a potent RNase inhibitor. The restriction of RNA synthesis obtained in these experiments was the same as in the absence of liver cytoplasm (see above). Thus it may be concluded that hepatoma cytoplasm contains a factor restricting transcription. This factor is absent in liver cytoplasm, and its nature remains unknown. Preliminary studies show that it is not a low molecular weight compound, since it is not lost by filtering the cytoplasm through Sephadex G-75. It is sensitive to heat (heating the cytoplasm at 50° for 10 minutes completely inactivates it) and to the freeze-thawing procedure.

Apart from the effect of the cytoplasm on transcription in the nuclei, we examined the permeability of the nuclear membranes to RNA on incubation of the nuclei in vitro.

Effect of the cytoplasm on the permeability of the nuclear membrane to D-RNA on in vitro incubation of the nuclei

This question was studied by hybridizing DNA with labeled nuclear RNA, using as the competitor either RNA isolated from a system containing nuclei and cytoplasm or RNA isolated from such a system after previous removal of the nuclei from it. The following data show that the permeability of the nuclear membranes of liver and hepatoma cells on incubation differs according to whether the incubation mixture contains liver or hepatoma cytoplasm in vitro.

Source of ^{14}C-RNA	Source of competitor RNA	Hybridization, %	Source of ^{14}C-RNA	Source of competitor RNA	Hybridization, %
N_L	$N_L^L \rightarrow N_C$	2.0 7.0	N_S	$N_S^S \rightarrow N_C$	2.3 3.3
N_L	$N_L^S \rightarrow N_C$	4.6 4.2	N_S	$N_S^L \rightarrow N_C$	2.0 6.8
N_L	$N_S^L \rightarrow N_C$	6.4 7.2	N_S	$N_L^S \rightarrow N_C$	3.0 4.4

Liver cytoplasm lowers the permeability of the nuclear membrane of hepatoma cells, thus impeding the release of synthesized D-RNA into the cytoplasm. On the other hand, in the presence of hepatoma cytoplasm the RNA synthesized in the tumor nuclei does pass into the cytoplasm.

We have shown that the in vitro incubation of isolated nuclei in cytoplasm provides both a system sufficiently homologous to intact cells, at least with respect to transcription, and a model for the study of nuclear-cytoplasmic relationships. Such a model system possesses considerable advantages for biochemical analysis, such as the availability and homogeneity of the material, the possibility of controlling the conditions in the medium surrounding the nuclei, etc.

The ability of tumor cytoplasm to restrict the transcription process, revealed in the present study, may prove to be an essential property of tumors. Our conclusion regarding the restriction of transcription is not contradicted by the published data, according to which tumors can synthesize proteins not generally found in normal tissues. As mentioned above, it seems that we are dealing not with structural genes but with the recurrent regulatory sequences of nucleotides in DNA. The nature of the factor responsible for this restriction is still obscure; it is not known whether its effect is confined to liver nuclei or whether it exists also in other tumor cells apart from the Seidel hepatoma. Further research may answer these and other questions.

BIBLIOGRAPHY

Anan'eva, L.N., Yu. V. Kozlov, A.P. Ryskov, and G.P. Georgiev. — Molekulyarnaya Biologiya **2**, No.5 (1968), 736.

Arion, V. Ya. and G.P. Georgiev. — Doklady AN SSSR **172**, No. 3 (1967), 716.

Britten, R.Y. and E.H. Davidson. — Science, Vol. 165, 1969.

Church, R.B. and B.J. McCarthy. — Biochem. Genet., Vol. 2 (1968), 55.

Georgiev, G.P. — In: "Khimiya i biokhimiya nukleinovykh kislot." Moscow, "Meditsina," 1968.

Georgiev, G.P. — Usp. Sovrem. Biol. 69, No. 3 (1970), 33.

Georgiev, G.P. and V.L. Mant'eva. — Biokhimiya, Vol. 27 (1962), 949.

Gribnau, A.A.M., J.G.G., Schoenmakers, M. van Kraaikamp, M. Hilak, and H. Bloemendal.— Biochim. biophys. Acta **224**, No. 1 (1970), 55.

Gurdon, J.B. and D.D. Brown. — J. Molec. Biol., Vol. 12 (1965), 27.

Harris, H. — In: "Nucleus and Cytoplasma," Oxford, 1968.

Lowry, H.O., T.N. Rosenbourg, G.A. Farr, and R. Randall.—J. Biol. Chem., Vol.193 (1951), 265.

Pogo, A.O., V.G. Alfrey, and A.E. Mirsky. — Proc. Natn. Acad. Sci. U.S.A. **56**, No. 2 (1966), 550.

Thompson, H.P. and R.J. McCarthy. — Biochem. Biophys. Res. Commun. 30, No. 2 (1968), 166.

UDC 575.24

A NEW TYPE OF SUPERSUPPRESSOR IN YEAST

S.G. INGE-VECHTOMOV and V.M. ANDRIANOVA
Leningrad State University

A functional test for allelism, carried out with a series of independently arising recessive suppressors, revealed the existence of two unlinked genes: S_1 and S. The existence of indicator genes which partly or fully neutralize the suppression was demonstrated. It is suggested that S_1 and S_2 control certain proteins participating in translation.

The study of the suppression phenomenon is of particular importance for understanding the genetic control of translation. Saccharomycetes yeasts are known to possess supersuppressors analogous to the nonsense suppressors of bacteria (Hawthorne and Mortimer, 1963). The supersuppressors decipher the nonsense codons UAA and UAG (the existence of the third nonsense codon has not been proved in yeasts) (Hawthorne, 1969). A characteristic feature of these suppressors is their dominance. The available knowledge on supersuppressors supports the hypothesis that the mechanism deciphering nonsense codons involves mutational alterations of tRNAs. Dominant suppressors of the same class of specificity, i. e. hindering the manifestation of the same set of mutant alleles, can arise by mutation in various (up to eight) genes. Moreover, all suppressors of a particular class translate the nonsense codon into the same amino acid residue. For example, class I supersuppressors restore the normal phenotype of the nonsense mutant with respect to cytochrome C by introducing a tyrosine residue into the polypeptide chain at the position corresponding to the nonsense codon (Sherman et al., 1968).

In bacteria and yeasts, the relationship between nonsense suppressors and nonsense codons is subject to the rules of codon-anticodon relationships as defined by the wobble hypothesis of Crick (1966). Moreover, yeasts contain classes of suppressors which are postulated by this hypothesis but have not yet been found in bacteria, in particular the so-called strict ochre suppressors (Hawthorn and Mortimer, 1968) and the missense-nonsense suppressors (Simarov et al., see p. 233 of this collection).

We describe here a hitherto unknown type of recessive supersuppressor in the yeast Saccharomyces cerevisiae.

The experimental material consisted of yeast strains from the Peterhof Genetic Collection, bearing mutations inducing a requirement for adenine, histidine, lysine and methionine as markers. The methods of obtaining these mutants and isolating multiple-marker yeast lines are described in a separate communication. The strains were cultured at 20°C.

Previously, we discovered recessive supersuppressors together with dominant ones in revertants with respect to adenine requirement. The recessive suppressors are sharply distinguished from the dominant ones by the pronounced depression of the strains carrying them (Inge-Vechtomov, 1964). We had two groups of strains which showed simultaneous requirements for adenine and histidine and could immediately revert to prototrophy for both characters. Some haploids carrying the ad_{1-6} and his_x mutations reverted on account of both dominant and recessive suppressors. This permits a comparative study of the two types of suppressors. Other haploids, carrying the ad_{1-14} and his_x mutations, reverted simultaneously with respect to both characters but strictly on account of recessive suppressors, which enabled us to obtain the reversions of interest at a probability of 1.

TABLE 1. Results of the tetrad analysis of diploids containing complementary and functionally allelic recessive suppressors

Diploid gentoype	Tetrads isolated	Number of viable spores in the tetrad		
		4	3	2
$\frac{a}{\alpha}\,\frac{s_{48}}{+}\,\frac{+}{s_{57}}\,\frac{+}{ad^{sb}_{1-6}}\,\frac{ad_{2-177}}{+}$	17	1	13	3
$\frac{a}{\alpha}\,\frac{s_{48}}{+}\,\frac{+}{s_{57}}\,\frac{+}{ad^{sb}_{1-6}}\,\frac{ad_{2-226}}{+}$	10	2	7	1
$\frac{a}{\alpha}\,\frac{s_{48}}{s_{70}}\,\frac{+}{ad^{sb}_{1-6}}\,\frac{ad_{2-226}}{+}$	8	7	1	—

Note. a and α are alleles controlling the affiliation to opposite types of pairing; ad) requirement for adenine; sb) suppressibility; s) recessive suppressor.

All the recessive suppressors isolated by us (over 300) were divided on the basis of the functional test for allelism into two complementary groups, apparently corresponding to two distinct genes. This assumption was tested by tetrad analysis of a diploid diheterozygous for complementary suppressors. The results obtained are given in Table 1. We found three groups of tetrads, differing in the viability of the ascospores in a nearly dihybrid ratio (1:4:1). At the same time the diploid containing allelic suppressors had nearly 100% viable ascospores. Table 2 presents the interpretation of the results obtained. It is clear from Tables 1 and 2 that the two genes, s_1 and s_2, representing the sites of the recessive suppressor mutations, are not linked together and that the mutation of both genes in a single haploid genome has lethal consequences.

While studying the comparative mutability of the recessive and dominant suppressors (Table 3) we isolated a series of revertants which were sensitive to high temperature and died at 36° (ts phenotype). Of the 68 revertants carrying recessive suppressors, 14 had the ts phenotype. At the same time, not a single temperature-sensitive form was detected among the 475 revertants carrying dominant suppressors. Thus, the appearance of ts revertants is characteristic only of the recessive supersuppressors.

TABLE 2. Interpretation of the results obtained on segregating a diploid diheterozygous for s_{48} and s_{57}

Ratio of tetrad types	Segregation into tetrads		
	by suppressors		by colony size
1P	s_{48}	+	Small
	s_{48}	+	"
	+	s_{57}	"
	+	s_{57}	"
4T	s_{48}	+	"
	s_{48}	s_{57}	None
	+	s_{57}	Small
	+	+	Large
1N	+	+	"
	+	+	"
	s_{48}	s_{57}	None
	s_{48}	s_{57}	"

N o t e. P) parental ditype; T) tetratype; N) nonparental ditype.

The recessive suppressors of the ts strains were subjected to a functional test for allelism with respect to the suppressor effect in a medium lacking adenine and histidine at 20° and with respect to temperature sensitivity in a complete medium at 36° (Table 4). In this experiment the prototrophic diploids bearing allelic suppressors had the ts phenotype, whereas the diploids possessing nonallelic suppressors and therefore auxotrophic for adenine and histidine proved to be temperature resistant. This means, firstly, that alleles of recessive suppressors control the ts phenotype and secondly, that the ts phenotype can arise as a result of mutations in each of the two s genes.

TABLE 3. Effect of selection conditions on the s:S ratio in mutants bearing the $ad^{sb}_{1-6}\ his^{sb}_{x}\ lys_{A12}$ and $ad^{sb}_{1-6}\ his^{sb}_{x}\ met_{A1}$ genotypes

Medium	s	S	s:S
Minimal	22	224	0.098 ± 0.018
Minimal + ad	79	230	0.343 ± 0.027
Minimal + his	21	174	0.116 ± 0.023

N o t e. The minimal medium contained methionine and lysine; ad, his, lys, and met are mutations inducing a requirement for adenine, histidine, lysine and methionine; sb) suppressibility; S) dominant suppressor; s) recessive suppressor.

In order to prove conclusively that the ts phenotype is due to a recessive suppressor we carried out a tetrad analysis of a diploid homozygous for the his mutation and for the recessive suppressor of the ts haploid. In 15 complete tetrads of this hybrid there was only a normal segregation (2:2) with respect to histidine requirement at 20°. All his haploids, i. e. those with a recessive suppressor, died at a temperature of 36° (see figure).

TABLE 4. Allelism of mutations to temperature sensitivity and s_1 and s_2 mutations

Strains of pairing type α	Strains of pairing type a			
	2-1A-P190 $(s_1\ ad^{sb}_{1-6}\ his^{sb}_{x}\ lys_{A_{12}})$		3-1A-P190 $(s_2\ ad^{sb}_{1-6}\ his^{sb}_{x}\ lys_{A_{12}})$	
	minimal 20°	complete 36°	minimal 20°	complete 36°
2-4B-P192 $(s_1\ ad^{sb}_{1-6}\ his^{sb}_{x}\ met_{A_1})$	+	−	−	+
24-4B-P192 $(s_1\ ad^{sb}_{1-6}\ his^{sb}_{x}\ met_{A_1})$	+	−	−	−
25-4B-P192 $(s_2\ ad^{sb}_{1-6}\ his^{sb}_{x}\ met_{A_1})$	−	+	+	−
31-4B-P192 $(s_2\ ad^{sb}_{1-6}\ his^{sb}_{x}\ met_{A_1})$	−	+	+	+
32-4B-P192 $(s_1\ ad^{sb}_{1-6}\ his^{sb}_{x}\ met_{A_1})$	+	−	−	+

The appearance of the ts phenotype as a result of mutations in the s_1 and s_2 genes in conjunction with the total recessiveness of the class of suppressors examined implies that we are dealing here with a mutational inactivation (probably incomplete, juding from the existence of ts phenotypes) of some of the proteins participating in translation. The code specificity of the recessive suppressors — namely, their ability to suppress two types of nonsense codons, apparently UAA and UAG (Inge-Vechtomov and Simarov, 1967), suggests that these mutations concern proteins involved in the termination of the synthesis of polypeptide chains.

If this is so, the mutations examined are not suppressors in the strict sense of the term but merely create suitable conditions for the manifestation of the potential polyvariability or ambiguity of translation and thus make the nonsense codon comprehensible. The latter assumption is supported by our previous finding that the manifestation of recessive suppressors depends on the genotypic background of the cell (Inge-Vechtomov and Andrianova, 1970). It also agrees with the fact that this class of supersuppressors has not been found so far in material of a different genetic origin, notably in the strains used by Hawthorne and Mortimer.

It has been shown by Phillips and his colleagues (1969) that nonsense codons can be translated as a result of disturbances in the termination apparatus, at least in bacteria. Our interpretation of the results obtained agrees also with the nature of the interaction of recessive and dominant suppressors. Thus, haploids bearing both types of nonsense suppressors show a markedly diminished growth rate even in comparison with haploids bearing recessive suppressors only. This phenomenon may result from the further disturbance of termination occurring with the appearance in the cell of a tRNA recognizing the nonsense codons and competing with the partly inactivated termination apparatus more efficiently than is the case in the wild type strains. These changes may lead to "fouling" of the ribosomes with polypeptides which cannot be detached and cross-linking of polypeptides controlled by neighboring translatable sectors of mRNA.

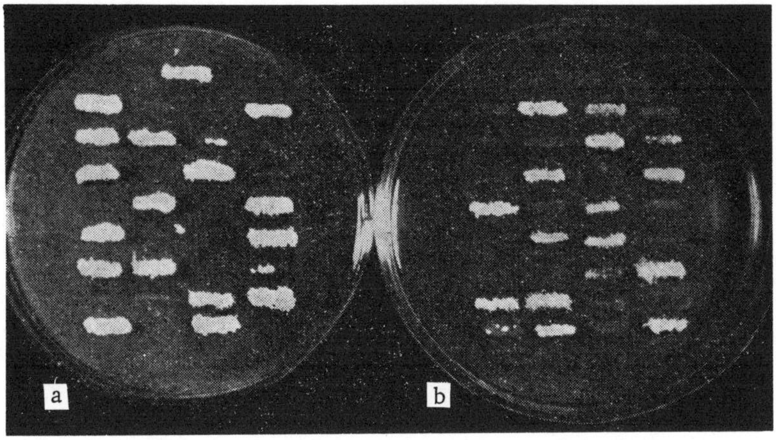

Segregation in eight tetrads with respect to temperature sensitivity in a complete medium — YEPD, 36° (a), and with respect to histidine requirement — C-hisS, 20° (b) in a diploid heterozygous with respect to the S^{ts} mutation. The complete genotype of the diploid is

$$\frac{a}{a}\ \frac{S^{ts}}{+}\ \frac{his_{x}^{sb}}{his_{x}^{sb}}\ \frac{ade_{1\text{-}14}^{sb}}{+}\ \frac{lys_{a12}}{+}\ \frac{+}{lis_{a21}}$$

BIBLIOGRAPHY

Crick, F. — J. Molec. Biol., Vol. 9 (1966), 457.

Hawthorne, D. — J. Molec. Biol., Vol. 3 (1969), 71.

Hawthorne, D. and R. Mortimer. — Genetics, Vol. 48 (1963), 618.

Hawthorne, D. and R. Mortimer. — Genetics, Vol. 60 (1968), 735.

Inge-Vechtomov, S.G. — Vest. LGU (Leningrad State University), Vol. 9 (1964), 112.

Inge-Vechtomov, S.G. and V.M. Andrianova. — Genetika 6, No. 11 (1970), 103.

Inge-Vechtomov, S.G. and B.V. Simarov. Investigations into Genetics, No. 3, p.127. Leningrad State University, 1967. (Russian)

Phillips, A., D. Schlessinger, and D. Apirion. — Genetics, Vol. 61 (1969), 47.

Sherman, F., J. Stewart, and R. Gilmore. — Biochim. biophys. Acta, Vol. 161 (1968), 270.

Simarov, B.V., L.N. Mironova, and S.G. Inge-Vechtomov. — This Collection, p.233.

UDC 575:577.1

NUCLEAR AND MITOCHONDRIAL GENETIC CONTROL OF PROTEIN BIOSYNTHESIS IN MITOCHONDRIA

V.S. GAITSKHOKI and O. I. KISELEV
Institute of Experimental Medicine
Academy of Medical Sciences of the USSR, Leningrad

On incubation with radioactive inorganic phosphate, isolated liver mitochondria synthesize a rapidly labeling RNA which effectively hybridizes with mitochondrial DNA and shows marked activity as a template in cell-free systems of protein synthesis. Model experiments show that RNA can in principle be transferred from the nuclei to the mitochondria and serve as a template for protein synthesis on mitochondrial ribosomes.

It is known that the structure and function of mitochondria are genetically controlled both by chromosomal genes and by autonomous mitochondrial genetic determinants; mitochondrial proteins are synthesized partly on cytoplasmic ribosomes and partly on mitochondrial ribosomes (Roodin and Wilkie, 1968). However, the interaction between the two genetic systems of the cell and the two systems of protein biosynthesis are still poorly understood.

The present communication deals with the role of mRNAs of nuclear and mitochondrial origin in the functioning of the mitochondrial system of protein synthesis.

MATERIALS AND METHODS

The mitochondria used in the experiments were isolated from rat liver by differential centrifugation and were freed from cytoplasmic impurities by repeated washing and RNase treatment. Mitochondrial RNA was extracted by the phenol method, mitochondrial DNA as described by Böttger et al. (1968), and nuclear DNA by the standard procedure involving phenol-detergent deproteinization and reprecipitation with isopropanol.

Radioactive RNA was obtained by injecting the animals with ^{32}P-orthophosphate (1μ curie/100 g intraperitoneally) or with ^{14}C-orotic acid (100 μ curie/100 g) 40 — 60 minutes before they were killed.

The method of incubation of isolated mitochondria for the preparation of pulse-labeled RNA was described earlier (Rakhimbekova and Gaitskhoki, 1969).

The radioactive RNA preparations were repeatedly reprecipitated in alcohol to a constant specific radioactivity. In some experiments the RNA isolated from the mitochondria was purified by gel filtration on Sephadex G-50 and in some experiments high molecular weight RNA precipitated in 1.5 M NaCl was used.

DNA-RNA hybridization was performed on HAWP nitrocellulose filters (Millipore, USA). The duration of annealing was 24—36 hours at a temperature of 66°.

RESULTS AND DISCUSSION

The rapidly labeling mitochondrial RNA isolated after injection of the animals with radioactive precursors contains nucleotide sequences closely homologous with those of mitochondrial DNA but also hybridizes with nuclear DNA (see below).

As can be seen from the results obtained, the rapidly labeling mitochondrial RNA contains approximately 2–3 times more nucleotide sequences hybridizable with mitochondrial DNA than sequences complementary to nuclear DNA. The percentage hybridization with nuclear DNA is nevertheless high, implying that some proportion of the constantly renewed mitochondrial RNA is of nuclear origin. The mitochondrial RNA hybridizing with nuclear DNA is probably mRNA of nuclear origin which participates in protein synthesis on mitochondrial ribosomes. This supposition is confirmed by the finding that the transfer of RNA from the nuclei to the mitochondria is accompanied by a stimulation of protein synthesis in the latter (Gaitskhoki and Kiselev, 1970).

The partial complementarity between mitochondrial RNA synthesized in vivo and nuclear DNA might be attributed to the existence of common nucleotide sequences in the mitochondrial DNA and in the nuclear genome; under such circumstances the RNA synthesized on these loci of the mitochondrial DNA would be homologous to the nuclear DNA. However, it was shown by direct DNA-DNA hybridization experiments (David and Wolsenholme, 1968) and by hybridizational analysis of the RNA synthesized on mitochondrial DNA in a RNA polymerase system (Borst and Aaij, 1969; Shmerling et al., 1970) that mitochondrial DNA does not contain sequences complementary to nuclear DNA. This conclusion agrees with our data on the hybridization of mitochondrial and nuclear DNA with pulse-labeled RNA synthesized in isolated mitochondria (see below).

It is evident that pulse-labeled mitochondrial RNA does not hybridize at all with nuclear DNA but is closely homologous with mitochondrial DNA. Figure 1 shows that the hybridization of the labeled RNA preparation exceeds 15% on "titration" of the synthesized RNA with increasing amounts of mitochondrial DNA immobilized on filters and the hybridization curve does not reach a plateau. The high hybridizability of mitochondrial RNA was recently described by the Borst group (Aaij et al., 1970).

Comparison of the data given on pp. 188–190 reveals clear-cut differences between the hybridization characteristics of preparations of mitochondrial RNAs synthesized in vivo and in vitro. Firstly, the RNA synthesized on isolated mitochondria is not at all complementary to nuclear DNA, whereas the radioactive mitochondrial RNA obtained after the introduction of labeled precursors into the animal organism shows a pronounced homology with nuclear DNA. This means that mitochondrial

DNA, or at least the part of it transcribed in vitro, lacks loci homologous to nucleotide sequences of nuclear DNA. Moreover, pulse-labeled RNA fror isolated mitochondria hybridizes with mitochondrial DNA to a much greater extent than does mitochondrial RNA synthesized within the organism.

DNA* on filter, μg	^{32}P-RNA in sample, pulses/min	Hybridization, %
	Experiment 1	
50	6,800	0.3
70	6,800	0.4
90	6,800	0.9
150	6,800	1.2
	Experiment 2	
100	7,200	1.35
100	14,400	0.58
100	36,000	0.41
	Experiment 3	
50	3,200	1.57
50	8,000	4.7
50	16,000	2.3
	Experiment 4	
200	80,000	6.2

* Experiments 1 and 2 — nuclear DNA; experiments 3 and 4 — mitochondrial DNA.

An increase in the hybridization rate of RNA can result either from a relative rise in the number of hybridizable molecules without any change in the qualitative composition of the RNA or from the appearance of new classes of hybridizable RNA, possibly by depression of certain loci of the genome. The simple hybridization data given above do not permit one to distinguish between these possibilities. We therefore compared the RNA preparations synthesized in vivo and in vitro by the competitive hybridization method, which reveals qualitative differences between the preparations compared. In these experiments in vitro labeled mitochondrial RNA was hybridized in the presence of three different preparations of nonradioactive mitochondrial RNA: 1) RNA from freshly isolated mitochondria; 2) RNA from mitochondria previously incubated under conditions allowing RNA synthesis; 3) RNA from mitochondria isolated from a tissue homogenate previously incubated under conditions allowing the synthesis of RNA. The result of these experiments (Figure 2) show that the RNA extracted from isolated mitochondria after their incubation competes effectively with in vitro labeled RNA but only partly competes with RNA from nonincubated mitochondria or from mitochondria incubated in a homogenate. The competition curves for the RNA from incubated mitochondria correspond to the model curves for

DNA on filter, μg	^{32}P-RNA in sample, pulses/min	Hybridization, %
	Experiment 1	
Mitochondrial, 50	66,000	12
Mitochondrial	130,000	3
Mitochondrial	380,000	1
Nuclear 100	130,000	0
	Experiment 2	
Mitochondrial, 60	6,250	13.6
Nuclear		
50	6,250	0.02
100	6,250	0.01
200	6,250	0.02
	Experiment 3	
Mitochondrial, 20	2,000	23.5

identical preparations, whereas the curves for the RNA from nonincubated mitochondria or from mitochondria incubated in a homogenate indicate a partial homology i. e. the absence in the competitor RNA of certain hybridizable sequences existing in the labeled preparation (Rachkus et al., 1969).

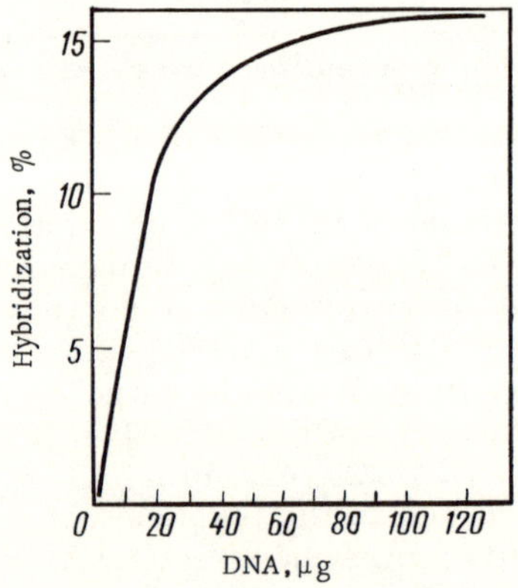

FIGURE 1. Hybridization of in vitro labeled mitochondrial RNA with mitochondrial DNA.

The amount of RNA in all samples is 100 μg (68,000 pulses/min). The time of annealing is 36 hours at 66°.

For a quantitative assessment of the degree of homology between the labeled mitochondrial RNA and the different types of unlabeled mitochondrial RNA we constructed graphs showing the relationship between g/c and g, where g is the concentration of the nonradioactive competing RNA and c is the relative amount of hybridizable labeled RNA prevented from hybridizing at the given concentration of the competitor (Melli and Bishop, 1969). The degree of homology between the labeled and unlabeled preparations is expressed by the tangent of the angle between the slope of the curve obtained and the abscissa, this value being the reciprocal (1/p) of the fraction of hybridizable sequences in the labeled RNA prevented from hybridizing by the presence of an infinitely increasing concentration of competitor. The magnitude of 1/p was determined by regression analysis. The data obtained (Figure 3) give the following 1/p values: for competitor RNA from incubated mitochondria 0.96; for competitor RNA from freshly isolated mitochondria 1.27; for competitor RNA from mitochondria incubated in a homogenate 1.4. Thus, the in vitro labeled RNA preparation contains respectively 100, 78 and 71% nucleotide sequences homologous to these three competing RNAs. This means that over 20% of the hybridizable sequences of the in vitro synthesized mitochondrial RNA consist of new classes of RNA which are absent in the mitochondria in situ. The appearance of these news RNAs may be attributed to a depression of certain loci of the mitochondrial DNA — a process which occurs in isolated mitochondria but not during the incubation of mitochondria in a homogenate, i. e. in the presence of nuclei and microsomal elements of the cellular fluid.

This assumption regarding the depression of some of the mitochondrial genes in vitro is further confirmed by the results of experiments on the competition of unlabeled RNAs from fresh and incubated mitochondria with identical labeled RNAs (synthesized in vivo and in vitro, respectively). Analysis of these data by the double reciprocals method (Shearer and McCarthy, 1967) showed (Figure 4) that in a state of saturation the proportion of mitochondrial DNA loci occupied by RNA synthesized in vitro (equal to the reciprocal of the ordinate intersect of the "straightened" hybridization curve) is roughly 2.5 times greater than the proportion of the mitochondrial genome which hybridizes with mitochondrial RNA synthesized in vivo.

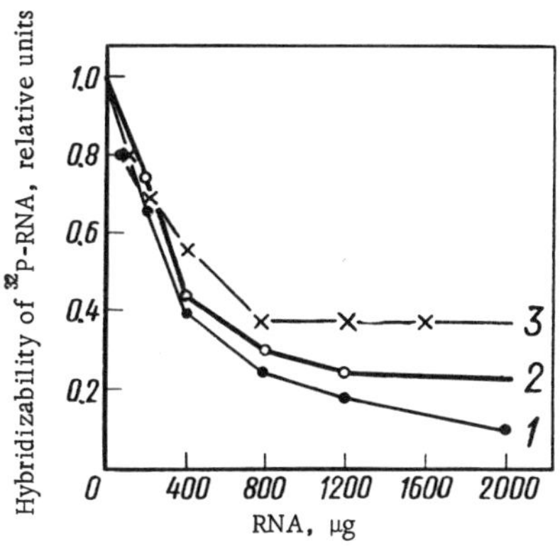

FIGURE 2. Competition between different preparations of mitochondrial RNA and in vitro labeled mitochondrial RNA.

Amount of DNA on each filter: 20 μg; ^{32}P-RNA in each sample 50,000 pulses/min. The hybridizability in the absence of competitor is taken as 1.0. 1) competitor RNA from incubated mitochondria; 2) competitor RNA from freshly isolated mitochondria; 3) competitor RNA from mitochondria incubated in a homogenate.

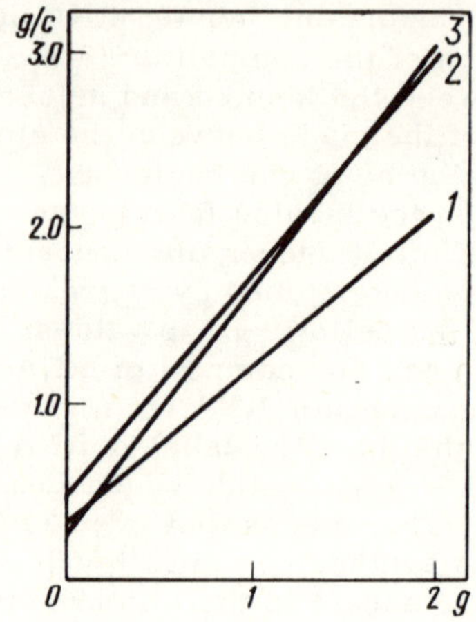

FIGURE 3. Determination of the degree of homology between synthesized mitochondrial RNA and various preparations of unlabeled mitochondrial RNA.

Experimental conditions and key as in Figure 2; see also text.

The results obtained suggest that nuclear control of the structure and function of the mitochondria is exerted in two ways: 1) by the "release" of genetic information in the form of mRNA molecules acting as templates for the synthesis of particular proteins on mitochondrial ribosomes; 2) by the repression of some genetic loci of the mitochondrial DNA. The detailed mechanisms of these processes are accessible to experimental research.

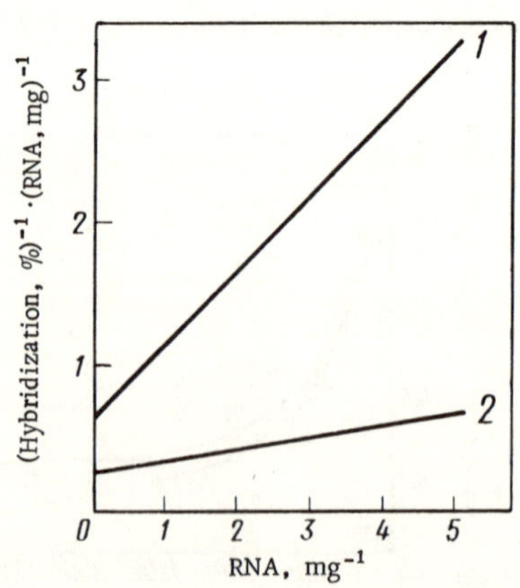

FIGURE 4. Determination of the proportion of mitochondrial DNA loci hybridizing with mitochondrial RNA at saturation:

1) competition between unlabeled RNA from freshly isolated mitochondria and labeled mitochondrial RNA synthesized in vivo; 2) competition between unlabeled RNA from incubated mitochondria and ^{32}P-RNA synthesized in isolated mitochondria.

BIBLIOGRAPHY

Aaij, C., C. Saccone, and M. Gadaleta. — Biochim. biophys. Acta **199**, No. 2 (1970), 279.

Borst, P. and C. Aaij. — Biochem. Biophys. Res. Commun. **34**, No. 3 (1969), 358.

Böttger, M., V. Wunderlich, M.Schütt, W. Förster, and A. Graffi. — Acta biol. med. germ. **21**, No. 5 (1968), 587.

David, J.B. and D.R. Wolsenholme. — Biophys. J. **8**, No. 1 (1968), 65.

Gaitskhoki, V.S. and O.I. Kiselev. — Proceedings 3rd All-Union Symposium on Cell Nucleus. Kiev, 1970. (Russian)

Melli, M. and J.O. Bishop. — J. Molec. Biol. **40**, No. 1 (1969), 117.

Rachkus, Yu. A., N.S. Kupriyanova, M. Ya. Timofeeva, and K.A. Kafiani. — Molekulyarnaya Biologiya **3**, No. 4 (1969), 617.

Rakhimbekova, L.S. and V.S. Gaitskhoki. — Molekulyarnaya Biologiya **3**, No. 3 (1969), 315.

Roodin, D. and D. Wilkie. Biogenesis of Mitochondria. London, 1968.

Shearer, R. and B. J. McCarthy. — Biochemistry **6**, No. 1 (1967), 283.

Shmerling, Zh. G., N. I. Borisova, and Ya.M.Rozovskii. — Molekulyarnaya Biologiya **4**, No.1 (1970), 144.

UDC 575:577.1

REGULATION OF THE TRANSCRIPTION OF MITOCHONDRIAL DNA

T.B. KAZAKOVA, M.M. GACHAVA and S.A. ERMEKOVA
Institute of Experimental Medicine
Academy of Medical Sciences of the USSR, Leningrad

The structural proteins of mitochondria do not inhibit the template activity of mitochondrial DNA but do inhibit the synthesis of RNA. A soluble DNP and specific basic proteins were isolated from rat liver mitochondria. The DNP lacks histone-type proteins. The specific basic mitochondrial protein, extracted by different concentrations of NaCl and HCl, inhibit the template synthesis of RNA on mitochondrial and nuclear DNA.

By now a sufficient number of experimental data on the specific molecular structure of mitochondrial DNA and its physicochemical distinctions from nuclear DNA has been accumulated to permit some conclusions about the mechanism controlling the genetic system of the mitochondria. Membrane structures constitute one of the levels of regulation of the sequence and direction of template synthesis on mitochondrial DNA. Studies of the metabolism of nucleic acids in bacteria have shown that the replication of DNA starts with the attachment of the chromosome to the cell membrane (Marvin, 1968), while the synthesis of mRNA requires the following structural complex: cell membrane-DNA-RNA polymerase — RNA (Tremblay et al., 1969). The attachment of DNA to the membrane may be necessary to give the template molecule a particular conformation ensuring the correct sequence of the stages of template synthesis and the control of this process by the cell.

To perform its biological functions, the mitochondrial DNA possibly has to acquire a particular molecular conformation by adhering to the lipoprotein structures of the mitochondrial membranes. The literature contains a few indications of a link between DNA and the inner membranes of mitochondria, mostly revealed by electron microscopy (Nass, 1969). We found earlier that the mitochondrial DNA is located not only in the matrix but also in the outer and inner membranes (Kazakova et al., 1969). The bond with the membranes helps to preserve the secondary structure of the DNA. DNA preparations from mitochondrial membranes are more active as templates in the RNA polymerase system of *Escherichia coli* than DNA isolated from the matrix. The purpose of this work was to determine the role of membrane structures in the regulation of the template activity of mitochondrial DNA.

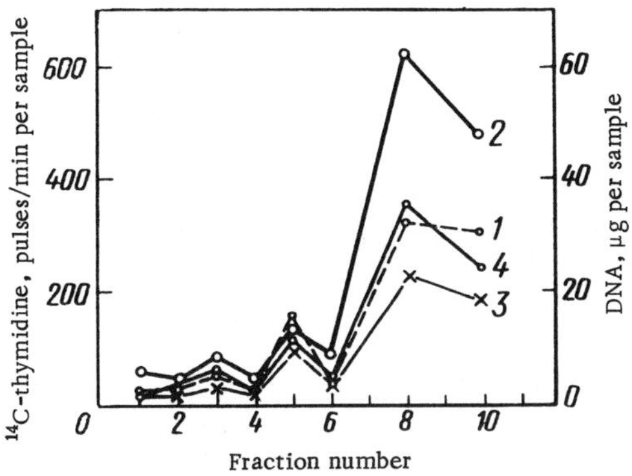

FIGURE 1. Distribution of ^{14}C-thymidine in structural elements of mitochondria before and after treatment with pronase and sodium deoxycholate:

1) DNA content of the sample, μg; 2) radioactivity of DNA, pulses/min per sample; 3) same after treatment of the membranes with pronase (200 μg/ml); 4) same after treatment of the membranes with sodium deoxycholate (1 mg/ml). Fractions 2—4 consisted of aggregated proteins and lipids, 4—6 of outer membranes, 6—10 of inner membranes of the mitochondria.

The experiments on the biosynthesis of DNA in the structural elements of the mitochondria were carried out as follows. Rats were injected intraperitoneally with ^{14}C-thymidine at a rate of 100 μcuries/100 g weight 24 hours before being killed. Membrane fractions of the rat liver mitochondria were obtained according to Parsons et al. (1966). The incorporation of the label was greatest in the fraction containing the inner mitochondrial membranes. By using various procedures for the destruction of the mitochondrial structure it was shown that mitochondrial DNA is bound more firmly with membrane proteins than with phospholipids; treatment of the membranes with pronase released greater amounts of DNA into the medium than treatment with sodium deoxycholate (Figure 1).

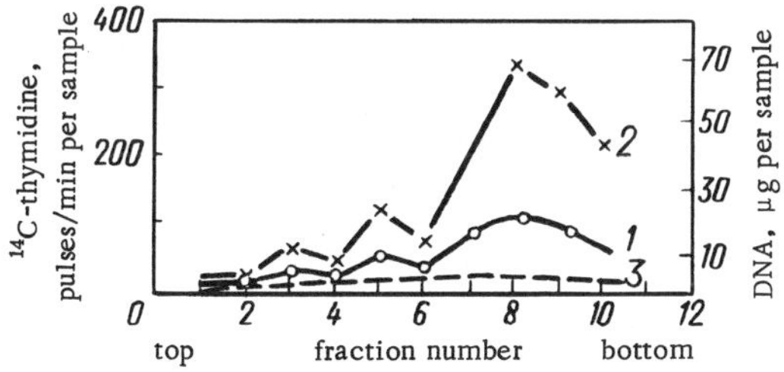

FIGURE 2. Distribution of ^{14}C-thymidine in the structural elements of mitochondria previously incubated in vitro with the label:

1) DNA content of sample, μg; 2) radioactivity of DNA, pulses/min per sample; 3) same after sonification of the membranes; fractions 1—4) aggregated proteins and lipids; fractions 4—6) outer membranes; fractions 6—11) inner membranes.

The bond between the DNA and the mitochondrial membranes facilitates the attachment of the nucleic acid to the membrane structures and its adoption of the exact spatial conformation required for directed syntheses on the DNA template. Maximum incorporation in the fraction containing the inner membranes was observed when the intact mitochondria were incubated in the presence of ^{14}C-thymidine in a medium with the following composition: Tris-HCl 20 μM, pH 7.4; KCl 33 μM; $MgCl_2$ 2μ M; GTP, CTP, and ATP 0.08 μM each (thymidine was added at the rate of 620,000 pulses/min). Disruption of the structural integrity of the membranes by ultrasonic vibrations (MSE ultrasonic generator, 1.8 A, 120 seconds) sharply reduced DNA synthesis in isolated mitochondria (Figure 2). A certain minimal structure of the mitochondrial membranes is apparently required for DNA replication.

The binding of the DNA with the membrane indicates its interaction with the lipid and protein components of this. It is well known that one of the main components of the membrane is a structural protein which reacts with the respiratory complexes and phospholipids of the membrane. We assumed that the structural protein can also react with molecules of mitochondrial DNA. In this case it would alter the conformation of the nucleic acid and in this way possibly affect its genetic activity. We examined the capacity of the structural proteins of the mitochondria to form a complex in vitro with DNA molecules and the propensity of this complex to carry out template synthesis of RNA in an artificial RNA polymerase system derived from E. coli. The structural proteins were isolated from the inner and outer membranes according to the method of Knippers and Hoffman-Berling (1966) and characterized according to their electrophoretic mobility in polyacrylamide gel using the method described by Takayama et al. (1964). The formation of a complex was assessed on the basis of the ability of actinomycin D to compete with the protein for binding sites on the DNA. The DNA preparations used were isolated from rat liver mitochondria according to the method of Böttger et al. (1968). The purity and specificity of the DNA preparation were assessed on the basis of the capacity of this compound to undergo renaturation (80—90%) and its degree of complementarity with mitochondrial RNA. In the latter case, RNA was synthesized on a mitochondrial DNA template in the presence of ^{14}C-ATP in a medium of defined composition (Kozlov and Georgiev, 1967) with added RNA polymerase, purified to the cellulose phosphate stage. Labeled ^{14}C-RNA was isolated and hybridized with mitochondrial DNA or with nuclear DNA as described by Gillespie and Spiegelman (1965) using 25 mm HAWP nitrocellulose filters manufactured by Millipore, USA. It is evident from the following data that the newly synthesized RNA is complementary to mitochondrial DNA but not to nuclear DNA.

	Mitochondrial DNA	Nuclear DNA
DNA, μg	27	100
RNA, pulses/min	1,200	1,200
Quantity of bound RNA,	906	40
pulses/min	75	3.3
Hybridization of DNA, %	11	0.13

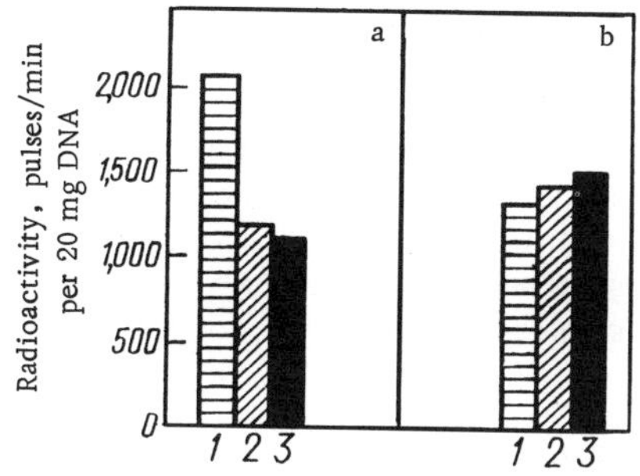

FIGURE 3. Incorporation of labeled ^{14}C-uridine into RNA synthesized on a template of mitochondrial or nuclear DNA:

a) nuclear DNA; b) mitochondrial DNA after purification on MAK; 1) free DNA; 2) DNA in complex with structural protein from outer mitochondrial membranes; 3) DNA in complex with structural protein from inner mitochondrial membranes.

The complexing of the structural proteins with DNA was carried out in 8 M urea with a subsequent dialysis against 4 M urea in 0.001 M pyrophosphate at pH 7.0, 2 M urea-pyrophosphate and four changes of 0.01 M pyrophosphate for 43 hours in the cold. It proved that with a DNA:actinomycin D ratio equal to 1:4.1, the DNA binds one molecule of the antibiotic per 2.8 nucleotide pairs. When structural protein isolated from the inner mitochondrial membranes is added to the DNA at a ratio of 0.7:1, the binding of actinomycin is reduced by 78% and saturation with the protein is achieved at a 1:1 ratio.

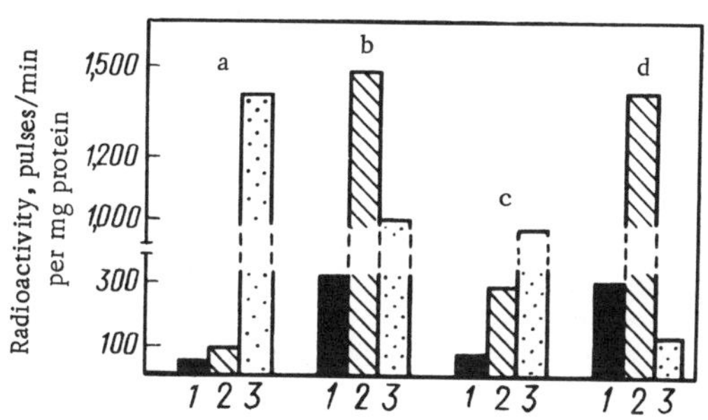

FIGURE 4. Incorporation of ^{14}C-amino acids into proteins of mitochondria incubated in vitro:

a) intact mitochondria; b) intact mitochondria +cytoplasm; c) mitochondria +cytoplasm +chloramphenicol; 1) protein extracted with NaCl; 2) protein extracted with HCl; 3) structural protein.

Incubation of the mitochondrial DNA — protein complex in the RNA polymerase system showed that the structural proteins do not exert any inhibitory effect on the template activity of DNA (Figure 3). This may be due to a change in the conformation of the DNA induced by the protein with a resulting increase of the frequency of hits of RNA polymerase on the molecule of DNA. The data obtained, however, do not enable any conclusion to be drawn as to the specific action of the structural proteins on mitochondrial DNA. The only proof of such specificity is the opposite behavior of nuclear DNA of the same origin with respect to complexing with structural proteins from mitochondrial membranes. Moreover, the mitochondria contain basic proteins which form complexes with mitochondrial DNA in vitro and thus inhibit its template activity. Thus, the successive extraction of mitochondrial proteins with different concentrations of NaCl and 0.25 M HCl yielded protein fractions which inhibited the synthesis of RNA on mitochondrial and nuclear DNA alike. These proteins were analogous as regards their action to the F_1 histone of calf thymus.

Analysis of the amino acid composition of the HCl extract of the mitochondria showed the absence of proteins rich in arginine or lysine, which implies that the proteins isolated from the mitochondria differ from the nuclear histones. Since mitochondria incubated in vitro synthesize mainly structural proteins, one might expect that the synthesis of basic proteins takes place outside the mitochondria.

In our experiments, isolated mitochondria were incubated in the presence of ^{14}C-amino acids (*Chlorella* protein hydrolysate) for 60 minutes at 30° in 5 ml of medium with the following composition: sucrose 0.15 M, Tris-HCl 35 mM, pH 7.4; KCl 25 mM; potassium phosphate 10mM; ATP 2mM; succinate 10 mM; ^{14}C-amino acids 0.5 — 1 μcurie/ml; mitochondria — 10 mg protein per ml; cytoplasm — 15 mg protein per ml; cycloheximide 500 μg/ml; chloramphenicol 50 μg/ml. In the absence of cytoplasm, the mitochondria synthesize structural protein but not proteins extractable with NaCl or HCl. Probably the basic proteins are synthesized on cytoplasmic ribosomes and then transferred to the mitochondria. Thus, proteins synthesized outside the mitochondria may also take part in the regulation of the gene activity of mitochondrial DNA (Figure 4).

To sum up, it appears that some fractions of structural and water-soluble proteins exert a regulatory effect on the template functions of mitochondrial DNA. The observed moderate stimulation of RNA synthesis on mitochondrial DNA templates under the influence of structural proteins can be attributed to a protein-induced alteration of the molecular conformation of the nucleic acid, similar to that induced when RNA polymerase fits into the DNA (Khesin, 1967). The change in the conformation of DNA in vitro may inhibit or activate the transcription process but it does not affect its course. Specificity of RNA synthesis can only be achieved in vivo, when the DNA molecule is attached to the membrane surface and its spatial orientation determines the direction and rate of the transcription process.

BIBLIOGRAPHY

Böttger, M., V. Wunderlich, M. Schütt, W. Förster, and A. Graffi. — Acta biol. med. germ. 21, No. 5 (1968), 587.

Gillespie, D. and S. Spiegelman. — J. Molec. Biol. 12, No. 3 (1965), 829.

Kazakova, T.B., M.M. Gachava, and K.A. Markosyan. — Biokhimiya 34, No. 5 (1969), 1000.

Knippers, R. and H. Hoffmann-Berling. — J. Molec. Biol. 21, No. 1 (1966), 292.

Kozlov, Yu. V. and G.P. Georgiev. — Molekulyarnaya Biologiya 1, No. 2 (1967), 190.

Marvin, D.A. — Nature 219, No. 5133 (1968), 485.

Nass, M.M. — J. Molec. Biol. 42, No. 3 (1969), 521.

Parsons, D.F., C.R. Williams, and B. Chance. — Ann. N.Y. Acad. Sci. 137, No. 2 (1966), 643.

Takayama, K., D.H. McLennan, A. Tzagoloff, and C.D. Stoner. — Archs. Biochem. Biophys. 114, No. 1 (1964), 223.

Tremblay, C.J., M. Y. Daniels, and M. Schaechter. — J. Molec. Biol. 40, No. 1 (1969), 65.

UDC 575:576.858.9

GENETIC DETERMINATION OF SPECIFIC mRNAs OF PHAGE T4

R. JAYARAMAN and E.B. GOLDBERG
Taft University, USA

A method for the quantitative assessment of the gene-specific mRNA of phage T4 has been developed. This method consists in the segregation of the complementary DNA strands, their hybridization with RNA, digestion of the nonhybridized sectors of DNA with endonuclease and determination of the protected genetic markers in the phage transformation system. It was shown that gene r II B is transcribed from the l strand and is an early gene, whereas gene 21 is transcribed later from the r strand.

The first major advance towards an understanding of the regulation of genetic transcription was the development of a method for the competitive hybridization of nucleic acids by Khesin and his colleagues (Khesin and Shemyakin, 1962; Khesin et al., 1962). By means of this method and improved versions of it (Bolle et al., 1968) phage mRNAs were divided into classes according to the time of their formation and disappearance in the infected bacterial cells. The next stage of transcription research was the determination of the kinetics of mRNA synthesis on individual genes. Studies of this kind have so far concerned only a few genes of phage T4 (Bautz et al., 1966) and a wider range of genes of phage lambda (Szybalski et al., 1969); deletion mutants were used in both cases. The method for segregating the DNA strands proposed by Guha and Szybalski (1968) also enabled determination of the polarity of transcription. We describe here a general methodology for the determination of mRNA at the same time enabling investigation of the kinetics and polarity of the transcription of individual genes of phage T4; the first results obtained by the use of this method are also given (see also Jayaraman and Goldberg, 1969; Wais and Goldberg, 1969).

The principle of the method is shown in Figure 1. The first stage consists in the segregation and purification of the complementary DNA strands of mature particles of phage T4. In the second stage, each DNA strand is hybridized with an unfractionated RNA preparation from phage-infected cells. The mixture containing the DNA-RNA hybrids is then treated with shark liver endonuclease (Trilling et al., 1970), which acts specifically on single-stranded DNA. This nuclease does not affect native DNA or DNA-RNA hybrids. Consequently, the genetic loci for which suitable complementary molecules are present in the mRNA preparation will be protected from the enzyme, whereas the loci not involved in transcription will be hydrolyzed.

Finally the enzyme is inactivated and the DNA-RNA hybrids denatured. The preparation of fragmented DNA obtained is used in the phage transformation system for an analysis of genetic markers (Veldhuisen and Goldberg, 1968). If the hybridization is carried out with sufficient amounts of DNA, the efficiency of transformation with respect to each marker will be proportional to the quantity of RNA complementary to the given locus. Figure 2 shows the results of such an experiment for two loci — the genes rIIB and 21. It is evident from Figure 2a that the marker r73 is not degraded by endonuclease and appears during transformation only when the RNA is hybridized with the l strand of the DNA. This is not surprising given that the mRNA of the rIIB gene is complementary to the l strand (Kasai et al., 1968). Hybridization of the RNA with the r strand shows only a background level of transformation with respect to the r73 locus. The results of transformation with respect to the tsN8 marker show that the mRNA of gene 21 is complementary to the r strand and is therefore transcribed with an opposite polarity.

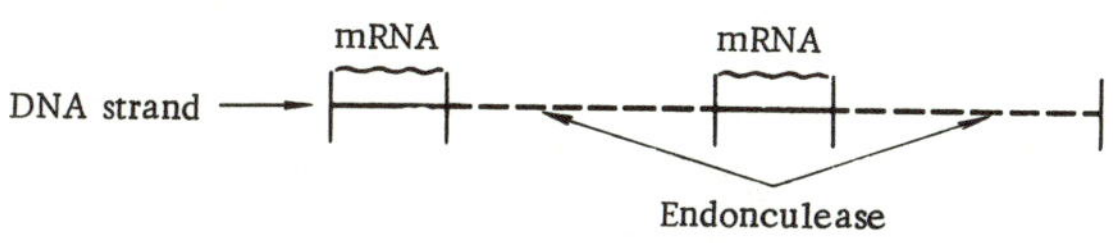

FIGURE 1. Determination of mRNA.

We must mention some limitations of the proposed method and also underline the risk of incorrect interpretation of the data obtained. As shown in Figure 2, a high RNA/DNA ratio lowers the efficiency of transformation. This reduction, although small, is quite consistent. The causes of this phenomenon remain obscure.

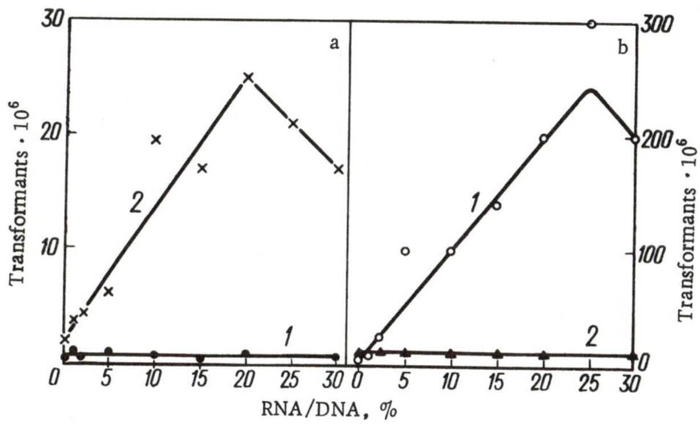

FIGURE 2. Efficiency of transformation of genes rIIB (a) and 21 (b):
1) r strand; 2) l strand.

The saturation ratio of RNA to DNA may vary greatly from one gene to another. For example, we were unable to protect gene 30 markers at the usual RNA/DNA ratio of 15. A normal level of protection required a fivefold increase in the relative concentration of RNA. Consequently, the data obtained for any gene may be too low if a saturation curve for it has not been plotted. The variability in the maximum level of mRNA (compare, for example, the data for the genes rIIB and 5 in Figure 3) may be due to differences in the average size of the mRNA of different genes. On hybridization, short mRNAs can prevent the nuclease from acting on correspondingly short DNA fragments with a molecular weight of less than $5 \cdot 10^5$ daltons and a low transforming activity (Zweerink and Goldberg, 1970). The observed reduction in the level of mRNA in cells 20 minutes after the start of infection can, in our view, be attributed to a disappearance of these particular mRNAs (Figure 3). However, this result can also be interpreted in terms of a decrease in the average size of the mRNA at this particular moment. Generally speaking, the size of the mRNA and consequently the observed level of mRNA may be influenced by the position of the marker involved in transformation relative to the promoter and terminator of the transcription and by the size of the transcribed sector of DNA. These topics are now being studied.

Figure 3 shows the results of our studies on the kinetics and polarity of transcription of certain genes (the kinetics and polarity of transcription of gene rIIB and gene e, coding for lysozyme, are given in Figure 3a). Our data agree with those obtained by the deletion method (Bautz et al., 1966). It is evident from Figure 3b that the transcription of gene 43 (DNA polymerase) proceeds in parallel to that of gene rIIB; the mRNAs of these genes appear in the infected cells 5 minutes after infection and begin to vanish after 10 minutes. The transcription of gene 30, coding for polynucleotide ligase (Figure 3, b), reaches a peak at five minutes and does not decline until 20 minutes after infection. Clearly, the transcription of the various early genes begins and ends synchronously, which is in agreement with previously published hypotheses (Salser et al., 1970).

The mRNAs of genes rIIB, e, 30 and 43 are complementary to the l strand, which means that these genes are transcribed counterclockwise. The mRNAs of genes 5, 21, 34 and 37 are complementary to the r strand and are therefore transcribed clockwise, i. e. in the same direction as the replication of DNA.

The data given in Figure 3c show that the transcription of genes 5 (baseplate) and 21 (membrane) reaches a peak at 15 minutes while that of gene 5 begins to decline from 20 minutes onwards (the level of transcription of gene 21 was determined only up to 20 minutes). Figure 3d illustrates the kinetics of the synthesis of the mRNA of genes 34 and 37, coding for structural components of the flagellar filaments. The mRNAs of these genes reach a peak at 20 minutes and then decrease in quantity.

The table opposite contains comparative data on the kinetics of transcription of the eight genes of phage T4 examined. The figures presented are fractions of the maximum content of the particular mRNA in the infected cells and are calculated on the basis of the data given in Figure 3.

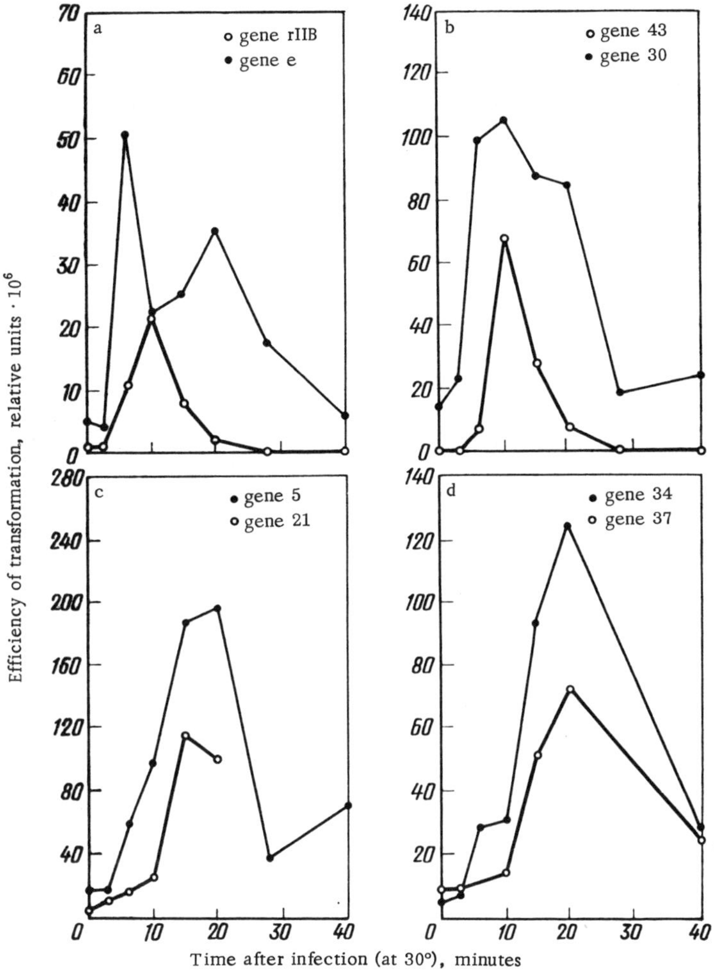

FIGURE 3. Kinetics and polarity of the transformation of genes rIIB, e, 30,43,5,21,34,and 37.

Relative amounts of mRNAs of different genes of phage T4 in infected cells of E. coli B

Minutes after infection	rIIB	DNA polymerase (43)	Ligase (30)	Lysozyme (e)	Flagellar filament (37)	Flagellar filament (34)	Baseplate (5)	Capsid (21)
3	0.0	0.0	0.10	0.0	0.0	0.02	0.0	0.04
6	0.45	0.09	0.92	1.00	0.0	0.20	0.24	0.09
10	1.00	1.00	1.00	0.14	0.07	0.22	0.45	0.18
15	0.31	0.40	0.79	0.43	0.67	0.74	0.94	1.00
20	0.03	0.09	0.77	0.67	1.00	1.00	1.00	0.85
28	0.0	0.0	0.05	0.33	—	—	0.10	—
40	0.0	0.0	0.10	0.02	0.25	0.20	0.29	—

At 3 minutes after infection the level of transcription does not exceed 10% of the maximal value for any of the genes. Transcription of the lysozyme and ligase genes reaches the maximum or near maximum level at 6 minutes, which is approximately the time that DNA synthesis in the cells begins. The transcription of most of the genes examined begins between 6 and 10 minutes after infection, when the functioning of the lysozyme gene slows down. Transcription of the gene of DNA polymerase and gene rIIB is switched off completely at 10 minutes, while the other genes continue to function until 20 minutes after infection. Thus, the eight genes examined have widely differing timetables for the start and end of transcription. It would be premature to assign these few genes to any of the known classes of T4 genes determined by the competitive hybridization method (Salser et al., 1970). We intend to classify the individual genes, compare the transcription and translation of certain genes in vivo, and determine the relationship between these two processes. We further hope to determine the size of the different mRNAs synthesized in vivo and establish the existence of polycistronic mRNAs. The proposed method is now being used for a comparison of transcription in vivo and in vitro in an effort to obtain a better understanding of the molecular mechanisms controlling transcription.

BIBLIOGRAPHY

Bautz, E.K. F., T. Kasai, E. Reilly, and F.A. Bautz. — Proc. Natn. Acad. Sci. U.S.A., Vol. 55 (1966), 1081.

Bolle, A., R. H. Epstein, W. Salser, and E.P. Geiduschek. — J. Molec. Biol., Vol. 31 (1968), 325.

Guha, A. and W. Szybalski. — Virology, Vol. 34 (1968), 608.

Jayaraman, R. and E.B. Goldberg. — Proc. Natn. Acad. Sci. U.S.A., Vol. 64 (1969), 198.

Kasai, T., E.K.F. Bautz, A. Guha, and W. Szybalsky. — J. Molec. Biol., Vol. 34 (1968), 709.

Khesin, R.B. and M.F. Shemyakin. — Biokhimiya, Vol. 27 (1962), 761.

Khesin, R.B., M.F. Shemyakin, A.M. Gorlenko, S.K. Bogdanova, and T.R. Afanas'eva. — Biokhimiya, Vol. 27 (1962), 1092.

Salser, W., A. Bolle, and R. Epstein. — J. Molec. Biol., Vol. 49 (1970), 271.

Szybalski, W., K. Bovre, M. Fiandt, A. Guha, Z. Hradecna, S. Kumor, H.A. Lozeron, V.M. Maher, H.J.J. Nijkamp, W.C. Summers, and K. Taylor. — J. Cell. Comp. Physiol., 74, Suppl. 1 (1969), 33.

Trilling, D.M., L.A. McNicol, B.D. Stollar, and E.B. Goldberg. — Biochim. biophys. Acta, Vol.217 (1970), 365.

Veldhuisen, G. and E. Goldberg. — In: "Methods in Enzymology," Vol. 12B, p.858. New York, Acad. Press, 1968.

Wais, A. and E.B. Goldberg. — Virology, Vol. 39 (1969), 153.

Zweerink, H. and E.B. Goldberg. — Virology, Vol. 41 (1970), 175.

UDC 575:576.858

IDENTIFICATION OF THE mRNA FOR THE CAPSID PROTEIN OF A PARAMYXOVIRUS

V.M.ZHDANOV, A.F. SITO, N.L. VARICH and M. I. PARFANOVICH
D.I. Ivanovskii Institute of Virology
Academy of Medical Sciences of the USSR,
Moscow

The replication of the virus causing Newcastle disease (NDV) involves the synthesis of several species of RNA, including an RNA with a sedimentation constant of 18S. This RNA, isolated in a sucrose density gradient, showed a marked template activity, inducing protein synthesis in a cell-free system on chick embryo ribosomes. Analysis of the products of this synthesis in polyacrylamide gel revealed a protein with a molecular weight of 59, 000 daltons (18S RNA has a molecular weight of 620,000 daltons). This protein was identified serologically as an S antigen or nucleocapsid protein.

The replication of paramyxoviruses involves not only the synthesis of virion RNA but also that of minus strands with sedimentation constants ranging from 10S to 37S (Zhdanov, 1968; Bratt and Robinson, 1967). The function of these strands is unknown; presumably they serve as mRNAs for the synthesis of the functional and structural proteins of the virus (Zhdanov, 1966).

The purpose of this study was to identify one of the species of RNA synthesized in cells infected with the Newcastle disease virus (NDV). The experiments were carried out with the Beaudette strain of NDV, cultured in monolayers of chick fibroblasts pretreated with typsin. The inoculum was 50 — 100 PFU (plaque-forming units) per cell. The culture was treated with actinomycin D at a concentration of 2 — 5 μg/ml to inhibit the synthesis of cellular RNAs and the newly synthesized RNA was labeled by introducing ^{3}H-uridine (specific activity 5 — 10 μcurie per ml) into the culture at various times after infection. RNA was extracted with phenol containing 0.5% sodium dodecylsulfate, precipitated in ethanol and stored in this form at —20° until further analysis.

Centrifugation of the RNA extracted from infected cells in a sucrose density gradient (10 — 30%) at 16,000 rpm for 16 hours in a Spinco L2 centrifuge (rotor SW 25.2) showed that it is heterogeneous, with sedimentation constants ranging from 10S to 48S. Figure 1 shows the sedimentation graphs of the RNA extracted 9 hours after infection. Of particular interest is the predominance of 18S RNA, the synthesis of which is most pronounced at a late stage of infection. This particular species of RNA was the object of our subsequent studies.

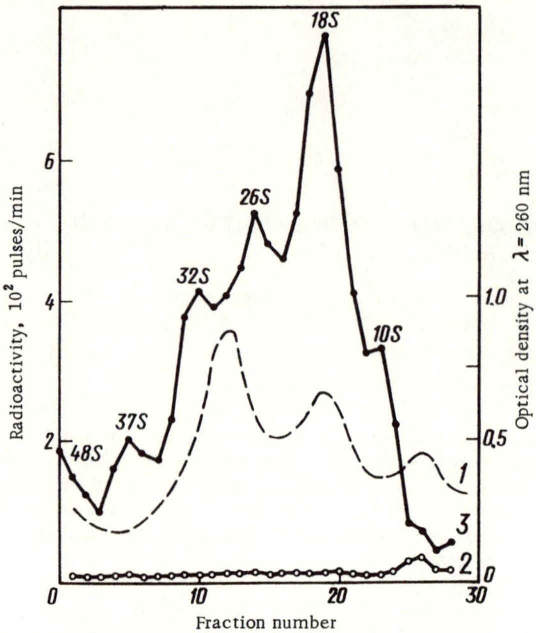

FIGURE 1. Sedimentation pattern of RNA extracted from noninfected cells (2) and from cells 9 hours after infection with NDV (3); 1) optical density of cellular RNAs at $\lambda = 260$ nm.

The homogeneity of the preparation was assessed by fractionation on a sucrose gradient. The RNA was again extracted with phenol, precipitated in alcohol and analyzed by electrophoresis in 2% polyacrylamide-agarose gel. As shown by the electrophoretic pattern in Figure 2, the 18S is homogeneous and free from other virus-specific RNAs. We determined the molecular weight of the 18S RNA by coelectrophoresis with ribosomal RNA markers labeled with ^{14}C-uridine. Comparison of the electrophoretic patterns obtained showed that the 18S RNA of NDV has a molecular weight of $6.2 \cdot 10^5$ daltons.

The complementarity between the 18S RNA and the virion RNA was assessed by determining the nucleotide composition of these RNA species, labeled with ^{32}P, by thin-layer chromatography (Randerath and Randerath, 1964), or by hybridization experiments (Kingsbury, 1966).

The nucleotide composition of the virion and 18S RNAs is shown below (%).

	Virion	18S
A	22.1	30.6
G	22.2	25.1
C	23.9	22.2
U	31.8	22.1

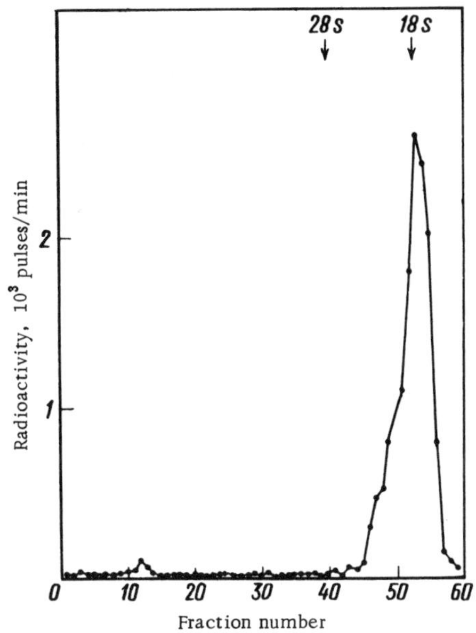

FIGURE 2. Electrophoresis of virus-induced 18S RNA of NDV in 2% polyacrylamide-agarose gel.

The position of the peaks of ribosomal RNA is indicated by arrows.

Clearly, the 18S RNA is largely complementary to the virion RNA in terms of its base composition. For the virion RNA, A + U = 54% and G+C = 46%. For the 18S RNA these values are respectively 53% and 47%.

Hybridization of the virus-induced 18S RNA with the virion RNA of NDV shows that the 18S RNA autohybridizes to the extent of 8.4% and hybridizes with the virion RNA 100%.

In a separate series of experiments, the 18S RNA and the products of its hybridization with the virion RNA were subjected to equilibrium centrifugation in $CsSO_4$ density gradients. The results of these experiments are given in Figure 3. It may be seen from this figure that 18S RNA has a buoyant density of 1.66 g/ml, which is typical of single-stranded RNA. The hybrid composed of 18S RNA and virion RNA has a buoyant density of 1.55 g/ml after treatment with pancreatic ribonuclease at a concentration of 10 μg/ml; such a density is typical of double-stranded RNA. The 18S RNA — virion RNA hybrid occupies an intermediate position: its buoyant density is 1.63 g/ml. By comparison of these data the proportion of double-stranded viral RNA can be calculated.

This calculation yields a value of slightly more than 20%. This can mean either that the virion RNA contains two sections complementary to the 18S RNA, since the virion RNA has a molecular weight of about $6.5 \cdot 10^6$ daltons, or that the virion RNA bears a single area complementary to the 18S RNA in view of the fact that the virion RNA is 10—12% autohybridized in the control.

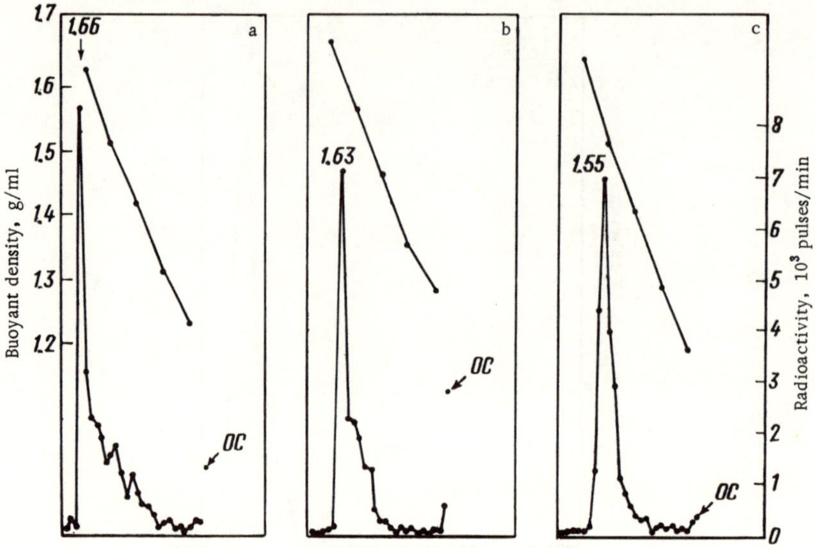

FIGURE 3. Density of the 18S RNA of NDV (a) and its hybridization products with virion RNA without treatment with pancreatic ribonuclease (b) and after such treatment (c).

Centrifugation in a $CsSO_4$ density gradient, SW 50 rotor, 40,000 rpm, for 72 hours.

We examined the template activity of the 18S RNA extracted from infected cells using chick embryo ribosomes; the endogenous templates were destroyed by incubating the ribosomes in a mixture containing the 100S fraction, amino acids and an energy-regenerating system, then the 18S was added and the mixture was incubated for different periods of time with a mixture of ^{14}C-amino acids (Zhdanov and Foster, 1969). Since the NDV 18S RNA centrifuged in a sucrose density gradient contained an admixture of ribosomal RNA, we made a parallel study of the template activity of 18S RNA extracted from noninfected cells. The template activity of NDV 18S RNA in a cell-free system of protein synthesis proved to exceed that of ribosomal RNA.

The products of protein synthesis were analyzed by electrophoresis in 5% polyacrylamide gel. Prior to electrophoresis the ribosomes were eliminated from the incubation mixture by centrifugation and the nonincorporated isotopes by dialysis. Figure 4 shows the electrophoretic pattern of the products of protein synthesis in a cell-free system in the presence of 18S RNA from infected cells. This system obviously produces proteins with similar electrophoretic mobility of which one predominates, forming a sharp peak. The molecular weight was determined by coelectrophoresis with bovine serum albumin. Calculation showed that the main peak of radioactivity is associated with a protein having a molecular weight of $5.9 \cdot 10^4$ daltons.

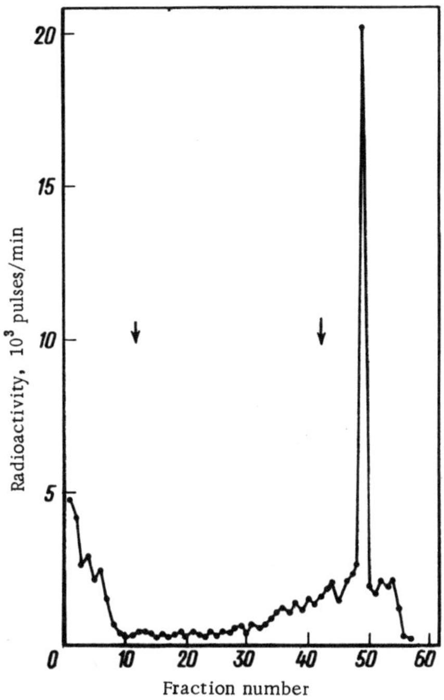

FIGURE 4. Electrophoresis of the protein synthesized in a cell-free system in the presence of NDV 18S RNA in 5% polyacrylamide gel.

The arrows show the position of the monomer and dimer peaks of bovine serum albumin.

To identify this protein serologically, we treated the preparations synthesized in the cell-free system with antisera obtained by fractionating the whole antiserum into S and V antibodies (Sito, 1970) and with normal rabbit serum. The results of one of these tests are given in the following table. The protein coded by the NDV 18S RNA is precipitated by S serum and practically does not react with normal or V sera.

Precipitation of proteins synthesized in a cell-free system with NDV-induced 18S RNA serving as template by virus-specific anti-S and anti-V sera

Serum	Radioactivity, pulses/min		
	total	in precipitate	percent in precipitate
V	15,368	659	4.3
V	18,273	651	3.0
S	22,081	14,172	64.2
S	20,690	13,263	66.1
Normal	19,376	812	2.2

The 18S RNA synthesized in NDV-infected cells thus contains the information for the synthesis of the capsid protein (S antigen). It is complementary to the corresponding gene on the virion RNA molecule and functions as a template in a cell-free system, inducing a specific synthesis according to the "one gene — one protein" principle.

BIBLIOGRAPHY

Bratt, M.A. and W.S. Robinson. — J. Molec. Biol., Vol. 23 (1967), 1.
Kingsbury, D.W. — J. Molec. Biol., Vol. 18 (1966), 204.
Randerath, K. and E. Randerath. — J. Chromat., Vol. 16 (1964), 111.
Sito, A.F. — Voprosy Virusologii, No.6 (1970), 736.
Zhdanov, V.M. — Lancet, Vol. 1 (1966), 1017.
Zhdanov, V.M. — Voprosy Virusologii, No.6 (1968), 686.
Zhdanov, V.M. and W.Foster. — Biokhimiya, Vol. 34 (1969), 1158.

UDC 575:576.858

EARLY AND LATE VIRUS-SPECIFIC POLYRIBOSOMES

A.G. BUKRINSKAYA, G.V. VORKUNOVA and O.G. NIKOLAEVA
D.I. Ivanovskii Institute of Virology
Academy of Medical Sciences of the USSR, Moscow

Early and late virus-specific polyribosomes were found in Ehrlich ascites carcinoma cells infected with Sendai virus. The early virus-specific polyribosomes arise on the parental viral ribonucleoprotein template and the late ones on newly synthesized virus-specific RNA template. Information in the early polyribosomes is read from the parental RNA existing in the form of a ribonucleoprotein. The protein element of this creates better conditions for translation than does the deproteinized viral RNA.

One of the peculiarities of paramyxoviruses is that cells infected with them synthesize RNA complementary to the parental viral RNA. The complementary RNA does not form virions; presumably it carries information for the synthesis of structural viral proteins (Kingsbury, 1966; Bratt and Robinson, 1967).

Study of the dynamics of polysome formation showed that infected cells produce polyribosomes of two types: associated with both the parental and newly synthesized viral RNA, or only with the newly synthesized RNA. The present communication is concerned with this phenomenon.

THE BUOYANT DENSITY OF POLYRIBOSOMES AT DIFFERENT STAGES OF INFECTION

The cellular polyribosomes were differentiated from the virus-specific ones by centrifugation in a CsCl density gradient (Figure 1). The washed and resuspended cells were infected with $^{32}PO_4$-labeled Sendai virus, and labeled amino acids (2μ curies/ml) were added to the cells for 5 minutes 1, 2 and 3 hours after infection.

The cytoplasmic extract was centrifuged in a stepped sucrose gradient (2—0.5M); the sediments were resuspended in 0.01 M phosphate buffer containing 0.01 M KCl and 0.003 M $MgCl_2$ and then fixed with 4% formaldehyde and centrifuged in a performed CsCl gradient (Superspeed 50 centrifuge, 3×5 rotor) for 12 hours at 35,000 rpm.

Fractions were collected by puncturing the tube and the buoyant density (ρ), the optical density and the radioactivity after precipitation of the acid-insoluble material with 5% trichloracetic acid were determined for every fifth fraction.

In the noninfected cells optically dense material was found in zones corresponding to a buoyant density of 1.55 and 1.52 g/ml, while the labeled amino acids were incorporated mainly in the 1.52 g/ml position. One hour after infection the radioactivity peak in the 1.52 g/ml zone decreased; at 2 hours incorporation of amino acids was detected in the 1.45 zone, and at 3 hours there was a large peak in the 1.49 — 1.52 g/ml zone. At 1 hour post-infection the parental label was detected in the 1.38 g/ml zone, corresponding to the density of partly deproteinized viral ribonucleoprotein (RNP) (Bukrinskaya and Zhdanov, 1970; Bukrinskaya, Zhdanov and Vorkunova, 1969) while by 2 hours post-infection this label had shifted to a higher density (1.41 — 1.45 g/ml), where it partly coincided with the incorporated amino acids. At no time after infection was there an appreciable quantity of parental label in the 1.52 zone (Figure 1).

It appears from these data that 2 hours after infection the cells produce polyribosomes with a buoyant density of 1.45 g/ml, unusual in noninfected cells, whereas at 3 hours polyribosomes identical to the cellular ones as regards density (1.49 — 1.52 g/ml) are formed. As will be shown below, the polyribosomes with a density of 1.45 and 1.49 — 1.52 g/ml are respectively early and late virus-specific polyribosomes.

Early virus-specific polyribosomes and their relation to the parental viral ribonucleoprotein

The following data indicate that the early polyribosomes are associated with an uncoiled parental ribonucleoprotein. 1) The 1.45 complex carries label not only from the parental RNA but also from the parental ribonucleoprotein after a preliminary labeling of the virus with a ^{14}C-hydrolysate of *Chlorella*. 2) Treatment of the 1.45 complex with EDTA causes its breakdown with the release of viral RNP in a partly deproteinated state with a buoyant density of 1.38 g/ml. 3) When treated with ribonuclease the 1.45 complex similarly breaks down, releasing fragments of native parental ribonucleoprotein with a buoyant density of 1.31 g/ml; in a sucrose gradient these fragments are distributed heterogeneously in the preribosomal, ribosomal and postribosomal regions; consequently, the viral RNA in the polyribosomes is not protected from the action of ribonuclease. 4) The 1.45 complex is not found in cell lines resistant to the virus. 5) Electron microscopy of the CsCl gradient fraction containing the 1.45 complex reveals polyribosomal structures associated with uncoiled viral ribonucleoprotein (Bukrinskaya and Zhdanov, 1970; Bukrinskaya Bykovsky and Zhdanov, 1969). Owing to the presence of viral structural protein in the early virus-specific polyribosomes we were able to separate them from the cellular polyribosomes.

For this purpose the cells were treated with actinomycin D, and ^{3}H-leucine and ^{3}H-phenylalanine (2 μ curie/ml) were added to them for 1 minute two hours after infection with the virus, labeled with $^{32}PO_4$. The cytoplasmic extract was divided in two parts. To one of these was added immune viral antiserum obtained according to Lief and Henle (1956) (HA (hemagglutinating activity) titer 1:1280; CF (complement fixation) titer with S antigen 1:160) to give a final dilution of 1:8. This mixture was then centrifuged in a sucrose density gradient for 2.5 hours at 25,000 rpm.

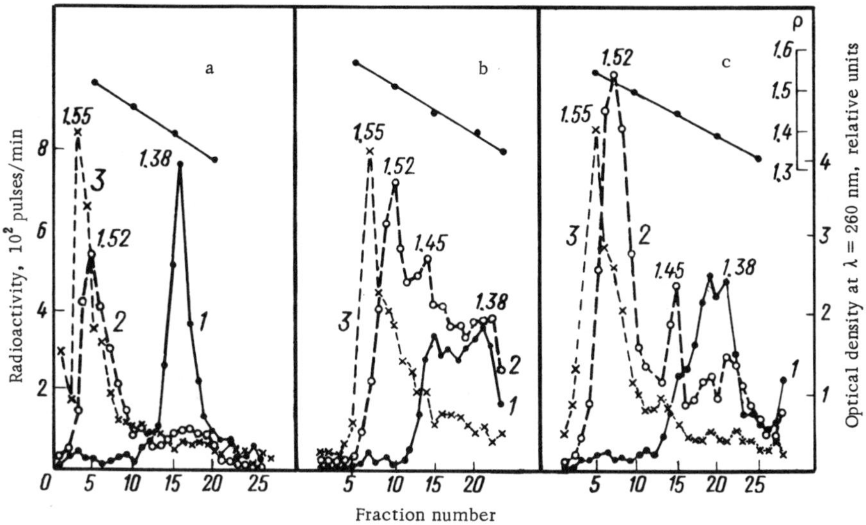

FIGURE 1. Changes in the buoyant density of polyribosomes in Ehrlich ascitic carcinoma cells infected with Sendai virus. ^{3}H-amino acids were added to the cells 1 hour (a), 2 hours (b) and 3 hours (c) after infection with the virus.

1) ^{32}P; 2) ^{3}H; 3) E_{260}.

The treatment with antiserum made the viral polysomes heavier (emerging in fraction 2 instead of 4). Subsequent centrifugation of the sucrose gradient fraction in a CsCl density gradient yielded a single component with a density of 1.43 g/ml, containing the parental viral label together with newly synthesized protein (Figure 2). The lowering of the buoyant density of the polyribosomes can be attributed to the presence of antiserum protein in them. Our data do not reveal on which type of viral RNA the early polysomes are formed, namely on parental RNA or newly synthesized RNA. The latter is present in a complex with the same buoyant density and is a product of the transcription of the parental genome by means of virion polymerase.

One of the hypotheses is that the parental viral RNA within the ribonucleoprotein functions as a template. To clarify this point we conducted experiments in a cell-free-protein-synthesizing system. It proved that the viral ribonucleoprotein (ρ =1.31 g/ml), after ribonuclease treatment and purification in a CsCl density gradient, can stimulate the incorporation of labeled amino acids into acid-insoluble material. This stimulation is absent if the incubation mixture contains ribonuclease and puromycin but lacks ATP or the 100S fraction (see table). Thus, the incorporation of labeled amino acids into the acid-insoluble material can be regarded as a specific process reflecting the synthesis of polypeptide chains. As shown in the table, the stimulatory effect of RNP on incorporation was invariably greater than that of viral RNA or polyuridylic acid at concentrations corresponding to their content in the RNP preparation.

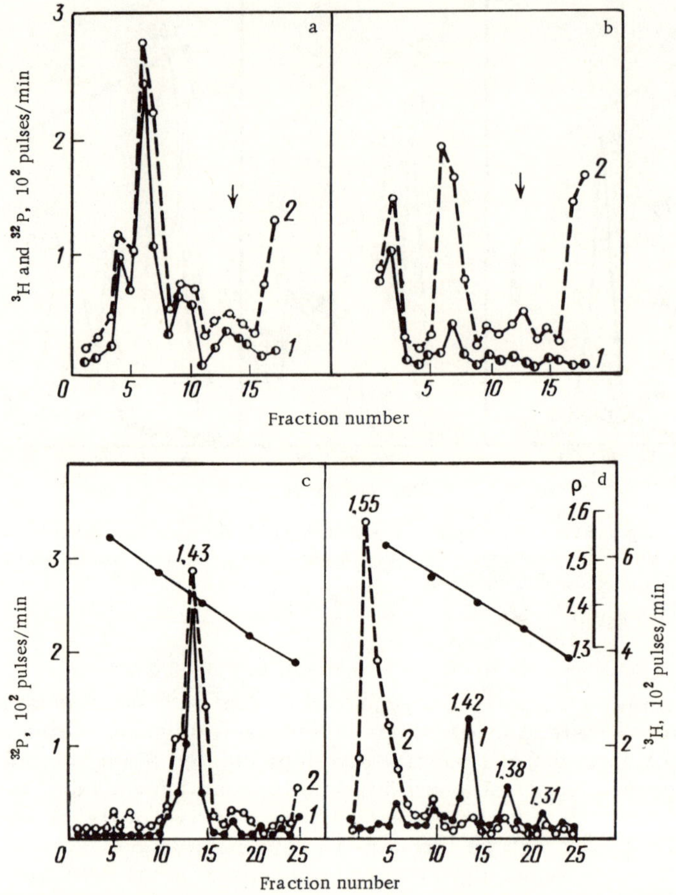

FIGURE 2. Effect of virus antiserum on the sedimentation and buoyant density of viral polyribosomes.

Radioactivity profile of a cytoplasmic extract treated (a) and not treated (b) with antiserum, centrifugation in a sucrose density gradient (15 — 60%).

Radioactivity profile of the 2nd (c) and 7th (d) fractions of the gradient shown in Figure 2b, on centrifugation in a CsCl density gradient.

After centrifugation in a sucrose density gradient some of the polyribosomes formed in the cell-free system were found on the bottom of the tube and some in the 180S position (Figure 3a). When the sucrose gradient fractions containing the radioactivity peak were centrifuged in a CsCl density gradient, after being fixed in formaldehyde, the polyribosomes were detected in the 1.43 — 1.45 g/ml density zone (Figure 3b). The labeled amino acids were found in the same complex after a one-minute pulse labeling (Figure 3c).

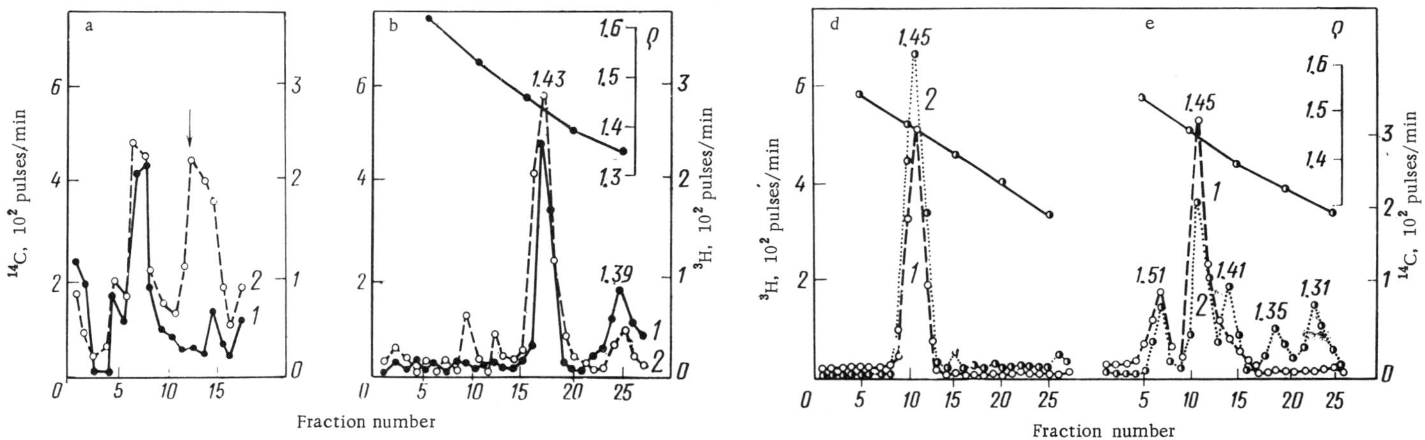

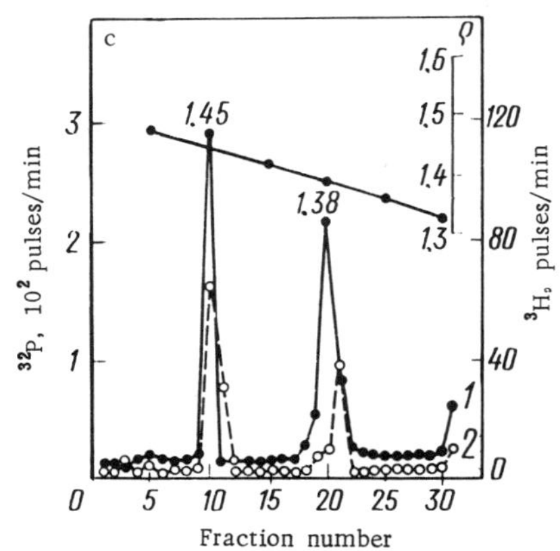

FIGURE 3. Characteristics of virus-specific polyribosomes in a cell-free protein-synthesizing system:

a) viral RNP labeled with a ^{14}C-hydrolysate of Chlorella was added to a cell-free system combining ^{3}H-uridine labeled ribosomes, and after 30 minutes' incubation the mixture was centrifuged in a 15–60% sucrose density gradient; the arrows indicate the position of monosomes in this and all other graphs; b) fractions 7 and 9 of the sucrose gradient were fixed with 4% formaldehyde and centrifuged in a CsCl density gradient; 1) ^{14}C; 2) ^{3}H; c) $^{32}PO_4$-labeled viral RNP was added to the cell-free system and after 25 minutes' incubation ^{3}H-leucine and ^{3}H-phenylalanine were added for 5 minutes at the rate of 2 μcurie each; the sediments obtained by centrifugation at 130,000 g were centrifuged in a CaCl density gradient; 1) ^{32}P; 2) ^{3}H; d) viral RNP was added to a cell-free system containing ^{3}H-uridine labeled ribosomes and after 25 minutes' incubation a ^{14}C-hydrolysate of Chlorella was added for 5 minutes (2 μcurie/ml). The incubation mixture was diluted with 0.01 M phosphate buffer containing 0.003 M $MgCl_2$, fixed and centrifuged in a CsCl density gradient; 1) ^{3}H; 2) ^{14}C; e) same as (d), but the addition of the ^{14}C-hydrolysate was followed by treatment with puromycin, 100 μg/ml, for 10 minutes; 1) ^{3}H; 2) ^{14}C.

When the incubation with labeled amino acids was followed by the addition of puromycin to the incubation medium for 5 minutes, about 50% of the protein pulse label vanished from the polyribosomes, and monoribosomes containing part of the protein label were split off. Another fraction of the protein label was found in association with the viral ribonucleoprotein in the 1.31, 1.35 and 1.45 g/ml positions (Figure 3d, e).

Template activity of Sendai virus RNP (pulses/min)

Components added to the incubation mixture	Ribosomes of Ehrlich ascites carcinoma cells		Chick embryo ribosomes
	^3H	^{32}P	^3H
None	360	—	242
Poly-U (200 μg/ml)	1,023	—	1,130
Viral RNA (200 μg/ml)	935	—	1,029
Viral RNP (200 μg RNA per ml)	1,544	3,282	1,284
Viral RNP + puromycin (100 μg/ml)	310	3,160	321
Viral RNP + RNase (5 μg/ml)	477	1,774	331
Viral RNP without ATP	—	—	192
Viral RNP without the 100S fraction	—	—	161

N o t e. $^{32}PO_4$-labeled RNP was added to the samples containing Ehrlich tumor cell ribosomes and unlabeled RNP to the samples containing chick embryo cell ribosomes.

^3H-amino acids were added simultaneously with the templates. Radioactivity in the acid-soluble fraction was determined after incubation for 30 minutes at 37°.

As was the case in vivo, the 1.43 — 1.45 complex formed in the cell-free system broke down after treatment with EDTA and ribonuclease and parental viral RNP with a density of 1.35 and 1.31 g/ml respectively was released. This means that in a cell-free system the virus-specific polyribosomes are formed on a template consisting of a partly deproteinized parental viral RNP with a density of 1.35 g/ml. Since the synthesis of polypeptide chains proceeds more rapidly if the system contains RNP than in the presence of the same amounts of viral RNA, we may conclude that the protein of the RNP creates optimal conditions for translation. However, in vivo and in vitro experiments cannot unconditionally be considered as fully analogous and the results obtained in a cell-free system cannot be construed as proof of the coding properties of the viral ribonucleoprotein.

Properties of the late virus-specific polyribosomes

Ascitic mice were injected intraperitoneally with Sendai virus; cells were collected 3 days after infection, washed, treated for one hour with actinomycin D (4μg/ml) and exposed to ^3H-uridine (5 μ curies/ml) for 5 hours. The cytoplasmic extract was analyzed in a sucrose density gradient (17 — 40%). Actinomycin D inhibited almost totally the incorporation of ^3H-uridine into the polyribosomal region of the noninfected control cells, whereas the polyribosomal region of the infected cells showed a marked radioactivity (Figure 4), indicating the incorporation of ^3H-uridine into virus-specific RNAs. The drop in the radioactivity on treatment of the cytoplasmic extracts with EDTA (0.02M) suggests that these virus-specific RNAs are associated with polyribosomes. To determine the buoyant density of the virus-specific polyribosomes, the sucrose gradient fractions corresponding to the radioactivity peaks were fixed with formaldehyde and recentrifuged in a CsCl density gradient.

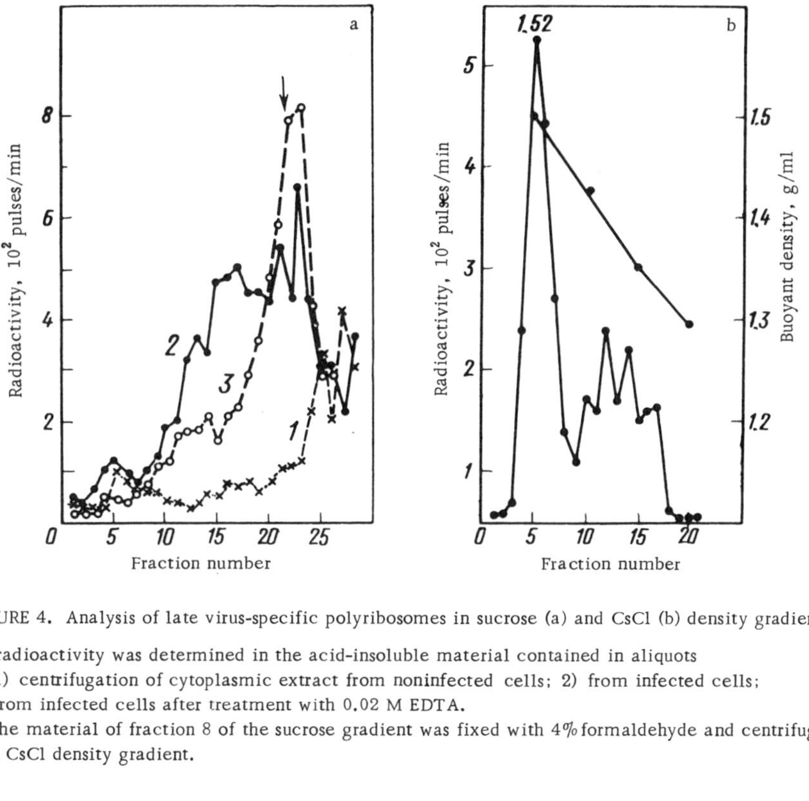

FIGURE 4. Analysis of late virus-specific polyribosomes in sucrose (a) and CsCl (b) density gradients. ^{3}H-radioactivity was determined in the acid-insoluble material contained in aliquots

a: 1) centrifugation of cytoplasmic extract from noninfected cells; 2) from infected cells; 3) from infected cells after treatment with 0.02 M EDTA.

b: the material of fraction 8 of the sucrose gradient was fixed with 4% formaldehyde and centrifuged in a CsCl density gradient.

The main peak of radioactivity appeared in the zone with a density of 1.52 g/ml. Smaller peaks were found in the 1.31, 1.35, and 1.38 g/ml zones. These components did not vanish upon treatment of the cytoplasmic extract with EDTA and probably represent native and partly deproteinized structures of newly synthesized viral RNP (Bukrinskaya, Zhdanov and Vorkunova, 1969). Polyribosomes with a density of 1.45 g/ml were not found at this stage of infection.

CONCLUSION

Ehrlich ascites carcinoma cells infected with Sendai virus produce early and late virus-specific polyribosomes. The early polyribosomes are associated with the parental viral ribonucleoprotein and have a buoyant density of 1.45 g/ml, whereas the late polyribosomes are formed on templates consisting of newly synthesized virus-specific RNA and have a buoyant density of 1.52 g/ml.

Since newly synthesized virus-specific RNA transcribed on the parental genome is present in the complex containing the early polysomes, the question of whether the early polysomes are formed on a template consisting of parental or newly synthesized RNA remains open.

BIBLIOGRAPHY

Bratt, M.A. and W.S. Robinson. — J. Molec. Biol., Vol. 23 (1967), 1.
Bukrinskaya, A.G., A. Ph. Bykovsky, and V.M. Zhdanov. — Virology, Vol. 39 (1969), 705.
Bukrinskaya, A.G., V.M. Zhdanov, and G.K. Vorkunova. — J. Virology, Vol.4 (1969), 141.
Bukrinskaya, A.G. and V.M. Zhdanov. — Molekulyarnaya Biologiya 4, No.2 (1970), 313.
Kingsbury, D.W. — J. Molec. Biol., Vol. 18 (1966), 204.
Lief, F.S. and W. Henle. — Virology, Vol. 2 (1956), 753.

UDC 575:577.1

REGULATION OF TRANSCRIPTION IN ANIMAL CELLS

R.I. SALGANIK
Institute of Cytology and Genetics
Academy of Sciences of the USSR,
Siberian Branch, Novosibirsk

Intensified transcription in cell nuclei is accompanied by an increase of the proportion of diffuse active chromatin. The chemical composition and other properties of active and repressed chromatin were examined. A study was made of the capacity of the chromatin of target cells to selectively bind inducers (hormones) acting specifically on these cells. For the most part the inducer hormone is bound to nonhistone proteins of the target cell chromatin. The phenomena of specific and nonspecific induction are discussed.

It has become evident in recent years that genetic induction is not a prerogative of microorganisms but represents a widespread mechanism for the control of transcription processes in the cells of multicellular organisms — animals and plants.

Many hormones act on target cells by switching on, during the process of transcription, the specific cistrons (DNA sectors) programing the synthesis of the enzymes responsible for the physiological function of the particular hormone. For example, hydrocortisone induces the synthesis of gluconeogenetic enzymes in liver cells (Rosen et al., 1959; Knox, 1960); insulin induces the synthesis of glycolytic enzymes in liver (Salas et al., 1963); gastrin induces the synthesis of histidine decarboxylase in the gastric mucosa (Salganik, 1969; Bersimbaev and Argutinskaya, 1969); DNA dependent synthesis of RNA and some proteins is induced by aldosterone in the kidneys, by testosterone in the prostate, and by estradiol in the uterus. Some food substances and metabolites also serve as inducers. Thus, protein hydrolysate induces the synthesis of various enzymes involved in amino acid metabolism in animal liver (histidase, serine hydratase, alanine transaminase, etc.); glucose inhibits the induction of these enzymes (Pestana, 1969). We showed not long ago that galactose induces the synthesis of galactose-1-phosphate-uridyltransferase in the liver of young rats (Salganik et al., 1970).

Hydrophobic chemical agents alien to the organism induce microsomal enzymes which hydroxylate them and thus facilitate their removal from the organism (Parke, 1968). Viral nucleic acids induce the synthesis of interferon and heterologous proteins the synthesis of antibodies. Thus, the mechanism of genetic induction performs not only regulatory but also crucial protective functions in the organism. Understandably, the genetic system of multicellular organisms differs markedly from that of unicellular ones. Genetic induction in differentiated cells obviously also differs from that in undifferentiated ones.

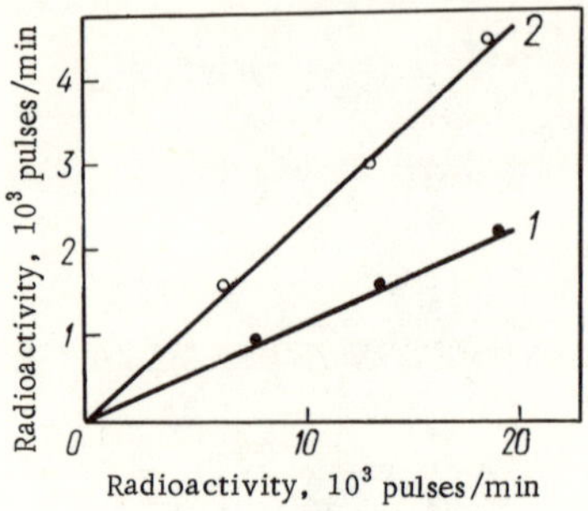

FIGURE 1. Interaction of the chromatins of rat kidney (1) and uterus (2) with 6,7-^{3}H-estradiol.

Concentration of chromatin: 0.2 mg protein per ml.

We shall discuss some aspects of genetic induction in animal cells with particular reference to hormones.

Induction in multicellular organisms clearly begins at the moment when the inducer "recognizes" the specialized target cells ready to perform specific functions. It was shown that the various hormones or inducers accumulate in their respective target cells, where they are selectively bound to the nuclei. Testosterone, for example, is bound selectively to the nuclei of the prostate (Anderson and Liao, 1968) and aldosterone to kidney nuclei (Fanestil and Edelman, 1966). We found that the selective binding of hormones is a property of the chromatin of the target cells. Using equilibrium dialysis, we were able to show that the amount of 6,7-^{3}H-estradiol bound to the chromatin of rat uterus is 3—5 times greater than that bound to kidney chromatin (Figure 1) while kidney chromatin binds 4—5 times more 1,2-^{3}H-aldosterone than liver chromatin (Figure 2); 1,2^{3}H-cortisone is bound almost equally by liver and kidney chromatins. What components of the chromatin bind the indicator hormones? Where are the receptor structures responsible for recognition located? We found previously that 1,2^{3}H-cortisone is bound almost exclusively by the non-histone proteins of the chromatin in vivo (Morozova and Salganik, 1969). DNA and histones practically do not bind labeled hormone *in vivo*. Experiments carried out in vitro similarly showed that ^{3}H-cortisone is bound predominantly by the non-histone proteins of the chromatin. In view of their capacity to bind different inducers, the chromatin proteins of diverse tissues must possess different physicochemical properties. The numerous attempts to detect differences in the fraction composition and properties of the histones of various tissues have met with little success (Busch, 1965). The heterogeneity of non-histone proteins of various tissues was examined in our laboratory by fractionation in polyacrylamide gel under conditions preventing the aggregation of proteins. With this method the non-histone proteins divided clearly into five fractions, but no differences were detected in the fraction composition of the non-histone proteins of rat liver, kidney, and thymus, or bird liver and erythrocytes. Possibly the methods employed are not sensitive enough, or the differences are such that they can only be revealed by other procedures.

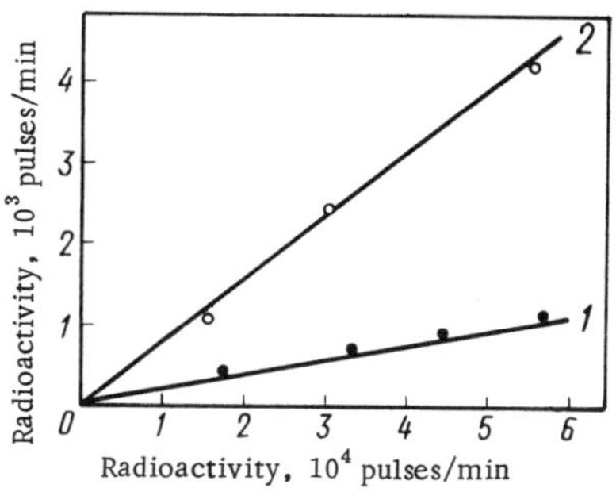

FIGURE 2. Interaction of liver (1) and kidney (2) chromatins with 1, 2-^{3}H-aldosterone.

Concentration of chromatin: 0.2 mg protein per ml.

Reaction with the inducer apparently alters the chromatin in such a way as to extend the transcription to certain DNA areas previously inaccessible to RNA polymerase. It is known that the template activity of chromatin increases markedly after induction. What changes in the state of the chromatin make certain parts of it available for transcription? Our experiments show that induction increases the amount of diffuse (active) chromatin and reduces the quantity of dense (repressed) chromatin.

The chromatin was divided into active and repressed fractions according to Frenster and his colleagues (1963). Active chromatin is distinguished from repressed chromatin by its high content of non-histone proteins and RNA and by the intensive incorporation of labeled precursors into the RNA in vivo (Arbuzova et al., 1968). The amount of active chromatin in the cell nuclei is proportional to the intensity of transcription. For example, the proportion of active chromatin in bird liver nuclei is about 16%, whereas in the nuclei of bird erythrocytes, where transcription is almost totally suppressed, active chromatin is practically absent. Injection of hydrocortisone into adrenalectomized rats increases the amount of active chromatin from 6—8% to 11—16% (Arbuzova et al., 1968). The proportion of active chromatin also increases in regnerating liver after hepatectomy, when several enzymes are induced. These transformations of the chromatin accompanying the induction of transcription are possibly associated with changes in the secondary structure of chromatin with an uncoiling of the areas where transcription is to begin.

Do these changes in the state of the chromatin result from disturbances of the bonds between DNA and proteins? Do the changes arise from local modifications of the chromatin proteins such as acetylation or phosphorylation? Is Frenster (1965) correct in assuming that we are dealing here with a competitive displacement of some DNA sections from DNA-protein complexes by other polyanions? Or does the inducer alter the conformation of the repressor proteins of the chromatin? For the time being it is hard to choose between these hypotheses. The intimate mechanism of control of the template activity of chromatin remains unknown; its disclosure is one of the most urgent tasks of molecular biology.

Many studies as well as our own have shown that induction in animal cells involves not only the synthesis of mRNA but also that of ribosomal proteins, tRNA, endoplasmic reticulum — in short, all the components of the protein-synthesizing apparatus (Gorski et al., 1965; Williams-Ashman, 1965; Salganik et al., 1967). The accumulation of these structures is obviously necessary to achieve a rapid intensification of the synthesis of adaptive proteins during a period of extreme conditions, when exposure to the inducer enables the organism to adapt to a changed environment. With the elimination of the inducer from the organism, however, the biosynthesis of the induced enzymes declines and the quantity of the endoplasmic reticulum and ribosomes decreases to the level observed in the control cells (Salganik et al., 1967). Such changes occur in liver cells after induction with hydrocortisone, for example. Recent studies in our laboratory have shown that the decline of induction is accompanied by a marked activation of lysosomal enzymes such as cathepsins, acid phosphatase, and ribonuclease (Figure 3). The activation of the lysosomal enzymes coincides in time with the depletion of the endoplasmic reticulum and ribosomes during the post-induction period. Conceivably, the activation of lysosomal enzymes brings about a disintegration of the cellular structures and adaptive enzymes that have become redundant with the fading of induction.

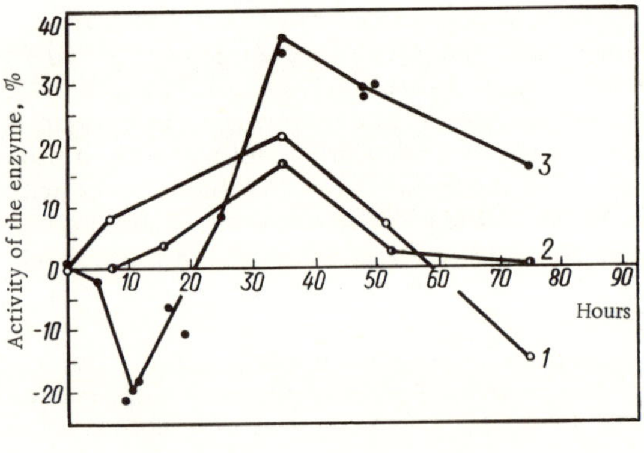

FIGURE 3. Change in the activity of some lysosomal enzymes at different times after the injection of hydrocortisone:

1) acid RNase; 2) acid phosphatase; 3) cathepsins.

In rats a three- to fourfold increase in the activity of tyrosine-α-keto-glutarate transaminase (TTA, Enzyme Catalogue No. 2.6.1.5) is observed as early as 5 hours after injection of hydrocortisone. Agar gel electrophoresis shows that rat liver contains two TTA fractions sharply differing in their mobility, one of them moving towards the anode (A fraction), the other toward the cathode (C fraction). Injection of hydrocortisone causes a sharp increase in the amount of the A fraction of TTA, whereas the concentration of the C fraction remains unchanged (Levin, 1966).

The stimulatory effect of hydrocortisone on transcription therefore appears to be confined to the cistron responsible for the A fraction of TTA — the inducible isozyme of this enzyme. The following table shows the activity of the two isozymes of liver tyrosine-α-ketoglutarate transaminase in control and hydrocortisone-treated rats and their sensitivity to Pronase (μ moles p-xyphenylpyruvate per sample per hour). Our experiments showed that only the inducible fraction of the enzyme is sensitive to the proteolytic action of Pronase. On incubation with Pronase, the activity of the A fraction decreases nearly tenfold while that of the C fraction decreases by less than 10%. Two processes are evidently crucial for regulation: firstly, the genetic apparatus of the cell must respond to the inducer by synthesizing adaptive enzymes and cellular structures; secondly, the cells must be capable of destroying the de novo synthesized structures and enzymes once these have become superfluous.

Experimental conditions	C fraction	A fraction
Normal	1.20 ± 0.04	3.22 ± 0.15
Induction with hydrocortisone	1.30 ± 0.09	8.33 ± 0.10
Induction in the presence of Pronase	1.14 ± 0.12	0.78 ± 0.05

This probably explains the activation of hydrolytic lysosomal enzymes during the post-induction period. The high protease sensitivity of the inducible TTA isozyme evidently brings about its selective disintegration after the disappearance of the critical conditions which induced the enzyme. Given the existence of standard monomers and programs, it is clearly more economic for the cell to assemble new structures and enzymes according to need and then break them down into standard monomers for further use as building units and energy sources than to conserve material and energy in idle structures.

BIBLIOGRAPHY

Andersen, K.M. and S. Liao. — Nature, Vol. 219 (1968), 277.

Arbuzova, G.A., I.M. Gryaznova, R.I. Salganik, and T.M. Morozova. — Molekulyarnaya Biologiya, Vol. 2 (1968), 308.

Bersimbaev, R.I. and S.V. Argutinskaya. — Thesis 2nd All-Union Biochem. Conference, p.37. Tashkent, 1969. (Russian)

Busch, H. Histones and Other Nuclear Proteins. — Academic Press, 1965.

Fanestil, D.D. and I.S. Edelman. — Proc. Natn. Acad. Sci. U.S.A., Vol. 56 (1966), 872.

Frenster, J.H. — Nature, Vol. 206 (1965), 1269.

Frenster, J.H., V.C. Allfrey, and A.E. Mirsky. — Proc. Natn. Acad. Sci. U.S.A., Vol. 50 (1963), 1026.

Gorsky, J., W.D. Noteboom, and J.A. Nicolette. — J. Cell. Comp. Physiol., Suppl., **66**, No.1 (1965), 77.

Knox, W.E. — In: "Synthesis of Molecular and Cellular Structure," p.13. New York, Ronald Press Co., 1960.

Levin, F.B. — Doklady AN SSSR **168**, No. 61 (1966), 1422.

Morozova, T.M. and R.I. Salganik. — Molekulyarnaya Biologiya, Vol. 3 (1969), 745.
Parke, D.V. The Biochemistry of Foreign Compounds. — Oxford, Pergamon Press, 1968.
Pestana, A. — Europ. J. Biochem., Vol. 11 (1969), 400.
Rosen, F., N.R. Roberts, L.E. Rudnick, and C. Nichol. — Endocrinology, Vol. 65 (1959), 256.
Salas, M., E. Vinuela, and M. Salas. — J. Biol. Chem., Vol. 238 (1963), 1165.
Salganik, R.I. — In: "Novye metody issledovaniya v gastroenterologii," p. 48. Novosibirsk, 1969.
Salganik, R.I., S.V. Argutinskaya, R.I. Bersimbaev, V.L. Knorre, N.M. Mertvetsov, and N.A. Solov'eva. — In: "Struktura i funktsiya kletochnogo yadra," p.23. Kiev, 1970.
Salganik, R.I., N.B. Khristolyubova, I.I. Kiknadze, T.M. Morozova, I.M. Gryaznova, and F.S. Valeeva, — In: "Struktura i funktsiya kletochnogo yadra," p.20. Moscow, "Nauka," 1967.
Williams-Ashman, H.G. — J. Cell. Comp. Physiol., Suppl., 66, No. 1 (1965), 71.

UDC 575:577.1

INTERACTION OF RIFAMPICIN AND RNA POLYMERASE

V.G. DEBABOV and G.I. KLIMOVA
Institute of Genetics and Selection of Industrial Microorganisms
Central Administration of the Microbiological Industry
Council of Ministers of the USSR,
Moscow

Acetylation of the antibiotic lowers its capacity to inhibit RNA polymerase by depriving it of the ability to form hydrogen bonds with the enzyme. The stepwise decrease in the affinity of rifampicin for the polymerase, occurring with modification, is regarded as an expression of the multifunctionality of the center binding rifampicin to the enzyme surface. This is confirmed by the existence of mutations leading to varying degrees of resistance of the enzyme to rifampicin.

The discovery that rifampicin and some of its semisynthetic analogs specifically inhibit bacterial RNA polymerase (Hartman et al., 1967) opened up new prospects for the study of this enzyme. Recent experiments in various laboratories (Toccini-Valentini et al., 1968; Igarashi and Yura, 1968; Wehrli, Knüsel and Staehelin, 1968), including our own (Alikhanian et al., 1970) yielded rifampicin-resistant mutants of Escherichia coli (rif-r) and showed that some of these mutants contain a modified RNA polymerase.

Today it is known that the enzyme binds rifampicin stoichiometrically (one molecule of the antibiotic per 13S monomer unit). High concentrations of urea disrupt the bond between rifampicin and the enzyme. Consequently, the bond is not covalent. Changes in the ionic strength do not affect the bond and ionic interactions apparently do not play an appreciable role (Wehrli, Knüsel, Schmid and Staehelin, 1968). The choice between hydrogen and hydrophobic bonds, however, is more difficult.

We acetylated rifampicin with acetic anhydride in pyridine at 0°. Judging from its UV spectra, the product obtained is free of the original compound in detectable amounts. We tested the ability of the product to inhibit RNA synthesis in vitro. As shown by the following data, acetylation sharply reduces the efficiency of inhibition. Acetylrifampicin competes very poorly with rifampicin for the enzyme, which means that the decreased inhibitory activity of the modified product results from its poor capacity for binding the enzyme.

Acetylation of the hydroxyl groups of rifampicin prevents them from forming hydrogen bonds and at the same time makes the molecule more hydrophobic. The results obtained could mean that hydrogen bonds are essential for stabilizing the enzyme-antibiotic complex, if it were possible to prove that the introduction of the acetoxy group does not have major steric consequences.

Rifampicin, μg/ml	Acetylrifampicin, μg/ml	Inhibition of RNA synthesis, %	Rifampicin, μg/ml	Acetylrifampicin, μg/ml	Inhibition of RNA synthesis, %
0.1	—	96	0.05	1.95	80
0.1	1.9	98	2.0	—	100
—	0.1	0	—	2.0	60
0.05	—	75			

Note. The reaction was carried out for 15 minutes at 30°. The sample contained 25μM Tris-HCl buffer pH 7.8; 7 μM Mg^{++}; μM KCl; 5 μM; β-mercaptoethanol, 40μM each of ATP, GTP, and CTP, 25 μg phage T2 DNA and 0.05 ml of an incubation mixture containing ^{14}C-UTP and 0.05 ml of enzyme in a final volume of 0.5 ml. Incorporation per sample about 4,000 pulses/min; background 100 pulses/min.

It is known that other modifications of the antibiotic, e.g. hydrogenation or oxidation of the macro-ring, or certain substitutions in the naphthalene nucleus, alter its affinity for RNA polymerase (Wehrli and Staehelin, 1969a). The complexity of the center binding rifampicin to RNA polymerase is supported by the following evidence: 1) the stepwise lowering of the affinity of rifamycins for polymerase with various modifications; 2) the existence of mutations leading to different degrees of resistance of the enzyme to rifampicin. It appears, therefore, that the rif-r mutations are spread over a rather large part of the genome and cause changes in one or several polypeptide chains of RNA polymerase. The position of only one rif-r mutation is known at present — in the β polypeptide (Rabussay and Zillig, 1969). Work on the mapping of various rif-r mutations is under way at our institute.

The adverse effect of high temperature on the bonding of RNA polymerase with rifampicin (Wehrli and Staehelin, 1970) agrees with the view that hydrogen bonds are involved in formation of the complex.

The latter property gives rise to some interesting effects of rifampicin in vivo. We examined the effect of low concentrations of rifampicin (1–10 μg/ml) on the growth of E. coli Hfr P13 in a complete medium at low and high temperatures (28 and 42°). Table 1 shows the results obtained.

TABLE 1. Growth of E. coli Hfr P13 at different concentrations of rifampicin

Concentration of rifampicin, μg/ml	Number of viable colonies per dish		Survival rate, % of the control	
	28°	42°	28°	42°
1	291	280	67	93
2	174	298	40	99
3	39	262	9	87
4	0	217	0	72
10	0	2	0	0
0	440	300	100	100

It is evident from Table 1 that rifampicin at a concentration of 4μg/ml causes 100% mortality of the cells at 28°, while the same concentration reduces the harvest by only 50% at 42°.

Previously we described rif-r mutants of E. coli Hfr P13 which behave similarly at a rifampicin concentration of 50 μg/ml, calling these mutants td (Alikhanian et al., 1970). These appear to be mutations with a slightly modified polymerase. The assignment of any mutation to td according to the phenotype depends on the concentration of antibiotic used, so that this classification hardly makes sense.

The search for synthetic compounds acting on the polymerase in the same way as rifampicin is of both theoretical and practical interest. From the theoretical point of view, stereochemical analysis of such compounds can shed light on the stereochemistry of the particular site on the enzyme surface binding the antibiotic.

Practically, rifampicin is one of the few compounds which can inhibit the replication of viruses in the cell (McAuslan, 1969). Compounds with a spectrum of action similar to that of rifampicin may prove to be effective antiviral agents.

We examined the antiviral agent oxalin. The structural formulas of oxalin and rifampicin are as follows:

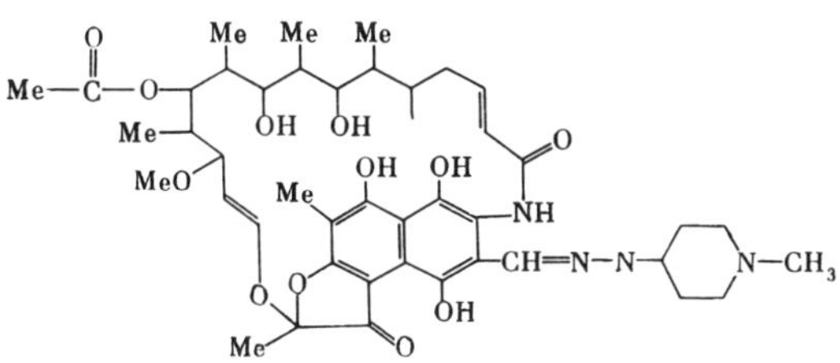

Rifampicin

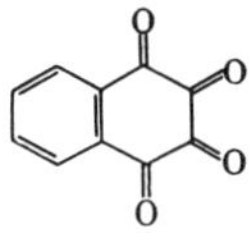

Oxalin

As shown in Table 2, the effect of oxalin on the RNA polymerase reaction in vitro resembles that of rifampicin, though the concentrations of oxalin used are much higher.

TABLE 2. Effect of oxalin on the RNA polymerase relation*

Presence or absence of the enzyme	Presence or absence of DNA	Oxalin, μg/ml	Incorporation of ^{14}C-UTP, pulses/min	RNA synthesis, %
+	+	—	3,179	100
+	—	—	101	0
+	+	200	1,105	35
+**	+	200	310	9
—	+†	200	1,053	36
—	+	200 ††	2,800	97

* For the conditions of the reaction see p.226.

** Preincubation of the enzyme with oxalin for 5 minutes.

† Preincubation of DNA with oxalin for 5 minutes.

†† Oxalin added 5 minutes after the beginning of the reaction.

BIBLIOGRAPHY

Alikhanian, S.I., V.G. Debabov, C.N. Tovmosian, S.T. Kopeickina, I.D. Nickiforova, T.A. Severina, E.S. Piruzian, and V.Z. Pogosov. — Proc. 1 Intern. Lepetit. Colloq., Florence, November, 1969, p.28. Amsterdam, North Holland Publ. Co., 1970.

Igarashi, K. and T. Yura. — J. Biochem. Soc. Japan, Vol. 40 (1968), 622.

Hartman, G., O. Honikel, F. Knüsel, and G. Nüesch. — Biochim. biophys. Acta, Vol.145 (1967), 843.

McAuslan. — Biochem. Biophys. Res. Commun., Vol. 37 (1969), 289.

Rabussay, D. and W. Zillig. — FEBS Letters, Vol. 5 (1969), 104.

Toccini-Valentini, G.P., P. Marino, and A. J. Colvill. — Nature, 220, No. 5164 (1968), 275.

Wehrli, W., F. Knüsel, M. Schmid, and A. Staehelin. — Proc. Natn. Acad. Sci. U.S.A., Vol. 61 (1968), 667.

Wehrli, W., F. Knüsel, and A. Staehelin. — Biochem. Biophys. Res. Commun., Vol. 32 (1968), 284.

Wehrli, W. and A. Staehelin. — Biochim. biophys. Acta, Vol. 182 (1969), 24.

Wehrli, W. and A. Staehelin. — Proc. 1 Intern. Lepetit Colloq. Florence, November, 1969. Amsterdam, North Holland Publ. Co., 1970.

UDC 575:576.858.9

TRANSCRIPTION OF THE PROPHAGE GENOME IN LYSOGENIC BACTERIA

E.I. GOLUB
Institute of General Genetics
Academy of Sciences of the USSR, Moscow

E. coli cells lysogenic for mutants of the UV-noninducible phages 18 and 299 lose their immunity to superinfection with phages genetically identical to the prophage at 42—46°. The loss of immunity is accompanied by depression of at least part of the prophage genome and also involves the functioning of one of the late phage genes. Nevertheless the prophage does not detach itself from the bacterial chromosome and consequently autonomous replication of the prophage and reproduction of phage particles do not occur in these cells.

Most of the data on the mechanism responsible for the lysogenic state were obtained in experiments with temperate inducible phages. The mechanism by which the lysogenic state is maintained in noninducible phages was largely obscure until recently. The reason for this was that any study of the mechanism responsible for the lysogenic state requires a means of disrupting this state by induction of the prophage. By definition, noninducible phages cannot be induced by any of the known agents acting on inducible phages.

In experiments with the noninducible phages 299 and 18 of E. coli (received from Professor F. Jacob) we were able to obtain mutants with a thermolabile repressor (Golub and Orlova, 1968; Golub and Reshetnikova, 1970). At temperatures above 42°, bacteria lysogenic for these mutants lose their immunity to superinfection with phage of the prophage type (Figure 1; see also Golub and Zvenigorodskii, 1969a; Golub and Reshetnikova, 1970).

Despite the loss of immunity, however, the prophage does not proceed to the stage of vegetative reproduction. The vast majority of bacteria neither reproduce phage nor die in the absence of superinfection (Golub and Zvenigorodskii, 1969b; Golub and Reshetnikova, 1970). Moreover, bacteria that are lysogenic for some mutants reproduce at 42—43° while remaining lysogenic (Figure 2; see also Golub and Zvenigorodsky, 1969).

Thus, the lysogenic state may persist in bacteria which have lost their immunity, i. e. when the repressor responsible for the immunity of the lysogenic bacteria to infection has been inactivated.

As might be expected, inactivation of the prophage repressor derepresses the prophage genome. DNA-RNA hybridization experiments have shown that bacteria lysogenic for the 299 ct11 phage mutant produce roughly 10 times more phage-specific RNA at 42° than at 37° (Golub and Zvenigorodskii, 1969a). In contrast to nonlysogenic bacteria or bacteria lysogenic for the wild type phage, the bacteria lysogenic for this mutant lose their ability to divide and produce filamentous forms when grown at 44—45° (Golub and Orlova, 1968, 1969). Such bacterial forms synthesize proteins and DNA at a normal rate while their nucleoids are distributed evenly and discretely throughout the length of the filiform cell, i. e. the process of cell division or, more accurately, cytokinesis, is selectively inhibited in such cells (Golub and Zvenigorodskii, 1970; Golub, Orlova and Reshetnikova, 1970).

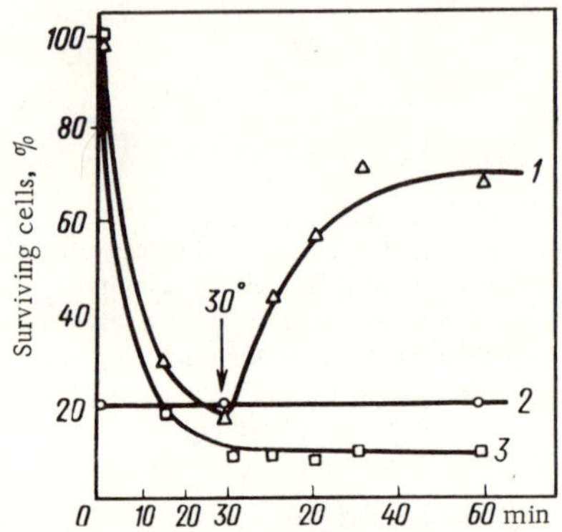

FIGURE 1. Loss of immunity to homoimmune superinfection in bacteria.

Bacteria in the stationary phase of growth were infected with phage 299 ct11 ct1 (a phage 299 mutant incapable of lysogenization) at a multiplicity of 2 particles per cell at 30° in medium A1 (0.03 M Na-K-phosphate buffer pH 7 + $1.5 \cdot 10^{-3}$ M $MgSO_4$ + $2 \cdot 10^{-3}$ M KCN). Unadsorbed phage were inactivated with antiserum, after which the bacteria were diluted 10^5 times with medium and incubated for 30 minutes at 42°. The bacteria were then diluted tenfold in the same medium and were incubated at 30°. Samples taken at regular intervals were seeded on dishes containing nutrient agar and incubated at 37°.
1) E.coli C (299 ct11); 2) E. coli C; 3) E. coli C (299 ct11 ct2).

We isolated 35 conditional lethal temperature-sensitive mutants of phage 299 ct11 unable to reproduce at 41°. Lysogenization of bacteria with these mutants revealed that in the presence of either of the two mutations — ts 30 or ts34 — in the prophage genome the cells start to reproduce normally at 44—45 without becoming filiform (Golub et al., 1971). Since these mutants are complementary, it is evident that at least two genes involved in phage reproduction are functional in the prophage state. The positive complementation test proves the derepression of the phage genome.

However, the ct11 mutation at the repressor locus described apparently does not totally derepress the prophage genome in such a way that all the genes of the prophage genome become functional. The introduction of a second mutation (ct1 or ct2) into the repressor locus of the mutant prophage 299 ct11 kills the lysogenic bacteria without allowing any reproduction of the phage (Golub et al., 1971). This is probably due to the activation of certain genes previously nonfunctional in the bacteria lysogenic for the 299 ct 11 mutant. Clearly, the introduction of a second mutation at the repressor locus into the prophage genome causes an additional depression of this.

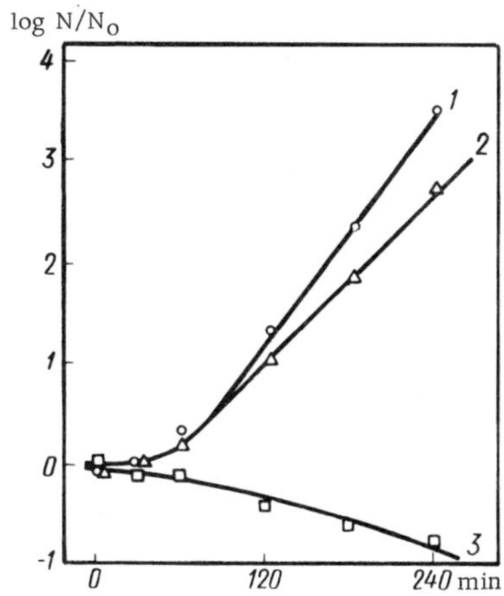

FIGURE 2. Bacterial reproduction.

Bacteria in the stationary phase of growth were diluted $10^5 - 10^6$ times with nutrient broth (Difco yeast extract 5 g; Bactopeptone 10 g; NaCl 10 g; glucose 1 g; H_2O to 1 liter, pH 7) and incubated at 42°;
1) E. coli C (299); 2) E. coli C (299 ct11); 3) E. coli C (299 ct11 ct2);
N_0 — initial titer of bacteria; N— titer of bacteria at the time indicated.

We therefore conclude that the ct11 mutation of the repressor locus causes only a partial derepression of the prophage genome. The ct11 mutation makes the lysogenic cell sensitive to phage infection which therefore culminates in the release of normal phage progeny. Consequently, the phage 299 ct11 repressor at high temperatures does not interfere with the development of the superinfecting phage but allows only partial derepression of the prophage genome. This implies that the repressor interacts differently with prophage DNA and with the DNA of the superinfecting phage.

The prophage repressor is currently regarded as a specialized product which does not perform any functions associated with phage reproduction. It was found, however, that bacteria lysogenic for phage 299 ct11, which has a temperature-sensitive repressor, as shown above, do not lose their immunity at high temperatures if the prophage bears the conditional lethal mutation ts16 (Golub et al., 1971). This mutation interrupts phage development 25 minutes after infection while the latent period of phage 299 ct11 under these conditions is about 30 minutes, i. e. it affects a "late" gene of the phage. Thus, the product of the late phage gene altered by the ts mutation can participate in the repression of the prophage.

As noted above, inactivation of the repressor of the noninducible phages 299 and 18 does not cause the prophage to pass into a vegetative stage of reproduction. The probable reason for this is that the prophage does not detach itself from the bacterial chromosome under such conditions (Golub and Zvenigorodskii, 1969a; Golub and Reshetnikova, 1970).

If bacteria which have lost their immunity are superinfected with phage of the same type as the prophage but bearing a genetic marker, the harvest obtained includes phage particles of both the superinfecting and the prophage type (Golub and Zvenigorodskii, 1969a). The superinfecting phage probably supplies the prophage with an enzyme necessary for the detachment of the prophage from the bacterial chromosomes. How is it that the superinfecting phage can direct the synthesis of the enzyme excising the prophage, while the prophage itself is unable to do so?

We have suggested that the incorporation of prophage in the bacterial chromosome damages the gene responsible for the synthesis of the enzyme detaching the prophage (Golub and Zvenigorodskii, 1969a). Recently Bertani (1970) reached a similar conclusion. Analysis of the behavior of the noninducible phage P2 showed that the gene controlling the detachment (and integration) of the prophage operates constitutively in the superinfecting phage, i. e. regardless of whether the prophage repressor is active or inactive. However, this gene ceases to function with the incorporation of the phageDNA into the bacterial chromosome.

In the case of noninducible phages it appears that the phage operon is active if the phage DNA is in a free state and inactive if the phage DNA is incorporated in the bacterial chromosome. Inactivation of the operon probably results from scission of the phageDNA during integration of the prophage.

Thus, there are two mechanisms of gene repression in temperate bacteriophages: by means of the phage repressor, and by mechanical lesion (scission) of the operon during integration of the prophage.

BIBLIOGRAPHY

Bertani, L.E. — Proc. Natn. Acad. Sci. U.S.A., Vol. 65 (1970), 331.

Golub, E.I. and G.G. Orlova. — Doklady AN SSSR 180, No. 4 (1968), 971.

Golub, E.I. and G.G. Orlova. — Mikrobiologiya, Vol. 38 (1969), 1054.

Golub, E.I., G.G. Orlova, and V.N. Reshetnikova. — Doklady AN SSSR, Vol. 192 (1970), 1149.

Golub, E.I., G.G Orlova, V.N. Reshetnikova, and V.I. Zvenigorodskii. — Molekulyarnaya Biologiya, Vol. 5 (1971), 543.

Golub, E.I. and V.N. Reshetnikova. — Mikrobiologiya, Vol. 39 (1970), 1057.

Golub, E.I. and V.I. Zvenigorodsky. — Virology, Vol. 39 (1969), 919.

Golub, E.I. and V. I. Zvenigorodskii. — Molekulyarnaya Biologiya, Vol. 3 (1969a), 869.

Golub, E.I. and V.I. Zvenigorodskii. — Doklady AN SSSR 188, No. 4 (1969b), 920.

Golub, E.I., V.I. Zvenigorodskii, I.I. Kurilova, G.G. Orlova, and Yu. K. Fomichev. — Mikrobiologiya, Vol. 39 (1970), 514.

UDC 575.24:576.8

CLASSIFICATION OF NONSENSE MUTATIONS IN THE ad_2 LOCUS OF YEAST

B.V. SIMAROV, L.N. MIRONOVA and S.G. INGE-VECHTOMOV
Leningrad State University

Two types of nonsense mutations found at the ad_2 locus were shown to be analogous to amber and ochre mutations (identified in phage-bacterium systems) with respect to their reaction to supersuppression. The supersuppressors examined were accordingly divided into two groups: those deciphering only one of the nonsense codons, and those deciphering both.

The phenotypic manifestation of certain mutant alleles in the yeast S. cerevisiae can be inhibited by supersuppressors (Hawthorne and Mortimer, 1963; Mortimer and Manney, 1964; Inge-Vechtomov, 1966). The position of these alleles on the complementation map and also the specificity of the suppressors examined confirm the view that the supersuppressed alleles bear nonsense codons and that their suppression involves mutational changes affecting the translation mechanism.

The principal genetic feature of the suppressors, enabling a more detailed discussion of the mechanism of their action, is their allelic specificity. We have shown, however, that this characterization is not universally applicable (Inge-Vechtomov and Simarov, 1967). In many cases a suppressor may decipher a nonsense codon but this will not be apparent at the haploid level because the amino acid involved in the translation of the nonsense codon is incompatible with the function of the enzyme. This type of deciphering of the nonsense codon can be detected in diploids by the change in the spectrum of interallelic complementation in the presence of the suppressor. Complementation will undergo a change in the presence of the suppressor because the cell controlled by the nonsense-bearing allele will synthesize not fragments but complete polypeptide chains capable of yielding active enzyme molecules by uniting with the mutant subunits produced by other mutant alleles.

Our studies of allelic specificity and the effect of various nonsense suppressors on complementation have shown that at least two types of nonsense mutations can occur at the ad_2 locus (Inge-Vechtomov and Simarov, 1967).

Here we present the results of our work on the allelic specificity and the effect of the dominant suppressor S_{509}, obtained by a reversion of the mutant p223 (type II nonsense mutation at the ad_2 locus), on interallelic complementation. We examined the specificity of this suppressor with respect to 22 alleleles of the ad_2 locus of which 13 contained a type I nonsense codon and 5 a type II nonsense codon. These 18 alleles were not mutually complementary; on the interallelic complementation map the functional lesions caused by them were polar or scattered throughout the map (Soidla et al., 1967). We also used 4 ad_2 mutants which bear slight nonpolar functional lesions and can therefore be regarded as missense mutants. The results of the analysis of the allelic specificity of the suppressor S_{509} are shown in the table. Since the tetrad analysis was carried out with hybrids heteroallelic with respect to the ad_2 locus and containing an allele known to be suppressible, the requirement for adenine always showed a $2^+:2^-$ segregation in the tetrads if the second allele was also suppressible. If the second allele was not suppressible, segregation gave tetrads with $3^-:1^+$ and $4^-:0^+$ ratios. In the table, all the alleles examined are divided into three groups: those bearing type I nonsense codons, type II nonsense codons, and minor functional lesions. It is evident from the table that the dominant suppressor S_{509} can overcome 6 of the 13 type I nonsense mutations and 4 of the 5 type II nonsense mutations.

Specificity of the S_{509} suppressor with respect to ad_2 alleles

Allele	Number of tetrads examined	Segregation in the tetrads			Suppressibility of S_{509}
		$3ad^-:1ad^+$	$2ad^+:2ad^-$	$4ad^-:0ad^-$	
16	19	5	4	10	—
75	11	5	6	0	—
91	9	7	1	1	—
99	5	1	3	1	—
105	11	0	11	0	+
113	9	7	2	0	—
120	10	6	3	1	—
133	6	0	6	0	+
144	13	0	13	0	+
148	7	6	1	0	—
190	10	0	10	0	+
199	15	0	15	0	+
201	13	0	13	0	+
72	7	0	7	0	+
137	11	0	11	0	+
177	9	0	9	0	+
223	This allele yielded the revertant R_{509}				+
226	12	8	3	1	—
47	9	0	9	0	+
192	7	0	7	0	+
163	11	8	3	0	—
209	10	9	0	1	—

Comparison of the specificity of the suppressor S_{509} with that of the suppressor S_5, which is concerned strictly with type I nonsense codons, leads to the conclusion that the type I nonsense codons are amber codons and the type II nonsense codons ochre codons. Thus, the suppressor S_5 deciphers only amber codons, whereas S_{509} deciphers both amber and ochre codons.

It is noteworthy that the suppressor S_{509} suppressed 2 alleles (of a total of 4 examined) containing minor functional lesions. This means that the suppressor S_{509} not only overcomes nonsense codons but also alters the sense of meaningful codons, thus acting as a missense suppressor. This hypothesis is confirmed by the observed effect of the suppressor S_{509} on interallelic complementation. These experiments were performed with monosporogenous clones isolated in the course of tetrad analysis and bearing various ad_2 alleles not suppressible by S_{509}. About 1,200 combinations of ad_2 alleles were examined in the presence of one or other of the suppressors. It was found that if the combination contains two unsuppressible alleles (U/U), the nonsense suppressor S_5 does not alter the complementation, while in a combination containing a suppressible and an unsuppressible allele (U/S) this same suppressor alters the complementation only slightly (91%). On the other hand, S_{509} causes in both cases suppressor-induced complementation (81% for U/U and 98.4% for U/S). Interallelic complementation induced by the suppressor S_{509} was also observed when the combination contained alleles with nonpolar functional lesions. These findings show that S_{509} can not only decipher the two types of nonsense codons but also alter the sense of meaningful codons.

What mechanism of supersuppression can explain the results obtained? It appears from the dominance and code specificity of the suppressor S_{509} that the mutation involved affects one of the types of tRNA; the existence of such a mutation was predicted on the basis of the Crick (1966) hypothesis of the interaction of mRNA codons and tRNA anticodons.

More specifically, the tRNA anticodon containing pseudouridine (Caskey et al., 1968) or 5-oxyacetyluridine (Murao et al., 1970) at position 1 must recognize three codons differing from one another by the presence of U, A, or G at position 3. In our case the mutant tRNA must recognize the codons UAA (ochre), UAG (amber) and UAU (tyr), accordingly causing the suppression of nonsense and missense mutations.

BIBLIOGRAPHY

Caskey, C.T., A. Beaudet, and M. Nirenberg. — J. Molec. Biol., Vol. 37 (1968), 99.

Crick, F. — J. Molec. Biol., Vol. 19 (1966), 457.

Hawthorne, D.C. and R.K. Mortimer. — Genetics, Vol. 48 (1963), 617.

Inge-Vechtomov, S.G. — Genetika, Vol. 7 (1966), 142.

Inge-Vechtomov, S.G. and B.V. Simarov. Studies in Genetics, No.3, p.127. Leningrad State University, 1967. (Russian)

Mortimer, R.K. and T.R. Manney. — Genetics, Vol. 50 (1964), 270.

Murao, K., M. Saneyoshi, F. Harada, and S. Nishimura. — Biochem. Biophys. Res. Commun., Vol. 38 (1970), 657.

Soidla, T.R., S.G. Inge-Vechtomov, and B.V. Simarov. — Studies in Genetics, No.3 (1967), 148. (Russian)

UDC 575:576.858

GENETIC INTERACTIONS IN POPULATIONS OF VENEZUELAN EQUINE ENCEPHALOMYELITIS VIRUS

Ya. Ya. TSILINSKII, B.V. GUSHCHIN, S.M. KLIMENKO and D.K. L'VOV
D.I. Ivanovskii Institute of Virology
Academy of Medical Sciences of the USSR, Moscow

Suspensions of uncloned wild type virus contain two types of virions: mononucleoid (accounting for 85% of the suspension) and di- and polynucleoid. Suspensions of clones isolated from the population consisted entirely of mononucleoid virions. Giant di- and polynucleoid virions reappeared in the viral progeny after mixed infection of the cells with two or several viral clones which separately contained and yielded strictly mononucleoid virions. It is suggested that the giant virions contain different genomes and are therefore heterozygotes; the biological significance of this phenomenon is discussed.

The populations of some arboviruses (Klimenko et al., 1965; Higashi et al., 1967; Bykovsky et al., 1969) and viruses of other taxons (Dahlberg and Simon, 1969) contain not only mononucleoid virions but also an admixture of di- and polynucleoid giant virions. Here we present evidence that the giant virions of Venezuelan equine encephalomyelitis virus (VEE) arise as a result of genetic interactions between the different viral genomes.

MATERIALS AND METHODS

The morphological composition of VEE virus populations isolated from a number of clones (Nos. 5, 6, 8, and 3/5), the morphological composition of temperature mutants (clones Nos. 7 and 14), and the morphological and genetic composition of a population obtained by mixed infection with different clones were examined. Passages of the virus were made in chick fibroblast (CF) cultures. The multiplicity of inoculation was one plaque-forming unit (PFU) per 500—1,000 cells. In the case of mixed infection with two or several clones, each of the viruses was introduced in the amounts indicated. Virus harvests were taken for morphological and genetic studies after 18—20 hours. Virus preparations for microscopy were made by the negative contrast method. The preparations were examined using a JEM-6-C electron microscope at a magnification of 500,000. The morphological studies were aimed at the detection and quantitative determination of giant virions in the viral suspensions. For this purpose at least 1,000 virions were examined in each preparation. Clonal analysis was used for determining the genetic properties of the virus.

The genetic characters of VEE virus and their phenotypic manifestation are designated as follows in the text and tables below:

S — plaque size in chick fibroblast cultures:
"+ +" — average diameter more than 8 mm;
"+ " — average diameter from 4 to 8 mm;
"+ —" — average diameter from 2 to 4 mm;
"— —" — average diameter from 1 to 2 mm;
"— — —" — average diameter less than 1 mm.

ia — resistance to the inhibitory action of agar polysaccharides:
"+" — resistant (the number and size of plaques under an agar overlay containing 0.6 mg/ml protamine sulfate are not increased);
"—" — sensitive (increased).

rct_{40} — ability to reproduce at 40°:
"—" — does not form plaques at 40°;
"+—" — the plaques formed at 40° are smaller than at 37°;
"+" — the plaques formed at 40° are equal in size and number to those formed at 37°.

T_{60} — resistance to heating at 60°:
"+" — inactivation index less than 1.5 log PFU after 10 minutes of heating;
"+ —" — inactivation index from 1.5 to 5.0 log PFU;
"—" — inactivation index more than 5.0 log PFU.

RESULTS AND DISCUSSION

The experiments showed that the population of VEE virus comprises virions of two types: 1) virions measuring 65—95 mμ (average length of the long axis of the virion) and containing a single nucleoid; 2) virions measuring more than 120 mμ and containing two or more nucleoids. These types of VEE virions are shown in the photomicrographs.

In contrast to the virus population, the clones isolated from it and the temperature mutants contained mononucleoid virions only. The morphological homogeneity of the clones persisted after passages in CF cultures.

TABLE 1. Percentage of giant virions in virus populations obtained 18 hours after the infection of CF cultures with mixtures of VEE virus clones

Number and characteristics of the clone	Clone number					
	5	6	7	8	14	3/5
5 (S + —, ia +, rct_{40} +, T_{60} —)	0	3,8	2.0	13.9	2.0	3.1
6 (S + —, ia —, rct_{40} +, T_{60} +)	3.8	0	4.0	9.2	9.0	—
7 (S —, ia +, rct_{40} —, T_{60} + —)	2.0	4.0	0	6.0	1.5	—
8 (S +, ia —, rct_{40} +, T_{60} +)	13.9	9.2	6.0	0	1.0	<0.1
14 (S — —, ia —, rct_{40} —, T_{60} —)	2.0	9.0	1.5	1.0	0	—
3/5 (S + +, ia —, rct_{40} +, T_{60} +)	3.1	—	—	<0.1	—	0

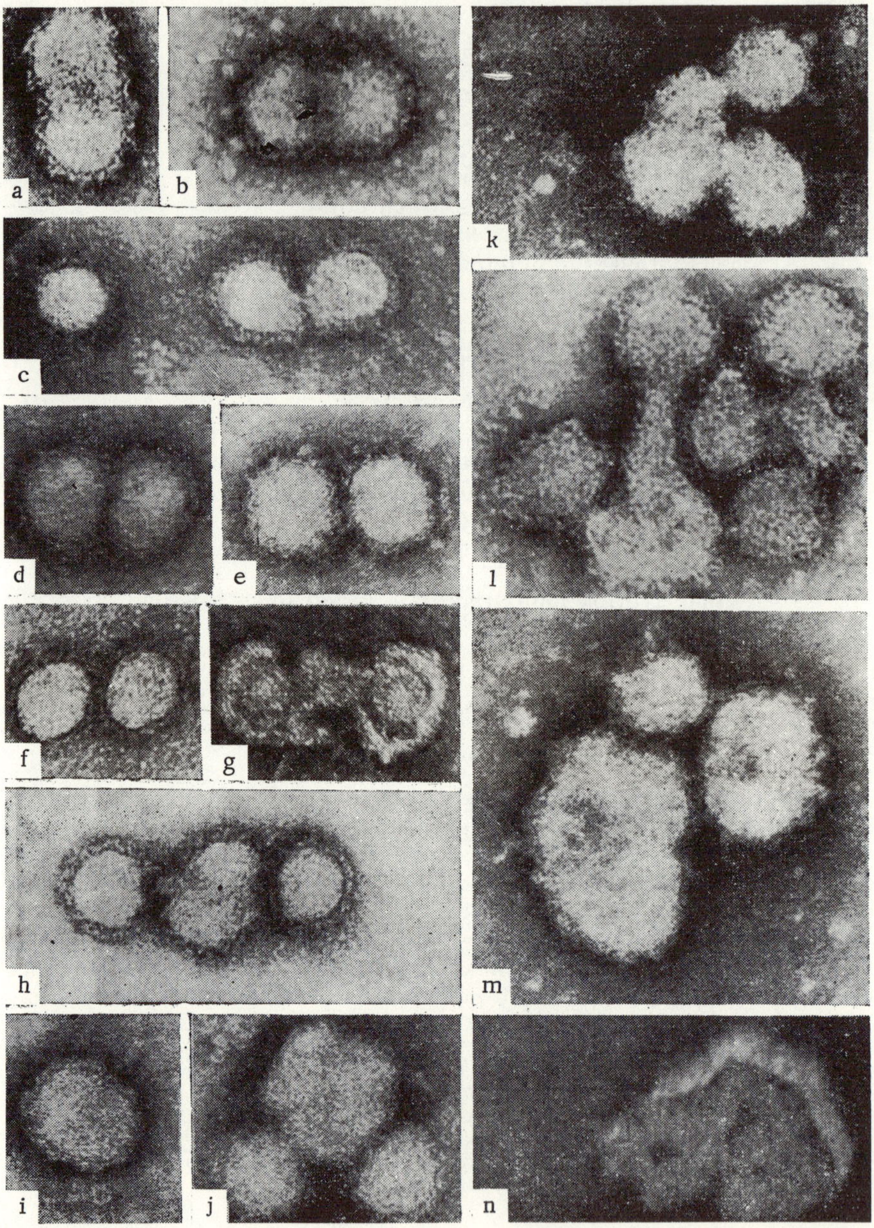

Different types of VEE virion:

a – n) giant virions; a, b, c, d, e, f, h, i, j) dinucleoid virions with different degrees of linkage between the nucleoids; k, l, m) polynucleoid virions, including irregularly shaped virions; g, n) disintegrated dinucleoid and polynucleoid virions. Remnants of the membrane and nucleoids of the virion are visible; c, h, j, m) mononucleoid virions are shown beside the giant ones. Contrasting agent 2% phosphotungstic acid. Magnification 200,000 ×.

Giant virions reappeared in the virus progeny on infection of the cell culture with two or several clones of the virus. The number of giant virions found with mixed infection depended on the properties of the clones participating in infection (Table 1). Certain combinations of clones failed to yield giant virions. The genetic constitution of the interacting clones was reflected in the size and structure of the giant virions. In the virus suspensions obtained by mixed infection with clones 5 and 8, the giant virions consisted mostly of polynucleoid forms, whereas dinucleoid virions predominated in the case of mixed infections with clones 7 and 8 and 6 and 14.

Analysis of the mixed infection caused by clones 5 and 8 and clones 7 and 8 showed that the formation of giant virions was accompanied by genetic interaction between the virus variants participating in the infection (Table 2).

The progeny of clones 5 and 8, differing in their resistance to agar inhibitors, contained not only particles with the properties of one parent but also a certain percentage of particles segregating into subclones which were resistant or sensitive to the agar inhibitors. Such particles were absent in the control tests, where each clone was transferred separately, suggesting that these clones were heterozygotes containing both parental genomes. The presence of heterozygotes in the mixed population was compatible with the existence of giant virions. Since the giant virions consisted of polynucleoid particles in this case, it appears that this type of giant virions is heterozygous in nature.

Apart from unchanged parental types, the progeny of clones 7 and 8 included particles which formed small plaques at 37° but reproduced rapidly at 40°, as well as large-plaque types which showed a lower capacity for reproduction at 40° (Table 2). These particles with altered properties were absent in the control tests, where each clone was grown separately. Thus, the joint replication of these variants of the virus led to a unilateral transfer of genetic information from the large-plaque clone, which can reproduce at high temperatures (donor), to the small-plaque cone, which lacks this property (recipient). The resulting recombinants were capable of reproducing at 40°. The donors contributing the genetic information associated with reproduction at high temperatures remained viable at 40° but their reproductive potential was lowered. The number of recombinants in the virus suspension exceeded by far that of giant virions. Clearly, the normal mononucleoid virions carried the recombinant genomes. The formation of the dinucleoid giant virions present in the virus suspension is probably associated with the recombination process.

On these grounds the formation of giant virions can be attributed to genetic interactions between the different viral genomes. Such genetic interactions require territorial matching of the replicative complexes of the different genomes. In view of the size differences between cells and virions it appears that the probability of such joint replication is extremely low given the random distribution of the sites at which infectious virions penetrate into the cell. Even in the case of such a large virus as the vaccinia virus, the frequency of recombination rises markedly when the cells are infected with virion clumps obtained by treatment of the virus suspension with a concentrated solution of magnesium chloride (Abel, 1962).

TABLE 2. Genetic composition of virus populations obtained by mixed infection with clones of VEE virus

Infecting clones	Number of subclones examined	Virus progeny				Percentage of giant virions
		Parental type	ia +, ia —	S—, rct_{40} +	S +, rct_{40} ±	
5+8	84	69	15	—	—	13.9
5 (ia +)	38	38	—	—	—	<0.1
8 (ia —)	41	41	—	—	—	<0.1
7+8	120	94	—	14	12	6.0
7 (S—, rct_{40} —)	42	42	—	—	—	<0.1
8 (S+, rct_{40} +)	82	82	—	—	—	<0.1

It follows from these data that the union of different viral genomes in giant virions is biologically advantageous. Such a union ensures the simultaneous penetration into the cell and joint replication of genetic material capable of recombination.

It appears, therefore, that the recombination of VEE virus takes place in two consecutive generations. In the first generation, the nucleoids formed in the cell by the joint replication of genomes capable of recombination and bearing unchanged parental genotypes unite into giant virions under the influence of some specific mechanism. The recombinational events occur in the next generation in the cells infected with the giant virions.

If the union of nucleoids into giant virions is mechanical in nature and represents a preliminary to the subsequent recombinations it is conceivable that there is some contact between the RNA of the partners united in the giant virion. This possibility is especially plausible with respect to the giant virions containing fused nucleoids. Such giant virions have been described by Higashi et al. (1967) for Chikungunya virus, which is related to VEE virus.

CONCLUSIONS

VEE virus can participate in various genetic interactions including the formation of heterozygotes and recombinations. The formation of ia —/ia+ heterozygotes was observed on culturing ia — and ia+ clones together. S—, rct_{40}+ recombinants were obtained by culturing the clones S+, rct_{40}+ (donor) and S—, rct_{40} — (recipient) together. On mixed function with different clones the genetic interactions were accompanied by the formation of di- and polynucleoid giant virions. The relationship between these phenomena is being investigated.

BIBLIOGRAPHY

Abel, P. — Virology, Vol. 17 (1962), 511.
Bykovsky, A.F., F.I. Yershov, and V.M. Zhdanov. — Virology, Vol. 4 (1969), 496.
Dahlberg, J. E. and E. H. Simon. — Virology, Vol. 38 (1969), 666.
Higashi, N., A. Matsumoto, K. Tobata, and J. Nagatomo. — Virology, Vol. 33 (1967), 55.
Klimenko, S.M., F. I. Yershov, J. P. Gofman, A.R. Nabatnikov, and V. M. Zhdanov. — Virology, Vol. 27 (1965), 125.

UDC 575.24:576.858

SYNTHESIS OF VIRUS-SPECIFIC STRUCTURES IN CELLS INFECTED WITH ts-MUTANTS OF POLIO VIRUS UNDER NONPERMISSIVE CONDITIONS

A.V. MIKHEEVA, E.A. YAKOBSON and Yu. Z. GENDON
Moscow Research Institute for Virus Preparations
Ministry of Health of the USSR

HeLa cells infected with ts-mutants of polio virus under nonpermissive conditions (40°) do not produce 150S virus particles or their 73S precursors. Only 14S and 5S particles are formed in such cells. The viral nature of the 14S and 5S structures was proved. It is suggested that the 5S structure is an early stage of the morphogenesis of polio virus. The arrest of morphogenesis at the 14S stage is due either to the absence of a specific factor required for the 14S→73S conversion or to mutational lesions in the proteins of the virus particles.

Sucrose gradient analysis of the cytoplasmic fractions obtained from HeLa cells infected with polio virus reveals four types of virus-specific structures with sedimentation constants of 5S, 14S, 73S, and 150S. At least three of these structures are known to be successive stages in the morphogenesis of polio virus: the 14S structures are precursors of the 73S procapsids, while these in turn unite with viral RNA to form mature 150S virions.

We examined the formation of such structures in HeLa cells infected with ts-mutants of polio virus under nonpermissive conditions (40°). The work was carried out with polio virus mutants of type I (ts2, ts5, and ts11) and type II (ts M/23), obtained respectively from the Mahoney and MEF_1 wild type strains.

None of the mutants obtained was capable of inducing the formation of virus particles (150S) or their precursors (73S) at 40°. Cells infected with the ts5 and ts M/23 strains produced 5S and 14S structures, while cells infected with the mutants ts2 and ts11 produced only 5S structures.

For the immunological characterization of these structures, extracts from cells infected with ts-mutants (before their transfer to the sucrose gradient) were treated with serum from rabbits immunized with a crude virus suspension obtained from cells infected with the wild type virus at 36°.

Such treatment of the cytoplasmic extract from cells infected with ts-mutants at 36° completely eliminates the three peaks of radioactivity corresponding to the 150S, 73S and 14S structures, as well as about 50% of the radioactivity of the 5S peak. Similar results with respect to the 14S and 5S peaks were obtained by treating cytoplasmic extracts from cells infected with ts-mutants at 40° with the antiserum.

These findings demonstrate the viral nature of the structures formed in cells infected with mutant virus under nonpermissive conditions. The arrest of morphogenesis at the 14S stage in the ts5 and ts M/23 mutants under nonpermissive conditions may result either from mutational lesions of the proteins constituting these particles or from the absence of the specific factor which, according to the literature, is required for the 14S → 73S stage of morphogenesis.

The discovery of mutants (ts2, ts11) forming only 5S structures at 40° can be regarded as an indication that these structures represent an early stage in the morphogenesis of polio virus.

MOLECULAR MECHANISMS OF GENETIC RECOMBINATION

UDC 575:576.8

DNA REPLICATION AND THE RECOMBINATION MECHANISM IN BACTERIA

S.E. BRESLER
A.F. Ioffe Physical and Technical Institute
Academy of Sciences of the USSR, Leningrad

In the course of conjugation, one of the two DNA strands of the donor cell is transferred to the recipient cell, where a complementary strand is immediately synthesized for it. The heterozygote has a double-stranded structure during conjugation. The average size of the integrated sector is $2 \cdot 10^8$ daltons. Transformation involves the integration of one of the two donor DNA strands into the chromosome of the recipient. Here the size of the integrated sector is 10^7 daltons. Each transformation event yields a molecular heterozygote in the recipient which is repaired by enzymes for the most part identical to those responsible for the repair of UV-induced lesions. The repair or correction of heterozygotes is allele specific. The same enzymes are involved in the process of genetic recombination at the level of double-stranded DNA.

The recombination of the genetic material of the two cells, donor and recipient, is conceived as the integration of DNA fragments from both parents into a single molecular chain. By means of electron microscopy and radioautography it has been possible to observe this process at the molecular level (see the article by M. I. Mosevitskii, p. 274 in this collection). We hope that this new technique will reveal many intimate details of this phenomenon. At this stage we can establish that the recombination process undoubtedly consists of both destructive enzymatic reactions, involving endonuclease and exonuclease, and synthetic reactions. The latter reactions involve DNA polymerase and ligase. Another complex process is the vegetative synthesis of DNA. Suffice it to say that the two DNA strands are replicated in opposite directions and that the synthesis apparently leads to comparatively short fragments (about 1,000 nucleotides long) which are joined together by ligase. The DNA polymerase which brings about the autoreplication of DNA has not yet been isolated and analyzed. For a long time it was assumed that the Kornberg enzyme is responsible for this function, but today it is well known that the Kornberg enzyme does not participate in vegetative autoreplication and apparently belongs to the category of repair enzymes (De Lucia and Cairns, 1969; Gross and Gross, 1969). The difficulty in isolating the vegetative DNA polymerase seems to be due to the integration of this enzyme into the bacterial membrane, which precludes its extraction under mild conditions. Ganesan and Lederberg (1965) have shown by pulse labeling of DNA precursors that DNA synthetase activity is concentrated in the membrane but they were unable to isolate the individual protein by disintegrating the membrane with agents such as detergents, urea, SH reagents, etc. We may well be dealing here not with a single protein but with a multienzyme complex comparable to those recently discovered in the majority of biochemical processes. Thus, one enzyme may initiate DNA synthesis, for example, by causing single-strand breaks in the template strand; a second enzyme may bring about elongation of the strands until a third enzyme stops the process and a fourth enzyme joins the pieces together, etc. A detailed study of this process necessitates obtaining mutants bearing lesions related to specific stages of the process. Thanks to the work of Bonhoeffer et al. (1967) we possess the valuable temperature-sensitive mutants ts43, which are defective for two functions simultaneously: they are unable, when grown at temperatures above 42°, to carry out vegetative autoreplication of DNA or a genetic recombination, i. e. they are of the rec⁻ type. We showed that these properties arise from a point mutation since the mutant reverts with a probability of the order of 10^{-8} per generation. The reversion eliminates both characters simultaneously. These rec⁻ mutants obviously differ in many ways from the recA, recB and recC mutants described by the Clark group (Willets and Clark, 1969). In all these cases the enzyme affected by the mutation was exonuclease. This could be shown in cell-free extracts. Here the mutations simultaneously affected two processes: genetic recombination and the repair of DNA lesions.

As for the Bonhoeffer mutants, it appears that the affected enzyme is synthetase or ligase. DNA synthesis accordingly does not occur in such cells at 43°; instead, there is an exonuclease-induced breakdown of DNA which can be assessed by the release of low molecular weight acid-soluble products. This breakdown takes place first and foremost near the replicative fork of the chromosome.

As regards the DNA of the ts43 mutant isolated by Bonhoeffer, it is reasonable to expect lesions of the DNA synthetase or ligase and also to expect that the same enzyme will participate in the vegetative replication of the chromosomes and in DNA synthesis during recombination. We shall begin the description of our observations with a comparison of the two processes. First, we carried out the conjugation of *Escherichia coli* K12 Hfr donor cells with F⁻ recipients bearing the ts43 mutation; secondly we induced sexduction into these cells from a number of F'lac strains containing different fragments of genetic substance in the episome. In the first case we looked for recombinants for the proximal markers thi and lac and in the second for the lac marker. Contraselection was carried out as usual in that the donor cells were streptomycin sensitive and the recipient cells streptomycin resistant. It is well known today that the conjugation process consists of a series of consecutive stages: 1) recognition of the donor cells by the recipients; 2) formation of a protoplasmic bridge or canal; 3) transfer of one DNA strand from the donor to the recipient cell with the concomitant synthesis of a DNA strand in the donor, though the transferred strand is of the pre-existing set (Bresler et al., 1969). The beginning and direction of the transfer as well as the selection of the strand to be transferred is entirely determined by the position and polarity of the F factor in the Hfr chromosome; 4) synthesis of a second strand for the fragment transferred to the recipient cell. This process has been demonstrated both with respect to conjugation between Hfr and F^- (Ohki and Tomizawa, 1968) and in sexduction (Eisenberg and Pardee, 1969); 5) genetic recombination proper between the fragment transferred and the chromosome of the recipient.

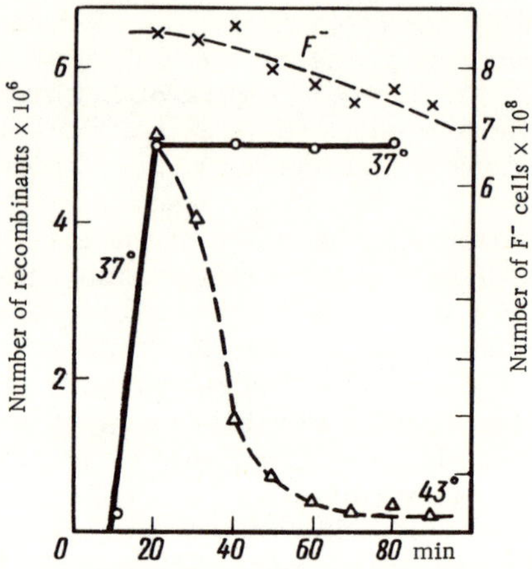

FIGURE 1. Heat shock of exconjugants.

The conjugation mixture HfrH × F^- ts43 was kept in contact for 20 minutes at 37°; the conjugating pairs were separated in a blender, after which the suspension was diluted 10^3 times with fresh medium and transferred to 43°, part of the suspension being left at 37°. Periodic controls were made for the number of $thr^+ leu^+$ str-r recombinants and viable F^- cells.

As mentioned above, the recipients bearing the ts43 mutation behaved in the same way as rec^- cells at 43°. Incubation at 43° of the merozygotes formed in the conjugation mixture at 37° caused the rapid and irreversible death of the potential recombinants (heat shock). The kinetics of this process is shown in Figure 1. The data in Figure 2 demonstrate that it is precisely the arrest of DNA synthesis in the recipient cell that has such lethal consequences on the yield of recombinants. In this experiment, 5-fluorodeoxyuridine (5-fluorouracildeoxyriboside or FUDR) was used to stop the synthesis of DNA in the merozygote. FUDR clearly has the same effect on the merozygotes — irreversible death of the potential recombinants. Previously we examined the sexduction of a strain with a short F'lac episome into a temperature-sensitive recipient (Bresler et al., 1968). It turned out that the yield of sexductants with respect to the lac gene was smaller by an order of magnitude at 43° than at 37° and that sexduction at 43° was markedly improved if carried out after a brief contact at 37° with subsequent strong dilution. We examined further the effect of temperature on sexduction in a number of donors with F' of different sizes: 200 PS (2.8% of the chromosome), E. coli ORF 206 (5.5%), and ORF 1(8%).

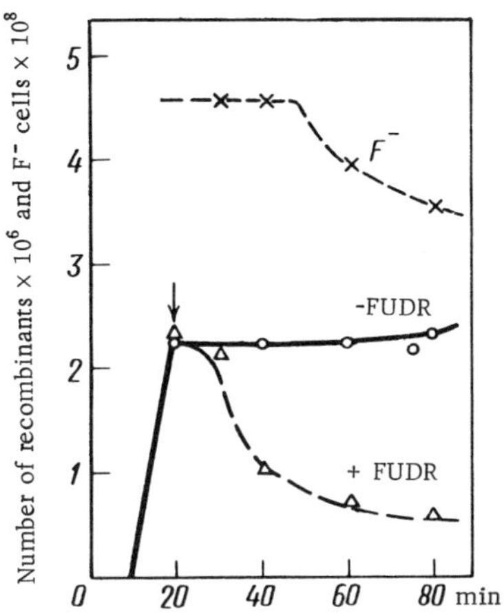

FIGURE 2. Kinetics of the yield of recombinants in the presence of 5-fluorouracil-deoxyriboside (FUDR).

The arrow indicates the moment when FUDR was added. All conditions are the same as in the preceding experiment, except for the addition of the antimetabolite to the exconjugants instead of their transfer to 43°.

In the controls ts43 → wild type revertants were used as the recipients. We found that sexduction in the controls is more efficient at 43° than at 37°. In all crosses of different F' with a temperature-sensitive recipient the yield of sexductants was much lower at 43° than at 37°. A definite relationship exists between episome length and the yield of sexductants at 37° and 43° — the shorter F', the smaller the difference between the maximal yields of sexductants at 43° and 37°. Figure 3 shows the kinetics of the yield of sexductants in crosses of F'lac (200 PS) and F'lac (206) with a temperature-sensitive recipient. The maximal yield of sexductants at 43° corresponds to 40% of the yield at 37° for F' (200 PS), and only 15% for the longer episome of F' (206). We also compared the behavior of the different sexductants in heat shock experiments similar to those described previously. Only the longest sexductants (ORF-1) yielded a product susceptible to irreversible death at 43° (Figure 4). This is understandable on the grounds that the long F' resemble Hfr donors: their markers can be transferred both in the form of episomes, i. e. with the F factor and as DNA fragments, without the F factor. Genetic recombination is indispensable in the latter case; hence the death of the cells at 43°, where recombination is blocked.

Let us examine the results obtained in terms of the mechanism of the conjugation process. Firstly, the transfer of genetic material into the temperature-sensitive recipient obviously proceeds without obstruction at 43°. Brief contact at 37° is absolutely insufficient for transfer of the episome. Such contact apparently helps during the initial stage — the formation of conjugation bridges. Our conclusion concerning the unobstructed transfer of genetic material into the ts43 strain has recently been

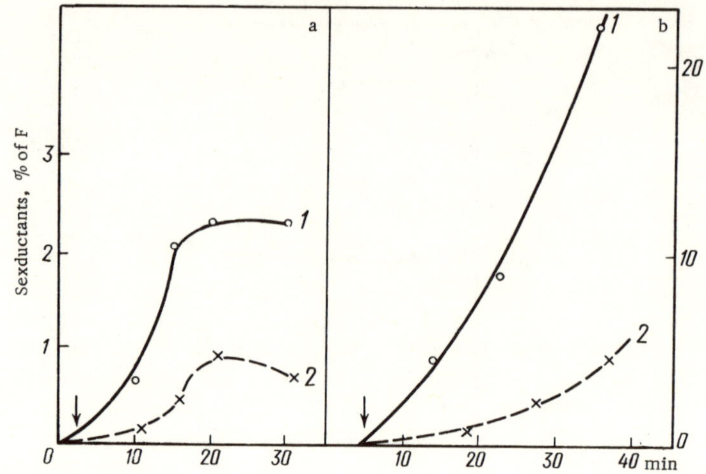

FIGURE 3. Kinetics of the yield of lac+ str-r sexductants at 37° (1) and 43° (2) from the following crosses:

F'lac (200PS) x F^- ts43 (a): F'lac (ORF 206) x F^- ts43 (b).
a: concentration of F'lac — 2.6×10^7 cells/ml, F^- — 2.4×10^8 cells/ml. The original contact at 37° lasted 2.5 minutes, after which the suspension was diluted 500-fold with medium warmed to 43° or 37°. The arrows in the graph indicate the times of sample collection, blender treatment and titration of sexductants;
b: concentration of F'lac — 1.6×10^7 cells/ml, F^- — 1.3×10^8 cells/ml. The procedure was the same except that the original contact lasted 5 minutes.

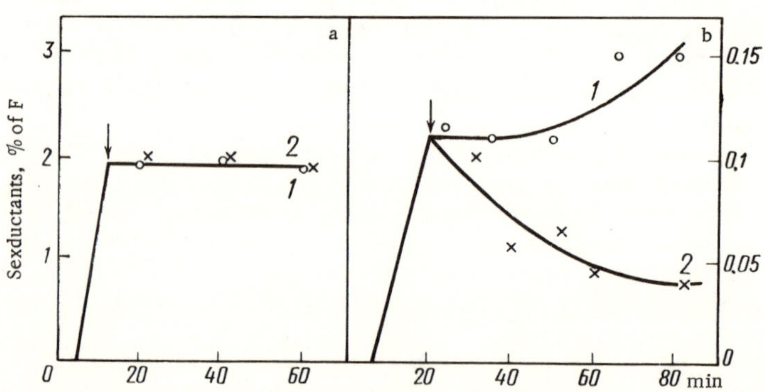

FIGURE 4. Heat shock of sexductants.

Kinetics of the yield of sexductants at 37° (1) and 43° (2) from the following crosses: F'lac (ORF 206) x F^- ts43 (a); F'lac (ORF 1) x F^- ts43 (b).
a: concentration of F'lac — 3×10^7 cells/ml, F^- — 3×10^8 cells/ml; after conjugation for 12 minutes the mixture was diluted 100-fold with fresh medium, treated in a blender to separate the pairs, and divided into two parts. Each part was further diluted tenfold with medium warmed to 43° and 37° respectively. The titers of the sexductants were determined at regular intervals;
b: concentration of cells — F'lac 4×10^7 cells/ml, F^- 2×10^8 cells/ml. The experimental procedure was the same except that conjugation lasted 20 minutes.

confirmed by the experiments of Stallion and Curtiss (1970), which were carried out in somewhat different conditions. Secondly, the DNA strand transferred to the temperature-sensitive recipient either undergoes degradation by exonuclease or survives by forming a stable structure resistant to the action of nucleases. These two processes are mutually competitive in the cell. The formation of a stable F' episome requires the synthesis of a second complementary DNA strand in the recipient and the closure of the whole fragment into a ring. The formation of a stable recombinant requires not only the synthesis of a complementary strand but also the integration of the fragment into the chromosome of the recipient. The latter process is rather prolonged (about 40 minutes). With the inhibition of recombination in the temperature-sensitive exconjugant at 43°, what remains of the two processes is simply a destructive event which leads to the observed death of the potential recombinants (Figures 1 and 4). We may now ask whether the death of the exconjugants accompanying heat shock results from the slowing down of the synthesis of the second DNA strand or from the impossibility of recombinational integration. The fact that the comparatively long sexductants ORF 206 (which transmit their lac marker for approximately the same time as that required for the donor to transmit the thr leu markers) are not susceptible to heat shock indicates that the shock blocks recombination in the recipient.

Let us consider the course of genetic recombination in bacteria. We examined the recombination process in two types of cross: transformation in *Bacillus subtilis* and conjugation in *E. coli* K12. Recombination follows different courses in these two cases.

Transformation always involves the integration of single-stranded DNA fragments about ten genes long or slightly longer. The single-strand nature of the integrated fragments was first demonstrated by us using genetic methods — by a clonal analysis of the progeny (Bresler et al., 1964). This provoked a long discussion until the existence of single-strand synthesis in transformation was conclusively demonstrated in various objects by both physicochemical and genetic methods in a number of laboratories.

In conjugation, the single-stranded DNA transferred from the donor to the recipient cell enters the genome as a double-stranded structure. The second strand is synthesized immediately in the recipient cell prior to recombination.

The immediate synthesis of the complementary DNA strand during conjugation has been proved by numerous experiments. We shall refer to our own work (Bresler, Lanzov, and Lukianez-Blinkova, 1968) involving the transfer of the lac marker from Hfr lac^+ to a lac^- rec^- recipient incapable of recombination. Under such conditions the cell begins to synthesize β-galactosidase (Figures 5 and 6), which is easy to trace. The effect obtained is the same whether the donor is Hfr H or Hfr C even though these transfer different DNA strands, one of them coding, the other noncoding. Consequently there is no doubt that the entry of one DNA strand into the cell immediately induces its complementary replication.

The difference between the recombination mechanisms in the two cases — transformation and conjugation — becomes quite evident if one considers the quantitative indexes of the processes involved. The most characteristic value is the physical scale of the gene map. This magnitude is defined as the distance between two adjacent breaks in the DNA strand or, in other words, the average size of the donor DNA fragment incorporated in the genome of the recipient. In the case of conjugation this magnitude corresponds to 1/15 of the chromosome of E. coli or $2 \cdot 10^8$ daltons according to the measurements of Wollmann, Jacob, Stent, and Fuerst (Fuerst, Jacob, and Wollmann, 1959; Stent and Fuerst, 1955).

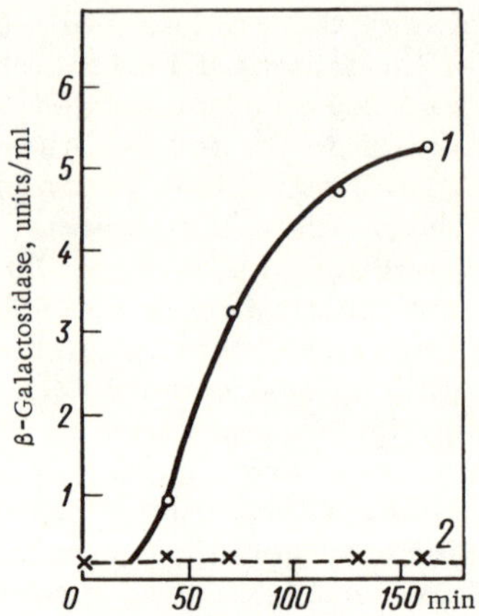

FIGURE 5. Kinetics of the synthesis of β-galactosidase after crossing Hfr H lac⁺ str-s with the recipient F⁻ lac⁻ str-r rec⁻ in the presence (1) and absence (2) of the inducer IPTG (isopropylthiogalactoside).

Concentration of Hfr — $9 \cdot 10^7$ cells/ml, F⁻ — $3.1 \cdot 10^8$ cells/ml. After conjugation for 5 minutes, streptomycin (1,000 μg/ml) was added to the medium and the samples were divided into two parts. Enzyme activity was determined at regular intervals according to the cleavage of β-nitrophenylgalactoside.

In the case of transformation, this value was determined by us and by the Lacks group (Guild, Cato and Lacks, 1968); in Bac. subtilis it is $1 \cdot 10^7$ daltons.

In the case of transformation, the fragments integrated are 20 times smaller and the recombination events per unit chromosome length 20 times as frequent. This is quite natural in view of the fact that the DNA sections integrated are single-stranded in transformation and double-stranded in conjugation.

Our experiments on the radioactive suicide of merozygotes provided direct proof of the double-stranded integration of DNA in conjugation (Bresler et al., 1968). Conjugation was carried out in a medium containing very hot ^{32}P (50 μcuries/mg). As a result, one of the two strands of the integrated DNA fragment was invariably radioactive. If the Hfr strain used for conjugation was previously grown in a radioactive medium with the same specific activity, both strands of the integrated fragments were equally radioactive.

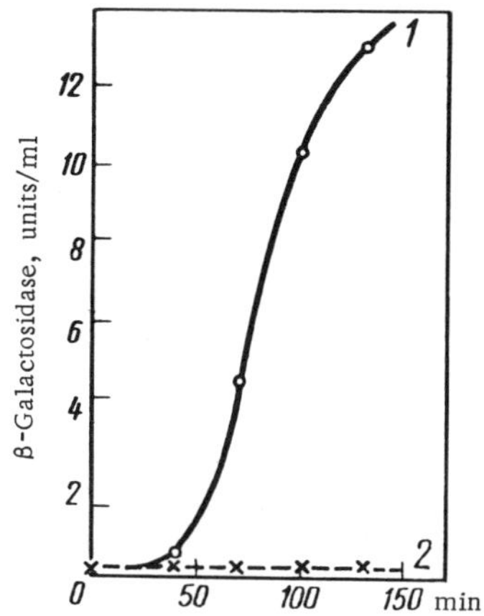

FIGURE 6. Kinetics of the synthesis of β-galactosidase after crossing Hfr C lac$^+$str-s with the recipient F$^-$ lac$^-$ str-r rec$^-$ in the presence (1) and absence (2) of the inducer IPTG.

Concentration of Hfr — $1.4 \cdot 10^8$ cells/ml, F$^-$ — $1.2 \cdot 10^8$ cells/ml. All other conditions as in Figure 5.

When the merozygotes are frozen and their radioactive suicide is observed, it is found that the ratio of the kinetic constants of death of the merozygotes is 2:1 (Figure 7).

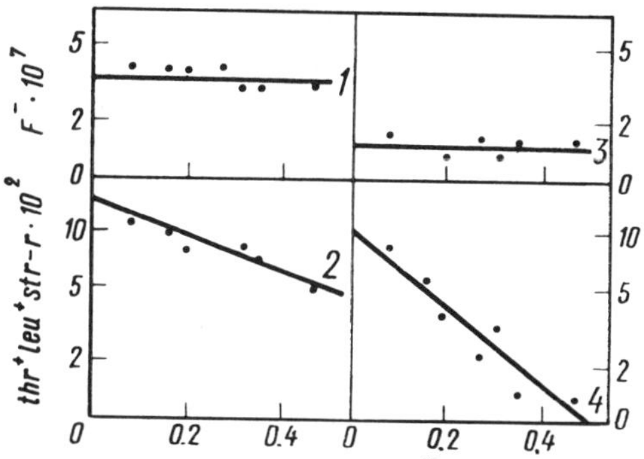

FIGURE 7. Kinetics of the radioactive suicide of exconjugants:

1,3) F$^-$ strain (control); 2) suicide of exconjugants grown in a nonradioactive medium but conjugated in a medium containing ^{32}P; 4) suicide of exconjugants grown in a medium containing ^{32}P and conjugated in the same medium.

This constitutes direct and quantitative proof of a double-stranded integration since such a ratio cannot be obtained in any other way.

Consequently the integration of DNA into the genome of the recipient during conjugation involves a double-stranded mechanism. A whole array of enzymes are engaged in this process. Endonucleases and exonucleases operate first, splitting the DNA into fragments and clearing single-stranded areas at the ends. The next stage is mutual recognition of the single-stranded DNA sectors according to the Watson-Crick principle of complementarity.

Finally, the gaps in the DNA must be patched by additional synthesis and the linear integrity of the entire chain restored by means of ligases (Figure 8). Conclusive proof of this hypothesis can only be obtained by constructing a system in which recombination can occur in vitro.

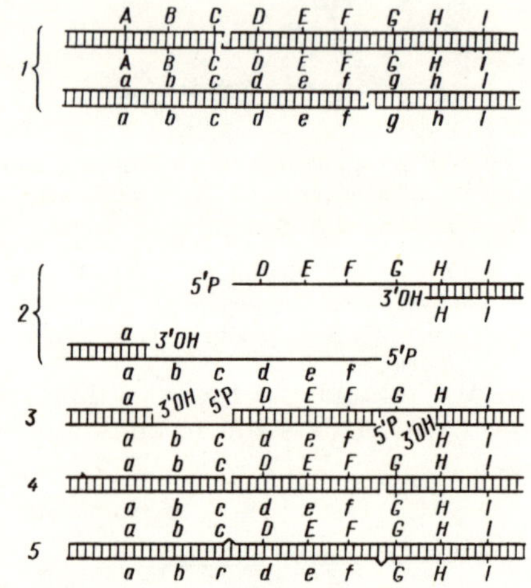

FIGURE 8. Possible scheme of genetic recombination:

1) splitting of DNA chains by endonuclease; 2) degradation of fragments from the 3' end by the action of exonuclease; 3) pairing of complementary chains; 4) patching of gaps in DNA by DNA polymerase; 5) complete repair of the lesions by ligase.

We shall consider some aspects of the recombination process on the basis of the results obtained.

One of the important factors in recombination is the prolonged action of the exonucleases, creating single-stranded areas at the ends of the DNA fragments and thus enabling recognition of the chains, synapsis and finally chemical union of the fragments. The length of the single-stranded terminal areas is of crucial importance. If this area contains a genetic marker in the form of two different alleles in the donor and recipient cells, synapsis yields a helix in which the two strands are not quite complementary. At the mutation site the strands do not correspond to one another since they originate from two allelic individuals. Such a double DNA helix can be termed a molecular heterozygote. Repair enzymes recognize such structures and correct them to the pattern of either the donor or the recipient DNA (Figure 9). In the final analysis, such a phenomenon may well be equivalent to recombination although it involves a totally different mechanism. In modern parlance, this is a genetic conversion.

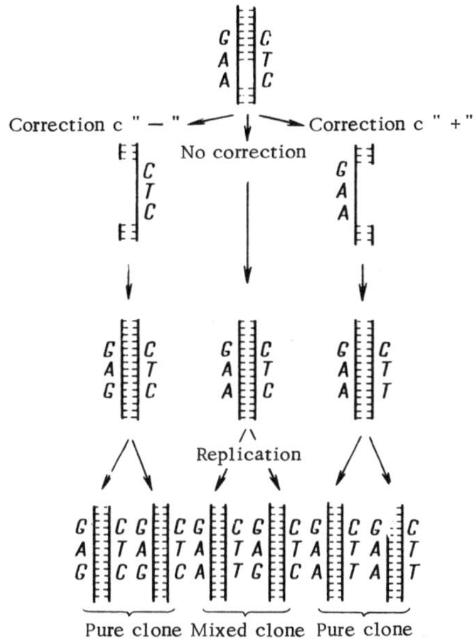

FIGURE 9. Scheme of transformation and subsequent correction: GAA/CTT — wild type (+); GAG/CTC — mutant type (−).

How can one distinguish between ordinary recombination and conversion? Is it possible to detect conversion experimentally? The answer appears to be yes — by examining the products of recombination. In an ordinary exchange of DNA fragments recombination is necessarily reciprocal since no allelic forms of genes can either vanish or reappear. It is difficult to establish reciprocity in bacteria since the zygote is incomplete, containing only a fragment of donor DNA, and moreover excess DNA breaks down and vanishes after the genetic exchange. Here too, recombination must proceed reciprocally, i. e. according to the rule of conservation of all allelic forms, but it is difficult to prove this point. It is possible, however, to select a special case in which the chromosome recombines with an episome bearing several genetic loci. Under such circumstances both recombination products — the new chromosome and the new episome — remain identifiable. Both structures can be studied, and the reciprocity of recombination is amenable to direct testing. This was done by Meselson (1967) and Herman (1968), who were able to show that long-range recombination, i. e. recombination between different genes, is practically always reciprocal, whereas the short-range recombination which takes place within a single gene is almost invariably nonreciprocal owing to the action of repair enzymes, i. e. conversion. Thus conversion can be defined as recombination in which the rule that the products must be reciprocal is violated. Conversion takes place on comparatively short sectors of the chromosome, within one or a few neighboring genes. At large distances conversion becomes an improbable event and only recombination in the previous sense of the term can occur. The two phenomena evidently take place at the same time, both being based on the incorporation of fragments of the donor chromosome into the recipient chromosome. The simplest model for studying conversion is transformation, which always yields a molecular heterozygote. Such a molecular heterozygote may be corrected towards the information of either the donor or the recipient, or it may remain uncorrected, yielding two types of progeny division (see Figure 9). We were able to determine the probability of the correction of the molecular heterozygote by studying the heterogeneity of the progeny of each transformed cell, i. e. by clonal analysis of the transformants (Bresler et al., 1968).

The correction must clearly be specific with respect to the nature of the unpaired DNA units, i. e. to the nature of the mutational lesion. This prediction was confirmed experimentally. We found a number of mutations located in a single gene for which the extent of correction ranged from 20% to 80%. It is furthermore known that some genetic markers are readily integrated and others only poorly. Markers in the latter category are repaired back to mutant DNA, i. e. they are virtually not subject to transformation. Louarn and Sicard (1968) have recently confirmed these findings in pneumococci. The study of transformants for two genetic markers at the same time has produced interesting results. For this purpose two mutational lesions must be combined in the same recipient cell. After subsequent transformation for both markers simultaneously it is possible to carry out a clonal analysis and determine the probability of the simultaneous repair of both lesions. If the two mutations are far apart, the probability of simultaneous repair equals the product of the probabilities of each separate repair. However, if the two mutations lie close to one another in the same DNA molecule they are repaired together. Adequate repair of one mutation inevitably results in the simultaneous repair of the neighboring mutation while the separate repair of this is a low probability event. Simultaneous repair becomes less and less probable as the distance between the two mutations increases, until the probabilities of the two repairs become completely independent of one another. This provides a quantitative estimate of the size of the DNA sector attacked by the repair enzymes (Figure 10). The value so obtained was $2-3 \cdot 10^6$ daltons, which corresponds to the length of 2—3 genes. This is a crucial parameter. Such extensive clearing of one of the DNA strands by exonuclease may seem puzzling. However, a wide scope of action appears to be typical of the repair or corrective enzymes. The same value can be calculated, for example, with respect to the repair of UV-induced lesions in bacteria.

About 2,000 radiation-induced pyrimidine dimers can be eliminated from a single cell of *E. coli* (Howard-Flanders and Boyce, 1966). Moreover, it is known that such massive doses of radiation induce degradation and resynthesis of about 70% of the total amount of DNA in the cell (Kaplan et al., 1969).

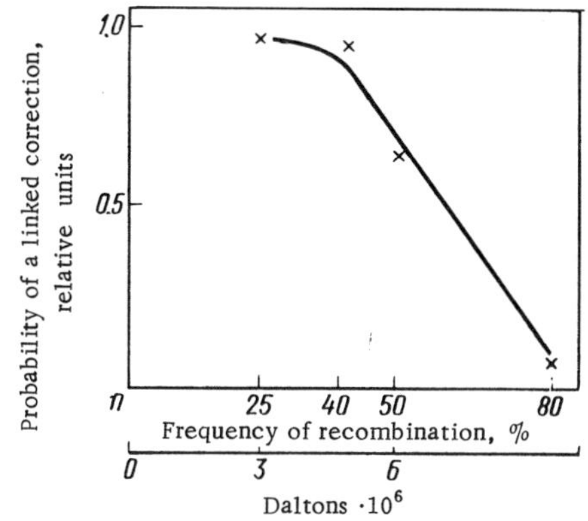

FIGURE 10. Effect of the distance between the markers on the probability of a linked correction.

Since the entire chromosome of E. coli weighs $3 \cdot 10^9$ daltons, it follows that 2,000 lesions correspond to the removal of $2 \cdot 10^9$ daltons or 10^6 daltons per lesion. Thus, the repair of UV-induced lesions leaves cleared areas as long as those created during the correction of molecular heterozygotes.

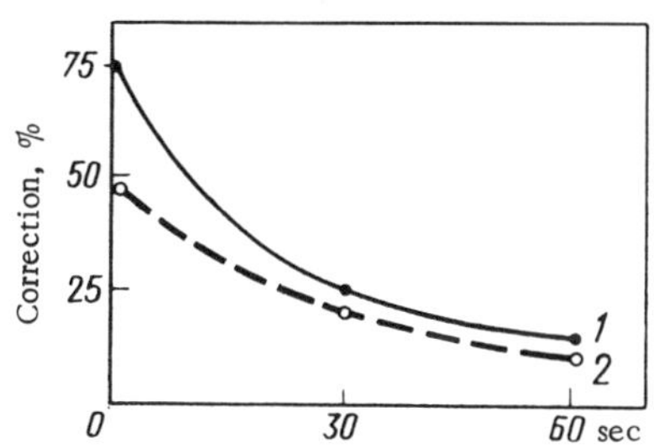

FIGURE 11. Effect of UV radiation on the efficiency of correction: 1) rib^0_{339}; 2) rib^0_{67}.

We demonstrated in the following way that the enzymes involved in the two cases are identical. Prior to transformation, the transforming DNA was exposed to large doses of UV radiation. This inflicted numerous lesions which occupied the repair enzymes. The correction of the transformation heterozygotes was immediately affected. Indeed, correction practically ceased, and almost all the transformed cells yielded mixed clones (Figure 11). The formation of molecular heterozygotes is very common in genetic processes and is apparently a concomitant both of any induced mutagenesis and of recombination. The correction of the molecular heterozygotes has crucial genetic consequences.

BIBLIOGRAPHY

Bonhoeffer, F., R. Hasselbarth, and K. Lehmann. — J. Molec. Biol., Vol. 29 (1967), 539.
Bresler, S., R. Kreneva, and M. Mosevitsky. — Z. VererbLehre, Vol. 95 (1964), 288.
Bresler, S., R. Kreneva, and V. Kushev. — Molec. and Gen. Genet., Vol. 102 (1968), 257.
Bresler, S., V. Lanzov, and A. Lukianez-Blinkova. — Molec. and Gen. Genet., Vol. 102 (1969), 269.
De Lucia, P. and J. Cairns. — Nature, Vol. 224 (1969), 1164.
Eisenberg, R. and A. Pardee. — J. Molec. Biol., Vol. 46 (1969), 355.
Fuerst, C., F. Jacob, and E. Wollmann. — C. r. hebd. Séanc. Acad. Sci., Paris, Vol. 243 (1956), 2162.
Ganesan, A. and J. Lederberg. — Biochem. Biophys. Res. Commun., Vol. 18 (1965), 824.
Gross, J. and M. Gross. — Nature, Vol. 224 (1969), 1166.
Guild, W., A. Cato, and S. Lacks. — Cold Spring Harbor Sympos. Quant. Biol., Vol. 33 (1968), 643.
Herman, R.K. — J. Bact., Vol. 96 (1968), 173.
Howard-Flanders, P. and R. Boyce. — Radiat. Res., Vol. 6, Suppl., p.156, 1966.
Kaplan, J., S. Kushner, and L. Grossman. — Proc. Natn. Acad. Sci. U.S.A., Vol. 63 (1969), 144.
Louarn, J. and A. Sicard. — Biochem. Biophys. Res. Commun., Vol. 30 (1968), 683.
Meselson, M. — J. Cell. Comp. Physiol., Vol. 70, Suppl.1, No.3, 1967.
Ohki, M. and J. Tomizawa. — Cold Spring Harbor Sympos. Quant. Biol., Vol. 33 (1968), 651.
Stallion, A.S. and R. Curtiss. — Genetics, Vol. 64, Suppl., No. 2, Pt. 2, p.62, 1970.
Stent, G. and C. Fuerst. — J. Gen. Physiol., Vol. 38 (1955), 441.
Willets, N. and A. Clark. — J. Bact., Vol. 100 (1969), 231.

UDC 575:576.8

THE RECOMBINANT YIELD STIMULANT (RYS) PRODUCED BY MALE STRAINS OF ESCHERICHIA COLI K12: ITS NATURE AND MECHANISM OF ACTION IN CONJUGATION

D.M. GOL'DFARB, G.I. GOL'DBERG, L.S. CHERNIN, L.A. GUKOVA,
I.D. AVDIENKO, V.N. KUZNETSOVA and I. Kh. KUSHNER
Institute of General Genetics
Academy of Sciences of the USSR, Moscow

Two distinct activities of filtrates of male strains — suppression of the aftereffect of nitrogen mustard and stimulation of the yield of recombinants in conjugation — are due to a single compound of a protein nature. E. coli HfrC RYS$^-$ mutants unable to synthesize this factor were isolated and it was established that these mutations are located in the episome. A method for the 300-fold purification of RYS is described. Evidence is presented that RYS acts at the surface of the recipient cell during conjugation and is manifested at the stages of conjugation following the transfer of genetic material from the donor to the recipient. A number of hypotheses are proposed with respect to the possible mechanism of action of RYS, which can be regarded as a special case of regulation of genetic interactions in bacteria at the intercellular level.

Earlier we reported that filtrates of male strains of E. coli K12 (Hfr, F$^+$, R$^+$) show three types of activity: 1) they increase the yield of recombinants in conjugation; 2) they inhibit the aftereffect of nitrogen mustard (ANM); 3) they partially but selectively inhibit DNA synthesis in F$^-$ cells (Gol'dfarb et al., 1969, 1970).

These three types of activity shown by filtrates of male strains of E. coli are sensitive to proteases, are inactivated at 75°, and do not diffuse through a dialysis membrane. It remained unclear, however, whether all these properties of the male filtrates are due to one or several protein-type compounds.

We found in a previous study (Gol'dfarb et al., 1970) that the recombinant yield stimulation (RYS) induced by male filtrates is due to their action on the recipient bacteria. However, these experiments did not reveal the exact stage of conjugation activated by the interaction of the male filtrate with the F$^-$ bacteria. Such an interaction could have the following consequences: 1) increase in the efficiency of formation of conjugant pairs; 2) enhancement of the stability of the pairs formed; 3) influence on the transfer of genetic material from the donor to the recipient. Finally, RYS may activate processes taking place within the recipient bacterium after completion of the transfer of genetic material, i.e. in the merozygote.

In view of the protein nature of RYS, established in an earlier study (Gol'dfarb et al., 1970), we suggested that this agent acts on the surface of the F^- cell since the probability of its penetration inside the bacterium was small. This focused attention on the possible interaction between RYS and the membrane of F^- bacteria. Finally, one of the unsolved topics concerning the genetic control of the synthesis of RYS by male cells is the functional role of the episomes, the presence of which is essential for the synthesis of this agent. This circumstance suggests that the episomal genome possibly controls the synthesis of RYS or derepresses a chromosomal gene.

We performed the following experiments with the aim of solving these problems.

MATERIALS AND METHODS

1. Bacteria. In addition to the strains described in our previous publications (Gol'dfarb et al., 1969, 1970) we used the following strains of E. coli K12: W1485 (met str-r); W485 F^- Rfi^+ (met str-r su-r cm-r tc-r), obtained by the cross W 1485 $F^- \times$E. coli K12 CSH-2 (222) F^- —Rfi^+ (met pro str-r su-r cm-r tc-r); HfrC7 and HfrC128 — mutants not releasing RYS into the culture medium; the other strains were derived from HfrC or from the mutants indicated and will be described below. The "male" phages f2 and MS2, the "female" phages φ II and T7, and the temperate phage P1kc were also used.

2. The media used for cultivating the bacteria and selecting recombinants have been described in previous publications (Gol'dfarb et al., 1969, 1970).

3. The preparation of filtrates and the determination of their capacity to inhibit ANM have been described previously (Gol'dfarb et al., 1969).

4. Selection of HfrC RYS^- mutants unable to synthesize RYS. E. coli HfrC, grown in meat peptone broth to a titer of $4-5 \cdot 10^8$ cells/ml, were centrifuged (4,000 rpm, 20 min), resuspended in 1/10 the original volume of 0.015 M phosphate buffer (pH 6.0), and treated with nitrosomethylurea (NMU) (2.5 mg/ml, 2 hours at 37°). The mutagen-treated culture was diluted 10^4 times with cold buffer pH 7.0 and seeded on Petri dishes containing medium M-9 supplemented with methionine (20 μg/ml). The dishes were incubated at 37°. The colonies obtained were replicated on a lawn of E. coli JC411, seeded on a medium for the selection of leu^+ str-r recombinants. The growth of the replicated colonies was determined after 36—48 hours. Donor colonies giving poor replicas in the form of areas of sparse recombinant growth were selected as probable mutants. Subclones of donor bacteria with a reduced capacity for recombination were tested for sensitivity towards phage f2 and the sensitive variants were selected for the subsequent experiments. Filtrates of cultures of selected probable mutants were tested for their capacity to increase the yield of recombinants during conjugation and to inhibit ANM in F^- bacteria.

To obtain revertants the mutant HfrC7 was treated with NMU and the donor colonies yielding normal replicas on a lawn of F^- cells of E. coli JC411 seeded on a medium for the selection of leu^+ str-r recombinants were selected as revertants.

5. Preparation of crude RYS preparation. An overnight culture of E. coli HfrC in medium M-9 containing methionine (20 μg/ml) was seeded in 20 liters of the same medium at an inoculum:medium ratio of 1:100. The bacteria were cultured at 37° with constant shaking to a density of $2-3 \cdot 10^8$ cells/ml. The culture fluid was separated from the cells in the cold by means of an S-44 separator. The liquid obtained was filtered through a No. 5 Chamberlain filter.

6. Hydroxyapatite chromatography (HA). HA was prepared as described by Levin (1962). In some cases the HA was boiled with double the quantity of 40% NaOH. The lowering of the ionic strength of the original filtrate, necessary for satisfactory adsorption on HA, was achieved by adding a saturating quantity of 0.5 M $CaCl_2$. To determine the quantity of $CaCl_2$ required for saturation 0.5 M $CaCl_2$ was added in small portions to aliquots of the filtrate (100 ml), the precipitate forming was removed by centrifugation and the next portion of $CaCl_2$ was added to the supernatant. This procedure was repeated until precipitation ceased. The total amount of $CaCl_2$ solution added was taken as saturating. The proportionate quantity of $CaCl_2$ was added drop by drop to the filtrate (20 liters) in the cold, mixing well. Then 50 g of a moist precipitate of HA was added to the suspension and the mixture was stirred for a further 20 minutes. After adsorption, the sediment was collected and eluted with 0.01 M sodium phosphate buffer pH 6.8 in the cold with stirring for 15 min. The latter procedure was repeated twice.

7. The material was concentrated by evaporation in an air current in a dialysis bag at room temperature, or by incomplete lyophilization. Concentration of the HA eluates for subsequent gel filtration was achieved by electrophoresis. Two electrodes were introduced in a test tube (5×25 cm) and the following solutions were layered consecutively: 0.2 M Tris-HCl buffer pH 7.5 containing 10% sucrose; the preparation, dissolved in 0.1 M Tris-glycine buffer pH 7.5 containing 5% sucrose; Tris-glycine buffer 0.2M, pH 7.5. The lower electrode served as the anode. A direct current (about 30 V/cm) was passed through for one hour. The concentrated substances gathered in a narrow zone which was then collected by means of a siphon.

8. Cellulose phosphate chromatography (CP). CP was treated with 0.5M solutions of NaOH and HCl in the sequence alkali-acid-alkali, and the ion exchanger was equilibrated with 0.01 M phosphate buffer pH 6.8. Chromatography was carried out in a $1.6 \cdot 10$ cm column at room temperature. HA eluate in a volume of 100—150 ml was introduced in the column at a rate of 40 ml/hour. The column was eluted in stepwise fashion with the same buffer containing ascending concentrations of NaCl (0.1—1M) at the same rate.

9. DEAE cellulose chromatography. DEAE cellulose was treated in the same way as CP (see paragraph 8). A column measuring 0.5×3 cm was used. The ion exchanger was equilibrated with 0.01 M Tris-HCl buffer pH 8.0. The column was eluted in a stepwise manner with the same buffer containing ascending concentrations of NaCl (0.1 — 1 M).

10. Gel filtration was carried out in a 0.9×80 cm column, filled with Biogel P-4 in distilled water. The preparation was applied to the column in a volume of 0.2 ml. The column was eluted with distilled water at a rate of 10 ml/hour. 2ml fractions were collected in an LKB fraction collector and registered by a "Uvikord" recorder.

The purification of the recombinant yield stimulant (RYS) from filtrates of E. coli Hfr male strains is outlined in the following scheme.

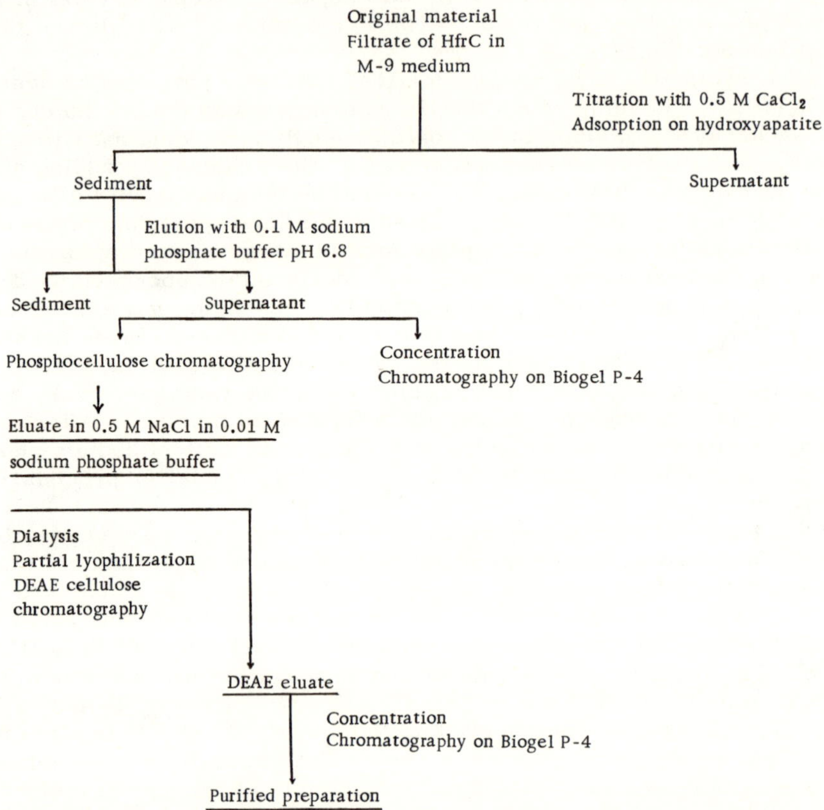

11. Examination of the capacity of the filtrates and fractions to stimulate the yield of recombinants in conjugation. A culture of E. coli K12 JC411, grown in meat peptone broth to a density of $2-3 \cdot 10^8$ cells/ml, was divided into 3 ml portions which were centrifuged (20 minutes, 4,000 rpm), resuspended in the same medium (control) or in the test material diluted 1:10 in the broth, and incubated for 30 minutes at 37°. The treated cells in a volume of $1.5-2$ ml were mixed with an equal volume of E. coli K12 HfrC ($1.5-2.5 \cdot 10^8$ cells/ml) previously washed twice in the broth (to remove the RYS). Crossing was carried out for 30 minutes, after which the mixture was diluted and seeded for the selection of leu^+ str-r recombinants.

12. Standard cross. 1.5—2 ml aliquots of cultures of JC411 F^- ($5 \cdot 10^8$ cells/ml) and HfrC ($2.5—5.0 \cdot 10^8$ cells/ml) were washed once in meat peptone broth, then mixed and incubated for 30 minutes at 37°. Next the mixture was diluted with phosphate buffer and seeded on a solid medium for the selection of leu^+ str-r recombinants.

13. The method of obtaining merozygotes and their treatment with filtrates of the HfrC strain have been described previously (Gol'dfarb et al., 1971).

14. Trypsin treatment of recipient bacteria "activated" by HfrC filtrate. E. coli K12 JC411 ($1 \cdot 10^8$ cells/ml) were washed by centrifugation and resuspended in the HfrC filtrate. The mixture was incubated for 60 minutes. The "activated" cells so obtained (E. coli K12 JC411 — "A") were washed twice and resuspended in broth containing trypsin (400 μg/ml). This mixture was incubated for 30 minutes. The enzyme action was stopped with soybean inhibitor (400 μg/ml), after which the bacteria were centrifuged and subsequently used as recipients in crosses with HfrC donors washed twice (to remove the RYS). Donors and recipients were mixed at a 1:1 ratio and 20 minutes were allowed for crossing to take place. Diluted samples were seeded in medium for the selection of leu^+ str-r recombinants.

15. Sodium dodecylsulfate (SDS) treatment of recipient bacteria "activated" with HfrC filtrate. E. coli K12 JC411 — "A" cells (see paragraph 14) were washed twice by centrifugation (4,000 rpm, 20 minutes) and resuspended in meat peptone broth containing 0.25% SDS. The cells were exposed to the detergent for a total of 30 minutes at 37°. Afterwards the bacteria were washed twice and subsequently used as recipients (see paragraph 14).

16. Determination of the capacity of bacteria subjected to various treatments to exhaust the active HfrC filtrate. E. coli K12 JC411 F^- cultures, grown to a density of $2 \cdot 10^8$ cells/ml, were treated by one of the methods indicated below.

a) Boiling for 15 minutes.

b) Treatment with a 2:1 chloroform-methanol mixture (Kanemasa et al., 1967).

c) Treatment with a 1:1:1 phenol-ethanol-water mixture (Barry and Goebel, 1951).

The treated and untreated (control) bacteria were washed three times in 0.85% NaCl and resuspended in a hundred times smaller volume of active HfrC filtrate. Exhaustion was carried out for 60 minutes, after which the bacteria were sedimented by centrifugation and the supernatant was tested for ability to stimulate the yield of recombinants in crosses between HfrC and JC411 F^- (see paragraph 11).

17. Determination of the capacity of phospholipids of E. coli JC411 F^- to inactivate a filtrate of the HfrC strain. Phospholipids were isolated according to Bauman et al. (1970). The active HfrC filtrate was mixed with an equal volume of a solution of phospholipids isolated from F^- bacteria, and the mixture was incubated for 60 minutes at 37°. The same filtrate not treated with phospholipids was used as the control. The samples were tested for their capacity to stimulate the yield of recombinants (see paragraph 11). The specificity of the effect of phospholipids was tested by experiments in which the phospholipids had been previously

destroyed by phospholipase A. Phospholipase A (100 μg/ml) was added to 0.5 ml of the phospholipid preparation and the mixture was incubated for 180 minutes at 37° in the presence of 10^{-2} M $MgSO_4$ and $5 \cdot 10^{-5}$ M $CaCl_2$. The phospholipase-treated phospholipids were tested for their ability to inactivate HfrC filtrates relative to that of intact phospholipids. The phospholipase A used in these experiments was isolated from cobra venom according to Edroards and Ball (1954).

RESULTS

The stimulation of the yield of recombinants and the inhibition of ANM are due to a single proteinlike substance present in filtrates of E. coli HfrC

The first proof that RYS and the ANM inhibitor are actually the same compound was obtained during a study of mutants (HfrC RYS^-) deficient for RYS. Two such mutants (HfrC7 RYS^- and HfrC128 RYS^-) were obtained from an HfrC strain treated with NMU (see Materials and Methods, paragraph 4). The mutation frequency was 10^{-5}. Both mutants were characterized by a reduced fertility; the number of recombinants obtained when they were crossed with E. coli JC411 F^- was much lower than that obtained when the same recipient was crossed with the HfrC strain (see below).

Type of cross	Frequency of leu^+ str-r recombinants
JC411 × HfrC7	$4.2 \pm 0.32 \cdot 10^{-2}$ (26)
JC411 × HfrC128	$3.8 \pm 0.53 \cdot 10^{-2}$ (5)
JC411 × HfrC	$4.5 \pm 0.39 \cdot 10^{-1}$ (16)

Note. Confidence limits of the average value at 0.05; the number of experiments is given in parentheses both here and in the following tables.

The reduction in the fertility of the mutants did not result from damage to the sex pili or disturbances in the process of formation of conjugant pairs. This conclusion is based on the following evidence: 1) the mutants adsorbed phage f2 to the same extent as did the HfrC strain (Table 1); 2) the mutants possessed intact sex pili, as demonstrated by electron microscopy (Figure 1).

At the same time the two mutants did not release RYS into the medium since their cell free filtrates did not increase the yield of recombinants when added to a conjugation mixture from which RYS had been previously removed.

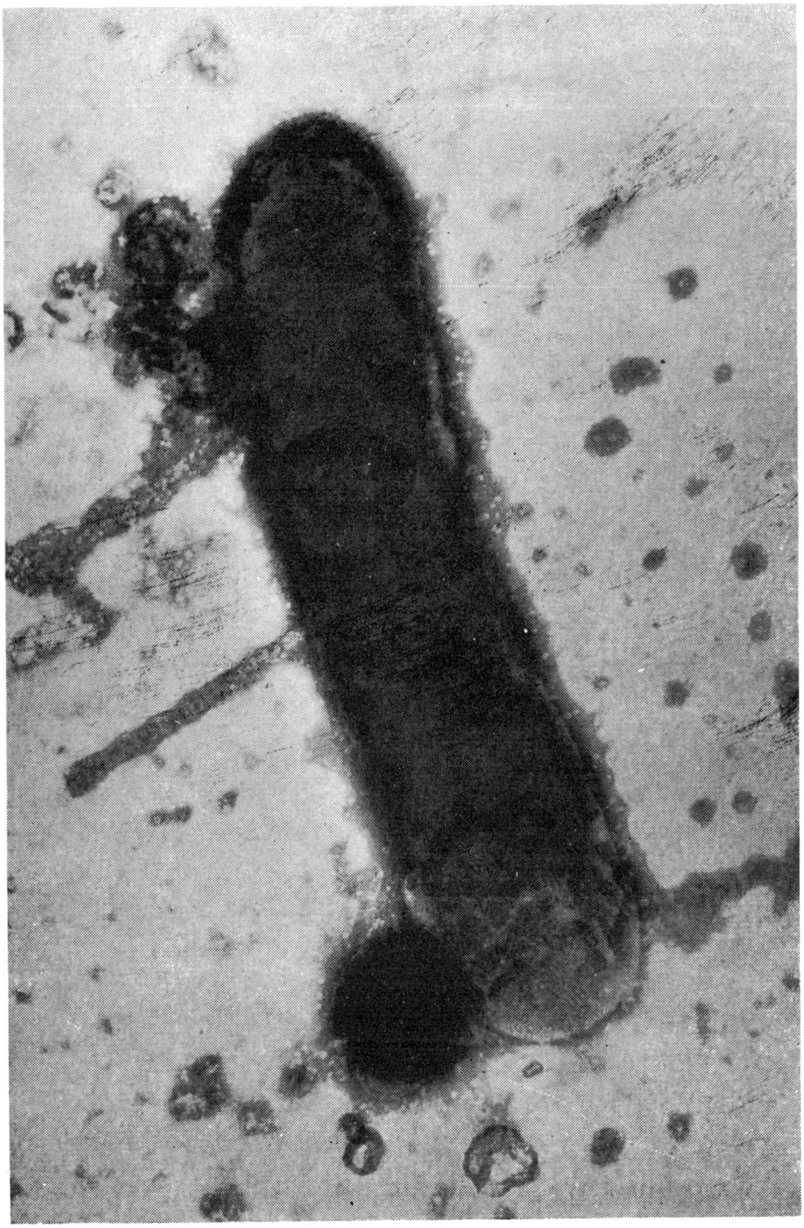

FIGURE 1. Electron micrograph of the mutant HfrC7 RYS$^-$ with adsorbed phage f2 (multiplicity of infection 200, magnification 17,000 x). Prepared by B.F. Poglazov.

TABLE 1. Adsorption of phage f2 on the HfrC strain and its mutants HfrC7 and HfrC128 (%)

Experiment	HfrC7	HfrC128	HfrC
1	74.4	81.5	77.2
2	78.4	83.1	75.2
3	63.8	80.9	61.0
Average	72.2	81.9	71.1

Note. The efficiency of adsorption was assessed from the titer of unadsorbed phage (Adams, 1959). Time of adsorption 10 minutes.

It is significant that these same filtrates did not inhibit ANM. In other words, the mutation to the RYS⁻ phenotype was simultaneously a mutation to the IANM (inhibition of ANM) phenotype. The reduced fertility characteristic of the RYS⁻ mutants can be partly counteracted by the addition of RYS-containing *E. coli* HfrC filtrates to the crossing mixture containing the mutant and a normal recipient.

Strain from which added filtrate derived	Number of leu^+ str-r recombinants, %	Survival rate of F^- cells incubated for 60 minutes after NM treatment, %
No filtrate added	100	62.4
HfrC7	112 (6)	62.3 (12)
HfrC128	105.9 (3)	68.6 (3)
RYS^+ revertant of the strain HfrC7	279.3 (3)	120.2 (3)
HfrC	240.3 (6)	101.7 (2)

Note. The number of colonies of *E. coli* JC411 F^- formed on seeding the culture immediately after a 5-minute treatment with nitrogen mustard (NM) (300 μg/ml) was taken as 100%. In the crossing experiments, the filtrates were added to JC411 F^- × HfrC cell mixtures from which RYS had been previously removed (the donor was twice washed free of culture liquid before the crossing).

In the following experiments we used the HfrC7 strain to obtain revertants (see Materials and Methods, paragraph 4) with a higher fertility in comparison with the mutant strain. Filtrates of cultures of the revertants stimulated the yield of recombinants and inhibited ANM (see above). These findings indicate that the production of RYS and of the factor inhibiting ANM is controlled by a single gene of the male strain of *E. coli* K12; mutation of this gene causes the loss of these activities, while its reversion restores them.

Type of crossing	Number of leu^+ str-r recombinants (%) in the presence of HfrC filtrate
JC411F^- × HfrC7	268.3 (20)
JC411F^- × HfrC128	343.7 (3)

Note. The number of recombinants obtained in crosses in meat peptone broth was taken as 100%.

The second proof that RYS and the IANM factor are identical was obtained by purifying RYS and assaying these two activities at different stages of the process. As shown by the following data, the original and diluted filtrates do not differ from one another as far as the RYS and IANM effects are concerned. Consequently, the activity of the preparations obtained was expressed quantitatively in terms of the reciprocal of the critical dilution at which the tested material exhibited the RYS and IANM effects.

Dilution of original filtrate	Stimulation of the yield of recombinants, %	Inhibition of ANM, %
Control (Meat peptone broth added)	28.6* (4)	61.7 ** (7)
1:20	95.0 (4)	122.3 (7)
1:200	82.9 (4)	121.3 (3)
1:500	38.6 (4)	—
1:1000	29.0 (4)	—
1:1600	—	122.4 (3)
1:3200	—	108.5 (5)
1:6400	—	62.7 (5)
1:12800	—	63.6 (3)

* The number of leu^+ str-r recombinants in the standard cross JC411 F^- × HfrC was taken as 100%.

** The number of E. coli JC411F^- colonies formed on seeding the culture immediately after a 5-minute treatment with nitrogen mustard (300 μg/ml) was taken as 100%.

It is evident from these data that filtrates of the HfrC strain inhibited ANM in F^- cells to a dilution of 1:3000 and stimulated the yield of recombinants to a dilution of 1:200, i.e., they contained, in arbitrary terms, 3,000 IANM units and 200 RYS units per ml. Both types of activity increased markedly with further purification of the material. As shown in Table 2, the HA eluate contained $4.8 \cdot 10^4$ IANM units per ml while its protein content was the same as in the original preparation. In one series of experiments the HA eluate was concentrated and subjected to gel filtration on a column containing Biogel P-4, while in other experiments further purification was carried out according to the diagram on p. 260.

The profile given by gel filtration of the concentrated HA eluate shows the presence of at least 4 components (Figure 2). The entire activity was concentrated in a single peak and amounted to $2 \cdot 10^5$ ANM units per ml and 10^4 RYS units per ml (see Table 2). Assuming that interactions between the separated components and the Biogel do not play a significant role under the experimental conditions used, it appears that a low molecular weight compound is responsible for the activity of the male-strain filtrates. Re-chromatography of the fractions corresponding to this first peak separates the mixture again into four components. In another series of experiments the HA eluate was subjected to chromatography on cellulose phosphate (see Materials and Methods, paragraph 8). The active component was eluted from the ion exchanger by increasing the molarity of the salt to 0.5. The active eluate obtained was dialysed against 0.01 M Tris-HCl buffer at pH 8 and concentrated by partial lyophilization.

TABLE 2. Purification of the recombinant yield stimulant (RYS) from filtrates of E. coli HfrC

Fraction	Protein, μg/ml	IANM units per ml	RYS units per ml	IANM/RYS activity ratio
Original filtrate	20	3,000	200	15
HA eluate	20	48,000	—	—
Concentrated HA eluate	500	400,000	16,000	25
Fractions corresponding to the first peak obtained by chromatography on Biogel P-4	150	200,000	10,000	20
Eluate after chromatography on cellulose phosphate (CP)	10	100,000	—	—
Concentrated CP eluate	40	400,000	16,000	25
DEAE eluate	10	400,000	—	—
Fractions corresponding to the peak obtained after gel filtration of the DEAE eluate	30	100,000	3,200	31.3

The preparation was chromatographed on a DEAE cellulose column, which adsorbed some of the contaminants while allowing the active component to pass through. Finally the DEAE eluate was concentrated by incomplete lyophilization and subjected to gel filtration on Biogel P-4, as in the case of the HA eluate. In this case instead of the four peaks obtained with the HA eluate,gel filtration revealed a single peak which contained both activities: RYS and IANM. Of particular interest is the coincidence between the first active peak detected by gel filtration of the HA eluate and the peak obtained after additional purification of the HA eluate (see Figure 2). However, we were unable to identify the protein in the purified fraction by acrylamide gel electrophoresis in either alkaline or acid conditions. The material from this peak gave a positive strain on paper with 1% ninhydrin in acidified acetone. The activity of the preparations obtained at the different stages of purification, like that of the original material, proved to be sensitive to temperature and protease. The degree of purification of the active agent may be roughly estimated as 200—300-fold on the basis of the data given in Table 2. The ratio between the two activities (RYS and IANM) does not change significantly in the course of purification of the HfrC filtrate and remains close to 20, as in the original material. This confirms our conclusion that both activities are due to the same component of the filtrate — a low molecular weight protein-like substance.

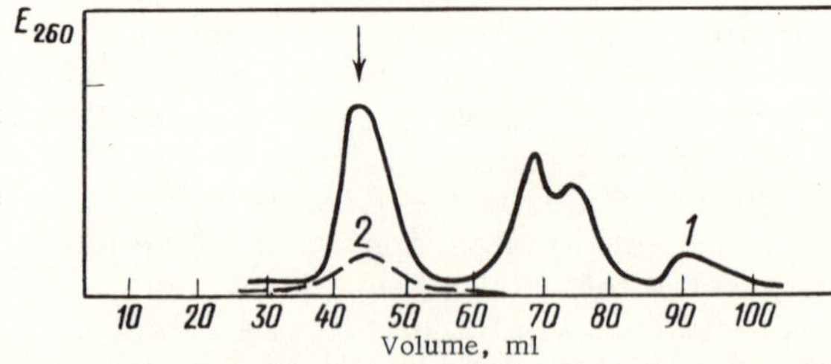

FIGURE 2. Column chromatography on Biogel P-4, 50 — 100 mesh (Bio-Rad Lab, U.S.A.). Column size 0.9 · 80 cm. Rate of elution 10 ml/hour. Fraction volume 2 ml. The arrow indicates the peak containing active RYS:

1) chromatography of the HA eluate; 2) chromatography of the DEAE eluate.

RYS acts after the transfer of genetic material from the donor to the recipient bacterium

The experiments described in this section were designed to determine the effect of HfrC filtrate on the yield of recombinants when applied to merozygotes isolated after conjugation, i. e. after the transfer of genetic material from the donor to the recipient bacterium has been completed and the donor has been removed from the crossing mixture. The efficiency of the method used by us for isolating merozygotes (see Materials and Methods, paragraph 13) was assessed in a special series of tests. We determined the number of recombinants for the proximal (leu^+) and distal (arg^+) markers by seeding samples of the merozygotes after a 15-minute cross (HfrC ×JC411) and samples from the standard cross taken at the same time. The efficiency of isolation of merozygotes was 99%. The subsequent experiments consisted in applying HfrC filtrate merozygotes isolated from crosses carried out without RYS. As shown by the following data, treatment of the merozygotes with HfrC filtrate steadily increased the yield of leu^+ str-r recombinants by a factor of three on average.

Filtrate added to the zygotes	leu^+ str-r recombinants, %
HfrC	292 (18)
JC411	120 (15)
None	100 (18)

The stimulatory effect of the filtrate depended on the "age" of the merozygotes and was obtained if the merozygotes were treated not later than 20 minutes after isolation (Figure 3). It is evident from these experiments that the RYS present in filtrates of *E. coli* HfrC acts after the completion of the transfer of the chromosomal fragment from the donor to the recipient and affects processes taking place in the merozygote.

RYS is fixed by receptors of F^- bacteria and acts through surface structures of the bacteria

We suggested in the introductory section of this work that RYS acts on the surface of the F^- cell, where it initially interacts with receptors. We tried to test this hypothesis in experiments designed to prove the following points.

1. RYS increases the yield of recombinants while it is located on the surface of the recipient cell.
2. Destruction of the RYS-bacterium complex stops the action of the stimulant, causing a decrease in the yield of recombinants.
3. Repeated treatment of such "inactivated" bacteria with active filtrate increases the yield of recombinants. These postulates were tested in experiments carried out according to the following scheme:

1 Treatment of bacteria with RYS	→	2 Destruction of the bacterium-RYS complex	→	3 Repeated treatment of inactivated bacteria
ACTIVATION		INACTIVATION		REACTIVATION

Inactivation of bacteria previously treated with RYS was achieved by trypsin or SDS (see Materials and Methods, paragraphs 14, 15).

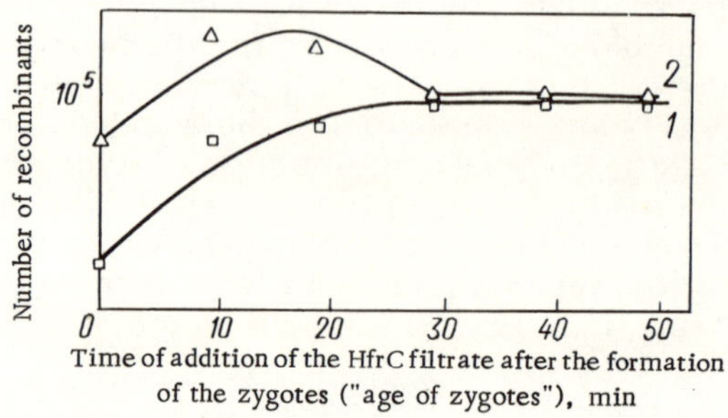

FIGURE 3. Yield of leu^+ str-r recombinants from zygotes of different age, treated with a filtrate of the HfrC strain.

Added to the merozygotes: 1) meat peptone broth; 2) filtrate of the HfrC strain.

The activation, inactivation and reactivation effects were assessed from the yield of recombinants in crosses of HfrC with F^- bacteria treated as indicated, after free RYS had previously been eliminated by double washing of the donor (see Materials and Methods, paragraph 11). The results (Table 3) show that activation of the cells with HfrC filtrates increased the yield of recombinants roughly 2.5 times relative to the control, in which the F^- cells were treated with broth. Treatment of the activated bacteria with trypsin or SDS lowered the number of recombinants to the control level (see Table 3). However, the reactivation of these bacteria by repeated treatment with HfrC filtrates proved effective only in the experiments involving SDS. We conclude that RYS is attached to the surface of the F^- cells, where it can be destroyed by trypsin or removed by SDS. The latter treatment frees the cell receptors and makes them available for repeated reaction with RYS.

Phospholipids of the F^- bacteria are the RYS receptors

The nature of the F^- receptors interacting with RYS was determined in experiments involving the exhaustion of HfrC filtrates with *E. coli* JC411 cells treated in various ways (see Materials and Methods, paragraph 16).

TABLE 3. Effect of the "activation," "inactivation" and "reactivation" of F^- cells on the yield of recombinants from JC411F^- × HfrC crosses

Activation	Inactivation	Reactivation	leu^+ str-r recombinants, %
Meat peptone broth	Meat peptone broth		100 (10)
Meat peptone broth	Trypsin		91.3 (10)
Meat peptone broth	Trypsin	Meat peptone broth	96.6 (6)
Meat peptone broth	SDS		140.3 (10)
Meat peptone broth	SDS	Meat peptone broth	143.7 (6)
RYS	Meat peptone broth		238.5 (10)
RYS	Meat peptone broth	RYS	219.6 (6)
RYS	Trypsin		126.5 (10)
RYS	Trypsin	RYS	93.7 (6)
RYS	SDS		134.0 (10)
RYS	SDS	RYS	314.0 (6)

As shown by the following data, the bacteria did not lose their capacity to exhaust HfrC filtrates when boiled or treated with phenol or methanol. Only treatment of F^- cells with a chloroform-methanol mixture produced this effect. Filtrates exhausted by bacteria previously treated with a chloroform-methanol mixture activated cells to the same extent as the original filtrates.

	leu^+ str-r recombinants, %
F^- cells "activated" with:	
HfrC filtrate	196.38
meat peptone broth	100
F^- cells "activated" by an HfrC filtrate previously exhausted by:	
intact F^- cells	98.37
F^- cells heated at 100° for 15 minutes	71.44
F^- cells treated with:	
chloroform-methanol mixture	175.88
methanol	90.31
phenol	90.19

Note. The HfrC donor was previously washed twice and crossed with a recipient "activated" with exhausted and nonexhausted filtrates.

This finding indicated the role of phospholipids in the receptor function of the cells. The phospholipid nature of the cell receptors interacting with RYS was confirmed in experiments with isolated phospholipids of *E. coli* JC411, which inactivated the stimulatory effect of HfrC filtrates.

F^- cells activated by:	leu^+ str-r recombinants, %
HfrC filtrates previously treated with phospholipids	97.32
HfrC filtrate	177.37
Meat peptone broth	100

The specificity of the observed inactivation was demonstrated in experiments with phospholipids previously destroyed by phospholipase A (see Materials and Methods, paragraph 17). The enzyme treatment deprived the phospholipids of their ability to inactivate RYS.

F^- cells activated by:	leu^+ str-r recombinants, %
HfrC filtrate + phospholipids previously treated with phospholipase A	283.91
HfrC filtrate + phospholipids	100
HfrC filtrate previously treated with phospholipase A	231.27
HfrC filtrate	236.24

N o t e. The number of recombinants obtained from the cross (JC411 F^- × HfrC) in which the F^- cells were treated with HfrC filtrate previously inactivated with phospholipids was taken as 100%. Average data of 8 experiments are given.

The 7 RYS^- mutation is located in the episome

In the introductory part of this article we stressed the need for data which would permit a choice between the following two hypotheses concerning the role of episomes in the induction of RYS synthesis by the cell:

1) the episome derepresses the structural gene of the chromosome responsible for RYS synthesis;

2) the episome contains the structural gene which controls RYS synthesis.

First of all we carried out experiments similar to ordinary complementation in which the Rfi^+ episome, which induces RYS synthesis in F^- cells, was transduced into the HfrC 7 RYS^- mutant. The capacity of the corresponding filtrates to inhibit ANM was used as the test for complementation in these experiments.

Both transductants of the type HfrC7 RYS^-Rfi^+ yielded filtrates which were unable to inhibit ANM. In other words, the transduced episome did not induce RYS production in the mutant. The functional potency of the R episome transduced into the mutant was proved by the finding that its transfer into the F^- strain by conjugation induced RYS synthesis in the latter. The results of these experiments can be interpreted in two ways: either the 7 mutation lies in the chromosome and therefore the derepressing effect of the episome is not expressed, or the mutant allele (RYS^-) dominates over the wild type allele (RYS^+). To test the second postulate it was necessary to remove the mutant episome from the HfrC7 RYS^- strain and replace it with a normal episome. If the activity of the filtrates obtained from such strains were restored this would prove the second postulate. From the strains HfrC and Hfr C7 RYS⁻ we selected variants possessing the F^- phenotype; these were designated respectively C/64F^- and C7/51F^-. These variants were selected by consecutive passages of the original cultures through meat peptone broth containing the phage MS2 (a description of the method will be published elsewhere).

Control:	
Survival rate of E. coli JC411 F^- incubated for 60 minutes after treatment with NM, %	57.4 ± 5.3
Experiment:	
% Survival rate of E. coli JC411 F^- treated with NM and incubated for 60 minutes with filtrates of E. coli K12 strains:	
HfrC	101.9 ± 9.0
HfrC7	59.6 ± 6.4
HfrC7R_1^+	58.2 ± 6.4
HfrC7R_2^+	63.7 ± 7.0
JC 411 F^-	62.0 ± 6.3
JC 411 $F^-R_1^+$	105.1 ± 5.1
JC 411 $F^-R_2^+$	100.2 ± 9.2

Note. The number of E. coli JC411F^- colonies formed on seeding the culture immediately after a 5-minute treatment with NM (300 μg/ml) was taken as 100%. The strains HfrC 7R_1^+ and HfrC 7R_2^+ were obtained by transduction of the episome with phage Pl_{kc} from the strain E. coli K12 CSH 2/222/F^-Rfi^+ to the mutant HfrC7 RYS^-. Transduction was performed according to Glover (1962). The strains JC411 $F^-R_1^+$ and JC411$F^-R_2^+$ were obtained by crossing JC411 with the strains HfrC7R_1^+ and HfrC7R_2^+ respectively.

Of the total of 200 subclones resistant to phage MS2 examined we selected variants possessing the following characteristics: 1) resistance to phages MS2 and f2; 2) sensitivity to phages T7 and φ II; 3) loss of donor properties in conjugation; 4) adoption of recipient role in conjugation with donor strains. Filtrates of cultures of the strains C/64F^- and C7/51F^- did not stimulate the yield of recombinants; however, the introduction of F or R episomes into these cultures leads to the synthesis of RYS (see below). These results support the second postulate and indicate that the gene controlling the synthesis of RYS lies in the episome.

Conditions of crossing	leu^+ str-r recombinants, %	Conditions of crossing	leu^+ str-r recombinants, %
Control 1:		HfrC	99.3 (6)
Standard crossing	100* (6)	C/64F^-	35.6 (6)
Control 2:		C/64F^+	121.9 (5)
RYS removed	32.8 (6)	C/64R^+	98.0 (5)
Experiment:		1485F^-	30.4 (5)
Addition of filtrates of the strains:		1485F^-R^+	121.4 (5)
HfrC7	35.6 (6)		
C7/51F^-	32.9 (3)		
C7/51F^+	116.4 (3)		
C7/51R^+	95.0 (3)		

* Frequency of recombinants $0.4 \cdot 10^{-1}$. Duration of the JC411F^- × HfrC cross 30 minutes.

Discussion

The results of these experiments together with the data published earlier lead to the following conclusions concerning the nature and mechanism of action of the RYS produced by male strains of *E. coli*. The episomes F and R control the synthesis of RYS — a protein-like substance of low molecular weight. Judging from the ultraviolet absorption pattern and the fact that we never managed to obtain a preparation containing more than $4 \cdot 10^5$ ANM units per ml, it is unlikely that RYS is a simple peptide. The lack of data on the chemical composition of the purified preparation precludes more definite conclusions. This compound is probably responsible for at least two of the three types of activity detected in filtrates of male strains of *E. coli* — namely, the effect on the yield of recombinants in conjugation and the suppression of ANM. Both effects are probably based on a single molecular mechanism in the cell, the nature of which is at present under investigation. The stimulation of the recombinant yield observed in our experiments is due to an effect on the recipient cell and takes place in the postconjugational merozygote, i. e. after completion of the transfer of the chromosomal fragment from the donor to the recipient. Since RYS increases the yield of recombinants only when applied to "young" merozygotes (see Figure 4), it appears that the RYS-induced stimulation of recombinations takes place in the merozygote before the donor fragment becomes completely integrated with the recipient chromosome. Further experiments are needed for a more thorough understanding of this topic.

We have also shown that RYS becomes bound to the cell surface and that its action requires the preservation of the RYS-bacterium complex. The RYS receptors in F^- cells are phospholipids. Thus, the biological activity of RYS appears to be based on its interaction with surface structures of the bacteria. As a result of this the secondary structure of the recipient DNA may be altered, thus facilitating its interaction with the fragment of donor DNA. Our findings widen the existing concepts of sexual differentiation in bacteria by showing that the male cell can synthesize RYS. We conclude that the genetic interactions of conjugating bacteria can be regulated at the intercellular level, in this case by the production of RYS on the part of the male cells. Furthermore, on the basis of its properties — protein nature, low molecular weight, action beyond the limits of the cells synthesizing it, on surface structures of recipient cells — the RYS factor may be assigned to the category of hormone-like compounds.

We are greatly indebted to V. O. Shpikiter and to A. M. Benenson and M. I. Karmanskii of the Protein Structure Laboratory of the Institute of Biological and Medical Chemistry of the USSR Academy of Medical Sciences for their advice and fruitful discussions, as well as to B. F. Poglazov of Moscow State University for the electron microscopy and to N. S. Gel'man (A. N. Bakh Institute of Biochemistry, USSR Academy of Sciences) for the phospholipase preparation and for valuable advice.

BIBLIOGRAPHY

Adams, M. Bacteriophages. — (Interscience) Wiley, 1959.
Barry, G. and W. Goebel. — J. Exp. Med. 94, No. 5 (1951), 587.
Baumann, L., W.C. Benz, A. Wright, and E.B. Goldberg. — Virology, Vol, 41 (1970), 356.
Edroards, S. and E. Ball. — J. Biol. Chem., Vol. 209 (1954), 619.
Glover, S.W. — Genet. Res., Vol. 3 (1962), 448.
Gol'dfarb, D.M., I.D. Avdienko, and L.S. Chernin. — Genetika 7, No.9 (1971), 133.
Gol'dfarb, D.M., L.S. Chernin, and Yu. K. Fomichev. — Genetika 5, No. 7 (1969), 74.
Gol'dfarb, D.M., L.S. Chernin, G.I. Gol'dberg, N.S. Akatova, and L.A. Gukova. — Genetika 6, No. 10 (1970), 107.
Kanemasa, L., S. Akamatsu, and S. Nojima. — Biochim. biophys. Acta 144, No. 2 (1967), 382.
Levin, O. — In: "Methods in Enzymology," S.P. Colowick and N.O. Kaplan (Eds.), Vol. 5, p.27. New York Academic Press, 1962.

UDC 575:576.858.9

STRUCTURAL TRANSFORMATIONS OF BACTERIOPHAGE DNA DURING REPLICATION AND RECOMBINATION PROCESSES

M.I. MOSEVITSKII
A.F. Ioffe Physical and Technical Institute
Academy of Sciences of the USSR, Leningrad

Circular forms of phage λ DNA participate in replication. Infection of E.coli BP33 with phage T2 causes a breakdown of the phage chromosome into independently replicating fragments with a molecular weight of $3 \cdot 10^7$. Electron microscopic autoradiography of intensively tritiated DNA molecules is proposed as a method for the study of structural transformations of nucleic acid molecules.

The molecular structure of the chromosome is an important factor in the mechanisms of DNA replication and recombination. Moreover, the chromosomal structure depends in turn on the replication and recombination processes. Viruses and bacteriophage are convenient models for the study of the relationship between chromosomal structure and function.

We shall begin by presenting data on the mechanism of the formation of the phage T2 chromosome. This work was done with the participation of L. E. Drabkina and A. L. Timkovskii. The genetic studies of Streisinger and his colleagues (1967), confirmed by electron microscopy (McHattie et al., 1967), have shown that the chromosome of T-even phages is a linear double helix with a molecular weight of $1.3 \cdot 10^8$ daltons. These chromosomes have uniform terminal sectors corresponding to $2-4 \cdot 10^6$ daltons and undergo cyclical permutation. This means that the chromosomes of T-even phages have a constant gene sequence but the position of the genes with respect to the ends varies. According to the model proposed by Frankel (1968) the structural peculiarities of the chromosomes of T-even phages result from the transition of the intracellular phage DNA into so-called catenomeres—giant DNA chains in which the recurring segments represent complete phage chromosomes. The union of phage chromosomes into catenomeres supposedly involves recombination at the redundant extremities. Subsequent inaccurate sectioning of the genomes of the phage progeny would lead to permutations and terminal redundancy. Our findings, however, do not confirm this model. In a study of the fate of ^{32}P-labeled chromosomes of phage T2 inside infected cells of *Escherichia coli* B we were unable to find catenomeres, but often observed the disintegration of chromosomes into fragments with an average molecular weight of $30 \cdot 10^6$ daltons. Such fragmentation was evident even before replication of the phage DNA began (Figure 1a) (Bresler et al., 1964). It is noteworthy that these fragments are capable of autonomous replication. This was established in another experiment in which the radioactive label was carried not by the parental phage DNA but by the phage DNA newly synthesized (Figure 1b) (Bresler et al., 1965). These results suggest a different hypothesis for the permutation and the terminal redundancy of the chromosomes of T-even phages. The only assumption required is that during the destruction of the parental chromosome, the breaks in the complementary chains are somewhat displaced, to an extent commensurate with the length of the redundant extremity. As a result, neighboring fragments possess overlapping terminal polynucleotide sequences which can be turned into double helices by repair synthesis.

According to the proposed model (Figure 2), the chromosomes of the phage progeny are built up of fragments and it is the recombinational union of these that creates the terminal overlapping. The cyclically permuted and terminally redundant phage chromosome can be explained on the basis that assembly can start with any fragment, which then determines the sequence of the remaining fragments (see Figure 2). Chromosomes with one missing or one excessive fragment are respectively incomplete or redundant. Particles with such chromosomes are invariably present in the phage population.

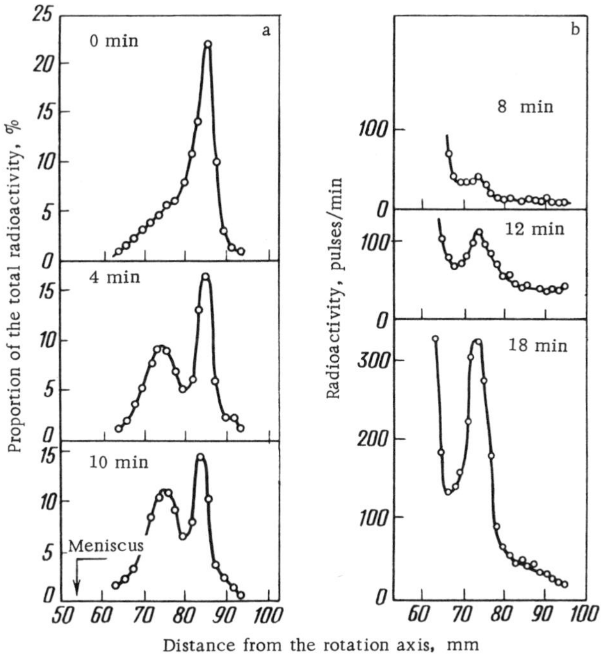

FIGURE 1. Sedimentation patterns illustrating fragmentation of the parental chromosome (a) and autonomous replication of the fragments (b) during the intracellular development of phage T2.

The time from the moment of infection of the cells to the collection of the respective sample is indicated on the diagrams.

Another study shows equally clearly the relationship between the structure and functions of the genetic material in phage λ. This work was carried out with the participation of L. E. Drabkina, L. V. Konevega and O. K. Legina. Examination of the fate of DNA of superinfecting phage λ revealed that proflavin at a concentration of 10 μg/ml or more completely inhibits the transition of phage λ DNA molecules from the linear form characterizing the mature phage to the curled or straightened rings commonly found inside infected cells (Drabkina and Mosevitskii, 1970). The same result was obtained on infection of sensitive cells. This confirmed the findings of other workers (Ravin and Andreeva, 1964; Bailey and Fray, 1969), who found that proflavin suppresses almost completely the replication of DNA and the production of phage progeny. It is known, however, that proflavin is not an efficient inhibitor of DNA synthesis in other cases. Thus, E. coli cells can reproduce for many hours in a proflavin-containing medium, though at a somewhat reduced rate. Proflavin does not inhibit the DNA replication in T-even phages, phage T1, etc. This means that the inhibitory effect of proflavin on DNA synthesis is specific to phage λ and is apparently associated with the suppression of DNA ring formation in infected cells. This conclusion can be confirmed experimentally: proflavin practically does not interfere with the normal development of the phage progeny if added 10—15 minutes after the start of intracellular phage development, when the rings of phage λ DNA are already complete, rather than at the very beginning. A number of different views on the structure of the replicative form of phage λ DNA have been published. Some workers maintain that phage λ DNA replicates in the form of linear catenomeres (Smith and Skalka, 1966), while others have found replicating rings of phage λ DNA (Ogawa et al., 1968). Our results show that the closure of phage DNA into a ring is a signal for the beginning of replication.

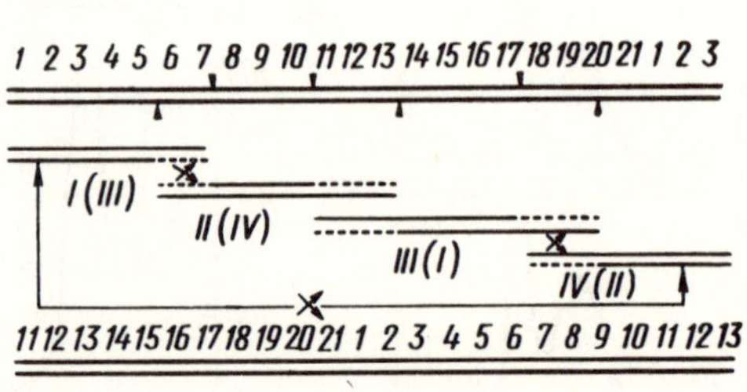

FIGURE 2. Possible mode of formation of cyclically permuted and terminally redundant linear chromosomes.

The inhibitory effect of proflavin on the formation of DNA rings undoubtedly deserves attention. This effect can be attributed to the increased rigidity of the DNA chain when complexed with proflavin, reducing the probability that the ends will meet.

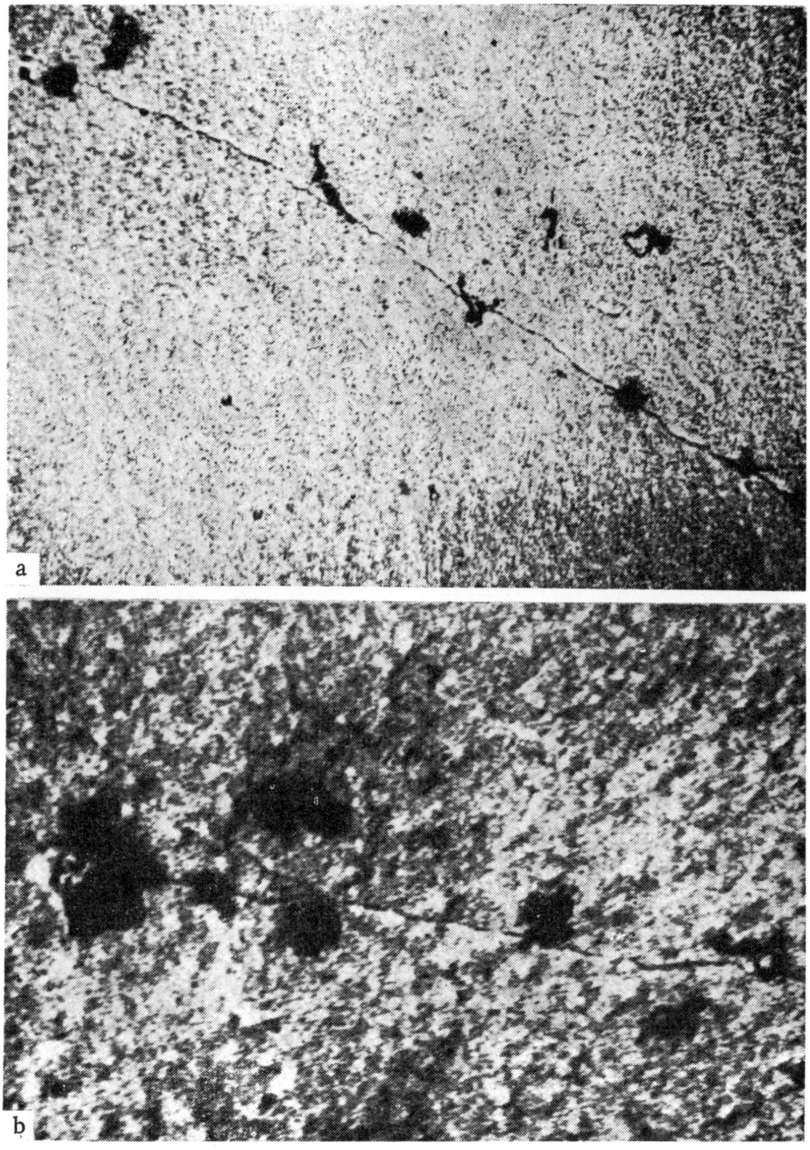

FIGURE 3. Electron microscopic autoradiographs of phage T1 DNA molecules.

Time of exposure two months; a) magnification 40,000 ×; b) magnification 50,000 ×.

Further research into the relationship between the structure and functions of the genetic material requires new laboratory methods. In order to obtain direct data on the structural changes in nucleic acids during replication and recombination we developed a method for the electron microscopic autoradiography of individual DNA molecules. This work was done with the participation of L. P. Dadivanyan. The new procedure enables the simultaneous observation of the DNA molecule and its autoradiograph. The action of the β-particles emitted from the labeled sections of the DNA molecule during exposure reveals the grains of the photoemulsion covering the preparation, thus producing an autoradiograph. As a model experiment we examined the recombination in phage T1. Particular attention was devoted to the sensitivity and resolving power of the method, i.e. to determining whether the distribution of the grains revealed reflects accurately the size and position of the labeled sectors of the DNA molecule.

Intensively tritiated phage T1 was obtained from infected E. coli B_3 thy^- by using ^{3}H-thymidine with a maximum specific activity of 20 curies/ /mmole as the thymine source. A mixture of ^{3}H-labeled and nonradioactive phage in the ratio 1:3 was used to infect E. coli B cells at a multiplicity of 3. The experiment was carried out under conditions allowing limited replication of DNA (in minimal medium containing 5μg/ml 5-fluorodeoxyuridine). Progeny phage particles purified from cell debris by a series of centrifugations and resuspended in 0.1 M NaCl to a concentration of 10^9 particles/ml were destroyed in such a way that the end of the DNA molecule released into the medium often remained attached to the protein coat.

To obtain straightened DNA molecules, preparations for electron microscopy were drawn out from the solution and shadowed with platinum at an angle of 6—8°. The shadowed specimens were coated with PR-9 nuclear photoemulsion according to the method of Caro and Van-Tubergen (1962) and exposed in tightly sealed metal cases at a temperature of 4°. The autoradiographs were developed with methol-hydroquinone and then placed for 3—6 hours in 0.05 N NaOH to remove the gelatin present in the emulsion. Figure 3 shows autoradiographs of DNA molecules labeled over their entire length and developed after a two-month exposure. The grains revealed obviously lie along the DNA strands and immediately below them. The center of the grain is not more than 0.2 microns away from the molecule, and the grains often overlap. This demonstrates the high resolving capacity of the method. A simple calculation shows that in the experiment described the totally labeled double-helical molecule of phage T1 DNA with a molecular weight of $30 \cdot 10^6$ emits an average of 1—2 β-particles per 24 hours of exposure. If the emulsion is applied to one side of the specimen it is struck by about half these β-particles. Only 10—20% of these are recorded, i. e. create latent centers of exposure. Because of its small energy, the β-particle emitted by tritium affects only one or rarely two grains of the emulsion. On this basis one would expect 6—12 visible grains in an emulsion developed after a two-month exposure over the DNA molecule of phage T1, and this is actually the case. Although adequate for the detection of labeled DNA molecules among unlabeled ones, this procedure does not reveal labeled and unlabeled sectors within the same DNA molecule.

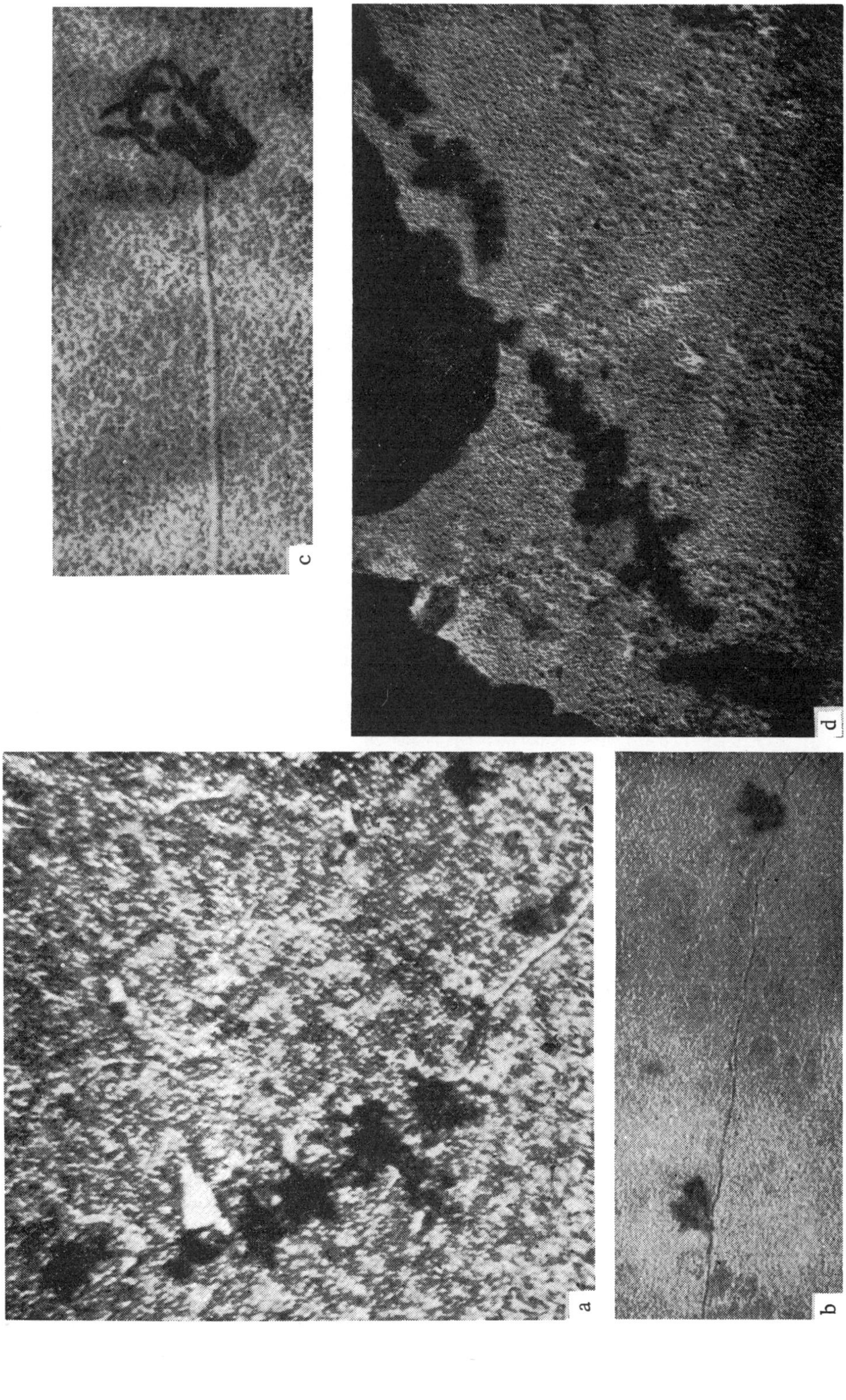

FIGURE 4. Electronmicroscopic autoradiograph of DNA molecules of phage T1.

Duration of exposure 6 – 7 months; a) magnification 50,000×; b) magnification 40,000×; c) magnification 80,000×; d) magnification 40,000× (sandwich).

For this purpose the exposure time of the autoradiographs must be increased ɔ 6—12 months. In some cases we placed the carbon film with the DNA preparation between two emulsion pellicles. Such autoradiographs are twice as sensitive as unilateral ones. Figure 4 shows autoradiographs of chromosomes of the recombinant phage progeny, developed after exposure for 6—7 months. It can be seen that the recombinant chromosomes may consist of extended labeled and unlabeled sectors (Figure 4a), but the incorporation of labeled material into unlabeled chains (Figure 4 b, c) and vice versa (Figure 4d) is negligible, being limited to fragments of between 1 and 2 microns ($2-4 \cdot 10^6$ daltons) or even less. This means that the recombination mechanism typical of phage T1 involves the incorporation of short fragments of DNA. According to the genetic data of Michalke (1967) the probability of crossing over in phage T1 is greater at the ends of the chromosome; we found that the exchanges are indeed more frequent near the ends of the molecule, but we do not yet have sufficient data to draw quantitative conclusions (Figure 4).

One would naturally like to know on the basis of the autoradiographic data whether one or both strands are integrated in the course of recombination. Although feasible in principle, such analysis requires a more highly perfected autoradiographic procedure enabling a more accurate count of the grains revealed.

BIBLIOGRAPHY

Bailey, A.D. and B.A. Fray. — Molec. and Gen. Genet., Vol. 104 (1969), 16.

Bresler, S.E., L.E. Drabkina, M.I. Mosevitskii, and A.L. Timkovskii. — Doklady AN SSSR, Vol. 156 (1964), 947.

Bresler, S.E., L.E. Drabkina, M.I. Mosevitskii, and A.L. Timkovskii. — Materials of the Symposium on the Molecular Nature of Viruses. Information Bull. 5th Scientific Council on Molecular Biology, No.4, 1965. (Russian)

Caro, L.G. and Van Tubergen. — J. Cell. Biol. 15 (1962), 173.

Drabkina, L.E. and M.I. Mosevitskii. — Molekulyarnaya Biologiya, Vol. 4 (1970), 591.

Frankel, E.R. — Proc. Natn. Acad. Sci. U.S.A., Vol. 59 (1968), 131.

MacHattie, L.A., D.A. Ritchie, and C.A. Thomas. — J. Molec. Biol., Vol. 23 (1967), 355.

Michalke, W. — Molec. and Gen. Genet., Vol. 99 (1967), 12.

Ogawa, T., J. Tomizawa, and M. Fuke. — Proc. Natn. Acad. Sci. U.S.A., Vol. 60 (1968), 861.

Ravin, V.K. and I.V. Andreeva. — Mikrobiologiya, Vol. 33 (1964), 819.

Smith, M.G. and A. Skalka. — J. Gen. Physiol., Vol. 49 (1966), 127.

Streisinger, G., J. Emrich, and F.W. Stahl. — Proc. Natn. Acad. Sci. U.S.A., Vol. 57 (1967), 292.

UDC 575:576.858.9

THE NATURE OF ULTRARARE RECOMBINATIONS

N.I. MATVIENKO
Institute of Molecular Biology
Academy of Sciences of the USSR, Moscow

On the basis of experiments with deletion mutants of phage T4 a hypothesis is proposed according to which rare recombinations result from the incorporation of a single-stranded fragment from the 196 parent into a single-stranded gap situated in the deletion zone of the 1299 parent.

It is known that the genetic map of phage T4 comprises about 2,000% recombinations and that its DNA molecule contains 200,000 nucleotide pairs. Consequently the average frequency of recombination between adjacent nucleotides must be about 0.01%. Such a frequency of recombination was observed by Benzer. He regarded it as the minimal value and called this primary unit of recombination the recon (Benzer, 1956). Tessman (1965) questioned this concept of a primary unit of genetic recombination. Determining the frequency of recombination between mutations located in a single hot spot, he obtained values much smaller than a recon. Previously we observed a recombination frequency close to the presumed magnitude of the recon (Matvienko, 1969). Crosses of the deletion mutants 1299 and 196 yielded wild type recombinants at a frequency of about $5 \cdot 10^{-5}\%$, which is less by two orders of magnitude than the presumed size of the recon.

FIGURE 1. Hypothetical mechanism of the formation of pseudowild recombinants.

The purpose of our work was to elucidate the nature of ultrarare recombinations with particular reference to the crossing of the 196 and 1299 deletion mutants of phage T4. The use of deletion mutants has considerable advantages in view of the lack of spontaneous revertants. It is known that the deletions 196 and 1299 overlap (Benzer, 1962). Hence the original assumption that the phage particles revealed in *Escherichia coli* K12 (λ) are not true recombinants but represent a pseudowild type arising from the suppression of the mutation 1299 by the mutation 196 when an unequal crossover brings both deletions into a single genome (Figure 1). The suppression hypothesis does not lend itself to direct verification. The existence of suppressions is usually established by the segregation of the original parental mutants among the recombinants obtained by crossing the pseudowild and wild types. However, recombination between the deletions used is so rare that the frequency of new r mutants from the wild type exceeds that of segregants by several orders of magnitude. The following experiments were made in order to test the hypothesis. The mutation W8-33, located to the right of deletion 1299, was introduced into the genome of one of the parents and the resulting change in the recombination frequency was determined. If the above hypothesis is correct, the introduction of a deletion into the genome of the 1299 parent should markedly reduce the frequency of recombinants; conversely, recombination should remain practically unchanged if the mutation is introduced into the 196 parent. It turned out that the recombination frequency decreased by half regardless of which parent received the W8-33 deletion (Figure 2). Thus, a marker located very close to the deletions 1299 and 196 proves to be unlinked. These results ruled out the hypothesis according to which all recombinants are of the pseudowild type. Next it was assumed that the pseudowild type contains only 50% recombinants and that the deletions are separated by a small interval within which reciprocal recombinants arise with a low and uniform probability. When the deletion W8-33 is introduced, one of the classes would then receive an additional W8-33 deletion, which would explain the 50% decrease in the recombination frequency. If this is actually so, the introduction of mutations C8 and C9 near the deletion 196 should similarly decrease the recombination frequency by half since true wild type recombinants will not appear in this case. However, the corresponding crosses did not yield any recombinants. Therefore we abandoned the hypothesis that the recombinants are of the pseudowild type as a result of suppression. The mutation C9 appears to be located in the hottest spot of gene rII and overlaps with deletion 196. If there is an interval between deletions 1299 and 196, therefore, it must be smaller than the C9-1299 sector. Translating the genetic distances into physical ones, this interval would correspond to only a few nucleotides. However, this estimate is inaccurate inasmuch as it is known that the genetic scale is reduced at the margins of deletions; this reduction apparently results from numerous factors which are difficult to assess.

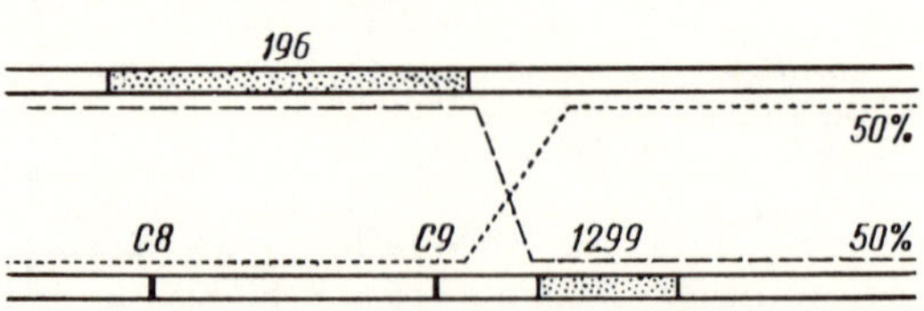

FIGURE 2. Hypothetical mechanism of recombination on the basis of data obtained by crossing the double mutants 196, W8-33 and 1299, W8-33.

In an attempt to obtain additional data we examined the inheritance of markers to the left of deletion 196 in the recombinants (Figure 3). For this purpose we formed the double mutants N3-196, C6-196, 1993-196 and 1368-1269. It turned out that the markers situated to the left of the deletion are normally linked and most recombinants receive the alleles of the 1299 parent, as should be expected.

In order to determine the inheritance of alleles to the right of deletion 1299 we prepared the double mutants 196-1519, 1299-1519, 196-ac_1^r, 1299-ac_1^r, 196-P219 and 1299-P219. The mutation ac_1^r, isolated by us from phage T4B, confers resistance to acriflavin at a concentration of 1 μg/ml in a solid medium. Like the mutations to acriflavin resistance in phage T4D, this mutation is located a short distance to the right of gene rII. The mutant P219, received from A. Maisuryan, lies in the gene controlling the synthesis of the flagellar filaments. All double-mutant crosses were performed in conjunction with 1299 × 196 control crosses in order to determine the extent to which the additional mutation lowers the frequency of wild type recombinants. The relevant data are shown in Figure 3.

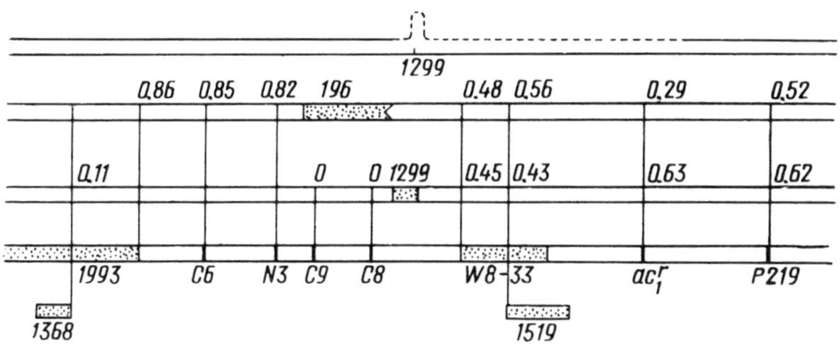

FIGURE 3. Pattern of the inheritance of adjacent markers.

The results were unexpected. In the presence of deletion 1519, the allele from the 1299 parent is inherited by the recombinants with a greater probability than is the allele from the 196 parent. In other words, a slight antilinkage is evident in mutation 1519. The mutation ac_1^r was tested in 13 experiments. All of them revealed a conspicuous antilinkage. The allele from the 1299 parent was passed on to the recombinants at a frequency twice that observed in the case of the allele from the 196 mutant. The P219 marker was inherited by the recombinants with roughly equal probability, probably because the mutation P219 is located rather far from the gene rII.

We conclude on the basis of these data that the rare recombinations result from the integration of a single-stranded fragment from the 196 parent into a single-strand gap situated in the deletion area of the 1299 parent. The existence of such fragments in the pool of intracellular phage DNA has been demonstrated beyond doubt. This hypothesis readily explains the antilinkage of the ac_1^r marker. Indeed, if the heterozygous region does not reach the ac_1^r marker, both strands of the recombinant will contain the marker from the 1299 parent; if the heterozygous region does reach the ac area, half the recombinants after segregation of the heterozygote will inherit the allele of gene ac from the 196 parent and half the allele from the 1299 parent. Accordingly inheritance of the ac allele from the 1299 parent must predominate. In other words, the antilinkage of the ac_1^r marker is due to the fact that the heterozygous region does not always reach the ac gene. Since the deletion 1519 lies much closer to the chiasma, fewer recombinants have a heterozygous region which does not reach this deletion and antilinkage will be less pronounced in this case. The lack of linkage of the W8-33 mutation may be due to a highly efficient repair of deletion heterozygotes and a free passage of information from one strand to the other. By now it has been proved that heterozygotes for long deletions do not appear in the phage progeny. This can be explained in two ways. Either these heterozygotes do not arise at all in the crossing process, or they are rapidly repaired. Our data indicate that such heterozygotes do appear and that their repair is not polar; in other words, such a deletion heterozygote can be repaired with equal probability to give a homozygote with mutant DNA in both strands or wild type DNA in both strands.

BIBLIOGRAPHY

Benzer, S. — In: "Chemical Basis of Heredity." Sympos. Chem. Basis of Heredity. Johns Hopkins Univ., 1956.

Benzer, S. — Scient. Am. 206, No.1 (1962), 70.

Matvienko, N.I. — Genetika 5, No. 4 (1969), 84.

Tessman, I. — Genetics 51, No. 1 (1965), 63.

UDC 575:576.8

INTERPRETATION OF INTRAGENIC RECOMBINATION PHENOMENA IN TERMS OF THE PRINCIPLE OF DIRECTIONAL CORRECTION

V.V. KUSHEV
A.F. Ioffe Physical and Technical Institute
Academy of Sciences of the USSR, Leningrad

A scheme of recombination developed on the basis of the Holliday model is proposed. It is shown that primary breaks in chromosomal strands induce local denaturation of the DNA of homologous chromatids. Heterozygous sites appearing in the recombining chromatids are corrected by a mechanism analogous to that engaged in repair. One of the DNA strands can be regarded as the donor, the other as the recipient. It is postulated that correction to the recipient text is more frequent than to the donor text.

When intragenic recombination was discovered it was assumed that the mechanism involved is identical to that of intergenic recombination. Further studies on fungi, however, showed that recombination between alleles can occur by a mechanism distinct from crossing over.

The four haploid products of meiosis can be readily observed in many fungi. Tetrad analysis has shown that any heterozygous locus usually yields the expected Mendelian segregation of 2:2 in the progeny. However, aberrant tetrads with a 3:1 segregation do appear, though infrequently. Aberrant segregation in yeasts was first demonstrated by Lindegren (1953), who termed it gene conversion following the lead of Winkler (1930).

This discovery was so unexpected that it failed to gain general recognition until Mitchell (1955) found a similar phenomenon in heteroallelic crosses of *Neurospora crassa*. This type of gene conversion was termed nonreciprocal recombination (Figure 1).

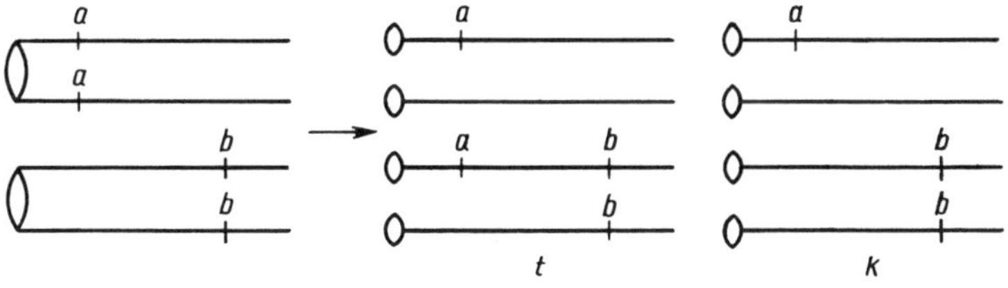

FIGURE 1. The main types of tetrads obtained from two-site (a × b) crosses.

Each line indicates a chromatid on the left-hand side of the diagram and a chromosome on the right-hand side. The oval indicates the centromere.

In the wake of this discovery studies of intragenic recombination by tetrad analysis were begun in a variety of fungi. It became clear that crosses of the majority of mutants yield mostly conversion-type k tetrads. Tetrads containing reciprocal recombination products (t tetrads) appear in such crosses at a frequency of not more than 10% (Table 1).

TABLE 1. Relative frequencies of k and t tetrads in interallelic crosses of fungi

Organism	Gene	k tetrads	t tetrads
Yeasts	is 1	52	–
		29	–
Yeasts	hi 1	980	101
Aspergillus	paba 1	373	20
Ascobolus	46	274	6
	y	308	18

N o t e. In the case of yeasts with the gene is 1 and *Aspergillus* – in incomplete tetrads. is 1 – synthesis of isoleucine; hi 1 – synthesis of histidine; paba 1 – synthesis of paraaminobenzoic acid; 46, y – coloration of ascospores.

However, the frequency of t tetrads increases sharply (to 50% or more) when the crossed mutants belong to a particular type which usually shows a tendency for postmeiotic segregation (Whitehouse, 1966). Thus, the frequency of t tetrads appears to be allele specific.

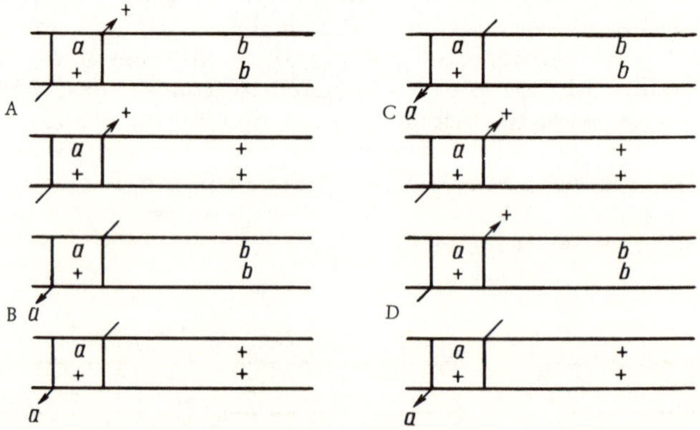

FIGURE 2. Scheme of the formation of different types of asci as a result of correction.

Two of the four chromatids are shown for each variant. The hybrid DNA occupies the site "a" in both recombining chromatids. The arrows show the direction of the correction.

The allelic specificity of intragenic recombination is also evident in the analysis of random samples of progeny: the homalleles often differ sharply from one another as regards their frequency of recombination with any other alleles of the same gene (Balbinder, 1962; Drapeau et al., 1968). Further research in this field revealed the phenomena of recombination polarity and high negative interference (Lissouba et al., 1962; Pritchard, 1960).

In an attempt to explain these peculiarities of intragenic recombination, Holliday (1964, 1968) and Whitehouse and Hastings (1965) proposed recombination models based on the hypothesis of the correction of molecular heterozygosis, put forward by Holliday (1962). Without entering into a detailed discussion of the specific structural details of these models (see Kushev, 1971) we shall point out their major shortcoming — namely, the inability to explain the interrelationship between reciprocal and nonreciprocal recombinations at the intragenic level. For example, four outcomes are possible according to Holliday since both homologous chromatids contain hybrid DNA (Figure 2). If correction proceeds in the same direction in both chromatids (towards + or towards —) the outcome will be a conversion, i.e. a nonreciprocal recombination (Figure 2a, b). If the correction takes place in opposite directions the result will be either the original (parental) condition (Figure 2c) or a t tetrad (Figure 2d).

The expected frequencies of the four possible products of correction are shown below.

Correction parameters in chromatid 2	Correction parameters in chromatid 1	
	p^+	p^-
p^+	3 + : 1 — (k)	2 +: 1 — (p)
p^-	2 + : 2 —	1 +: 3 — (k)

The frequency of correction to the wild type is designated p^+, and to the mutant type p^-. Since p^+ usually does not differ much from p^-, the frequencies of t and k tetrads should be roughly comparable. In the latest publication of Holliday (1968) no attempt has been made to explain this fundamental discrepancy.

In our view, the most probable cause of the nonreciprocity is directional correction.

Let us assume that the parameters of the correction of heterozygous sites in chromatids 1 and 2 (Figure 2) can differ sharply from one another. Let us assume further that the direction of correction depends mainly on the polynucleotide strand in the hybrid region. In this region, one of the strands (distinguishable by its free end) can be regarded as the donor and the other as the recipient (until the hemichiasma turns into a complete chiasma). It is known from transformation data that the correction of heterozygotes to the recipient text is many times more frequent than to the donor text. Such is the case at least with the LE mutations of pneumococci (Gray and Ephrussi-Taylor, 1967). The existence of directional correction is evident from the results of clonal analysis of transformants (Bresler et al., 1968; Kushev et al., 1970) and *Bac. subtilis* phages (Trautner and Spatz, cited in Whitehouse, 1970), as well as from the data of Lacks (1970), who isolated hex mutants of pneumococci capable of integrating LE mutations with an efficiency typical of HE mutations.

Thus, if the donor allele in chromatid 1 is "+ " and that in chromatid 2 is "-" and if the directionality factor of the correction is 10 — 20 (Gray and Ephrussi-Taylor, 1967), it follows that t tetrads will appear 10 times less frequently than k tetrads (Table 2).

Some types of mutation (HE) can be corrected also toward the donor text, as a result of which the correction parameters in the two recombining chromatids tend to become equal. The outcome is a high frequency of t tetrads and a high recombination frequency.

The proposed model differs further from that of Holliday in the following important aspects.

1. The probability of conversion of the hemichiasma into a complete chiasma is taken to be much lower than 0.5 (for example, 0.1 — 0.05 for Sordaria fimicola (Kushev, 1971).

2. The secondary breaks occurring during conversion of the hemichiasma into a complete chiasma do not necessarily arise at strictly homologous sites. Hence the need for repair enzymes for the completion of the crossover process (and not merely for conversion).

TABLE 2. Probability of appearance of different types of tetrads in interallelic crosses involving LE mutants, calculated on the basis of directional correction

Correction parameters of chromatid 2	Correction parameters of chromatid 1	
	$p_1^+ = 0.09$	$p_1^- = 0.91$
$p_2^+ = 0.91$	0.0819 6 + : 2 − (k)	0.8281 4 + : 4 − (p)
$p_2^- = 0.09$	0.0081 4 + : 4 − (t)	0.0819 2 + : 6 − (k)

3. The recombinators are never located in the middle of genes but are terminal, like the operators in transcription.

The proposed model is free of most of the shortcomings of the theories of Whitehouse and Holliday. It also provides an interpretation of other phenomena associated with intragenic recombination — namely, polarity, map expansion, and both types of high negative interference.

BIBLIOGRAPHY

Balbinder, E. — Genetics, Vol. 47 (1962), 545.

Bresler, S.E., R.A. Kreneva, and V.V. Kushev. — Molec. and Gen. Genet., Vol. 102 (1968), 257.

Drapeau, G.R., W.J. Brammar, and C. Yanofsky. — J. Molec. Biol., Vol. 35 (1968), 357.

Gray, T.C. and H. Ephrussi-Taylor. — Genetics, Vol. 57 (1967), 125.

Holliday, R. — Genet. Res., Vol. 3 (1962), 472.

Holliday, R. — Genet. Res., Vol. 5 (1964), 282.

Holliday, R. — In: "Replication and Recombination of the Genetic Material." Canberra, 1968.

Kushev, V.V. Mechanisms of Genetic Recombination. — Moscow-Leningrad, "Nauka," 1971. (Russian)

Kushev, V.V., S.E. Bresler, and R.A. Kreneva. — In: Sintez, struktura i svoistva polimerov, p.245. Leningrad, 1970.

Lacks, S. — J. Bact. 101, No. 2 (1970), 373.

Lindegren, C.C. — Genetics, Vol. 51 (1953), 625.

Lissouba, P., J. Mousseau, G. Rizet, and J.L. Rossignol. — Adv. Genet., Vol. 11 (1962), 343.

Mitchell, M.B. — Proc. Natn. Acad. Sci. U.S.A., Vol. 41 (1955), 215.

Pritchard, R.H. — Genet. Res., Vol. 1 (1960), 1.

Whitehouse, H.L.K. — Nature, Vol. 211 (1965), 708.

Whitehouse, H.L.K. — Biol. Rev., Vol. 45 (1970), 265.

Whitehouse, H.L.K. and P.J. Hastings. — Genet. Res., Vol. 6 (1965), 27.

Winkler, H. Die Konversion der Gene. Jena, 1930.

UDC 575:576.858.9

PRIMARY RECOMBINATIONAL EVENTS DURING THE TRANSFORMATION OF PHAGE T4B

Yu.P. VINETSKII
Institute of General Genetics
Academy of Sciences of the USSR, Moscow

Clonal analysis shows that every recombinational event in the area of integration of the donor DNA includes the formation of a molecular heterozygote as an intermediate stage of recombination. The heterozygous area is 400 − 600 nucleotides long. When the donor molecules consist of a single helix, a donor mutation can be corrected to the opposite chain of the recipient in the heterozygous area. In a system with terminal markers it was shown that the mechanism of the integration of donor DNA consists of two independent exchanges between donor and recipient.

This study is concerned with the modes of integration of the donor DNA of phage T4 into the recipient genome and the primary reactions involved in the pairing of the donor fragment with recipient phage DNA during transformation.

The development of phage T4 in bacteria involves the synthesis of numerous replicas of phage DNA. Recombinational events take place mainly between daughter replicas, the total number of which can reach several hundred. For this reason one of the most probable mechanisms of double crossover consists of consecutive genetic exchanges between different replicas of phage DNA. Three original phage chromosomes are required in this case for the formation of a primary recombinant molecule. At the same time it is thought that the mechanism in bacterial transformation consists in the insertion of a donor DNA fragment into the recipient chromosome. We tried to make a choice between these models in the case of the recombination occurring during the transformation of phage T4.

Our work was based on an analysis of the transformant genotypes for the conditional lethal markers of the recipient phages. For this purpose bacterial spheroplasts were infected with two recipient phages bearing deletions in the rII region and two conditional lethal rII^{ts} mutations to the right and left of the central deletion marker, and with donor DNA bearing two limiting deletions which overlapped with the marginal markers of the recipients. The scheme of the experiment is shown in Figure 1. The double mutants ts820, 1589 and ts113, 196 were used as the recipient phages. The deletions 1589 and 196 overlap, but complementation occurs with respect to the rII function (Champe and Benzer, 1962). Transformation was carried out in *Escherichia coli* K12 (λ) spheroplasts, where the recipient phages do not reproduce with a single infection.

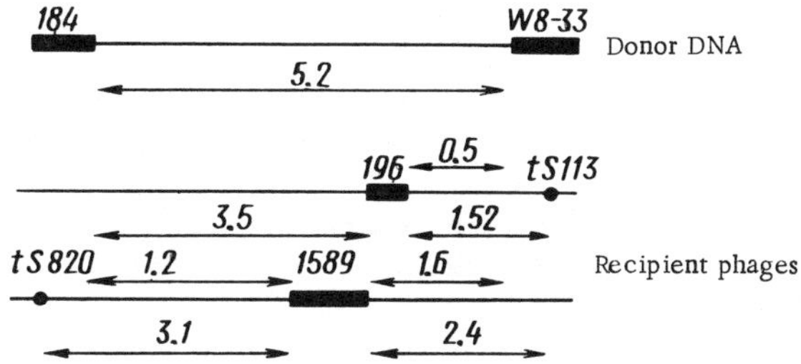

FIGURE 1. Scheme of crosses in the rII region during the transformation of phage T4.

Analysis of the transformant genotypes enables selection between the two models of integration. In the case of independent genetic exchanges the transformants must obviously contain both recipient (parental) and recombinant alleles of the ts mutation (i.e. ts^+, ts^+ and ts, ts). On the other hand, if the donor fragment is incorporated into a single replica of the recipient genome (or into the original genome), the outcome must be various combinations of parental alleles with no recombinant configurations. The distortions caused by genetic exchanges between recipients prior to their interaction with the donor DNA are negligible because of the very short genetic distances between the recipient markers.

The distribution of transformant genotypes by class is as follows:

	Donor DNA	
	native	denatured
Total number of plaques counted	257	412
Proportions of transformant genotypes		
820^+, 113^+	0.53 (1.0)	0.42 (0.79)
820, 113^+	0.2 (0.91)	0.25 (1.14)
820^+, 113	0.195 (1.08)	0.30 (1.65)
820, 113	0.075 (1.07)	0.03 (0.43)

Note. The figures in parentheses show the ratio between the experimentally determined proportion of transformants of each possible genotype and the value expected theoretically on the basis of the genetic distances between the markers, taking into account back crosses between the recipients.

It is evident from the above data that the transformants included all the possible combinations of the marginal mutations and their wild type alleles in accordance with the theoretically expected values. No marked differences were observed between the experiments using native and denatured donor DNA in this respect.

Thus, genetic analysis of the transformants shows that integration of the fragment of transforming DNA comprises two separate and consecutive instances of recombination occurring to the right and left of the central deletion marker.

Let us consider a single instance of recombination. It is generally assumed that one of the first stages is the pairing of the DNA molecules. If one of the participating molecules bears a point mutation, the recombinant molecule becomes heterozygous in the pairing region (molecular or internal heterozygotes).

Analysis of the progeny of the primary heterozygous recombinant molecule reveals a segregation of characters and the formation of a mixed clone arising as a result of the replication of the primary recombinants. Guerrini and Fox (1968) performed such a clonal analysis in the case of bacterial transformation. However, processes resembling the repair of DNA can also take place in the pairing region. These processes convert the primary recombinant from a heterozygous to a homozygous state before the first replication begins. It appears that this phenomenon is the result of two consecutive events: enzymatic degradation of one DNA chain in the unpaired area, and repair synthesis restoring homozygosis with respect to one of the parental characters (correction of heterozygotes). The correction of heterozygotes can be detected in a clonal analysis by the presence of pure clones. This fact was observed by Bresler and his colleagues (1968) in the case of bacterial transformation.

Knowledge of the specificity of the correction of molecular heterozygotes is essential for an understanding of the primary processes involved in recombination. We investigated two topics: the dependence of repair on the molecular state of the transforming DNA, and the directionality of correction processes. The experiment was based on the genetic analysis of the recombinant particles formed in a linked transformation for two markers — a deletion and a conditional lethal mutation, both located in the rII region. Recombinants were selected on the basis of the characteristics of the wild type allele of the genetic marker (rII^+), and the genetic analysis was based on the presence of the conditional lethal mutation. Conditional lethal mutations located in different parts of the rII region were obtained by means of various mutagens: am N6 AZad (nitrous acid), ts 820A2h3 (2-aminopurine), and ts113 B9-B10 (hydroxylamine).

Figure 2 shows the relationship between the proportion of transformants with the am^+ genotype and the genetic distance from the near end of the deletion marker to the N6 mutation. In this experiment the transforming DNA contained the N6 mutation, and various deletion mutants of phage T4 were used as the recipients. The denatured DNA preparation was freed from traces of double-stranded molecules by passing through methylated albumin (MAK) columns.

As the distance increases, the proportion of am^+ transformants rises to a plateau and then rises further. The proportion of am^+ transformants is roughly twice as high when denatured DNA is used as the donor as in the case of native DNA. The am^+:am ratio in the transformant population is 0.4:0.6 for native DNA and 0.8:0.2 for denatured DNA. These data can be interpreted as follows. If the mutation lies not far from the rII^+ marker, it enters the genomes of the recombinant phages at a high frequency. As the distance increases there is a greater probability that the mutation will fall into the heterozygous region; hence the plateau. The existence of the plateau indicates that the heterozygous region is rather large (about 3.3% recombinations or 400—500 base pairs in this case).

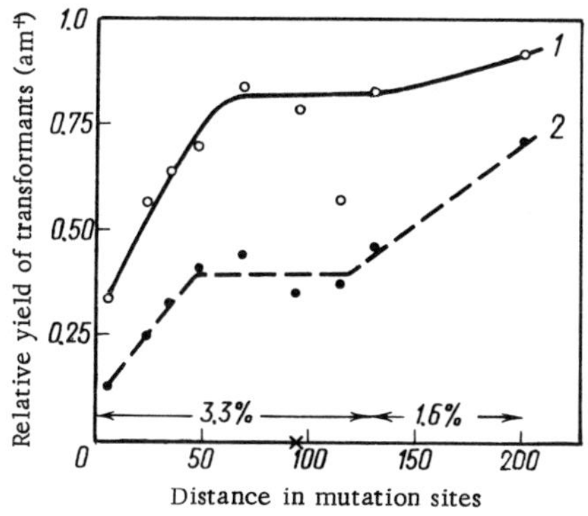

FIGURE 2. Proportion of am$^+$ transformants as a function of the distance between the N6 mutation in the donor DNA and the rII marker in the recipient:

1) denaturated DNA; 2) native DNA.

The fact that the proportion of am$^+$ type recombinants arising from transformation with DNA is four times greater than the mutant type (a 1:1 ratio can be expected in the case of an equally probable segregation) can be interpreted as the result of a correction of the donor mutation N6 to the recipient (am$^+$) type. Correction in the AZad segment apparently requires the crossing of denatured single-stranded denatured DNA with double-stranded DNA. There is no appreciable correction if the cross involves two double-stranded DNAs (as in the experiments involving transformations by native preparations).

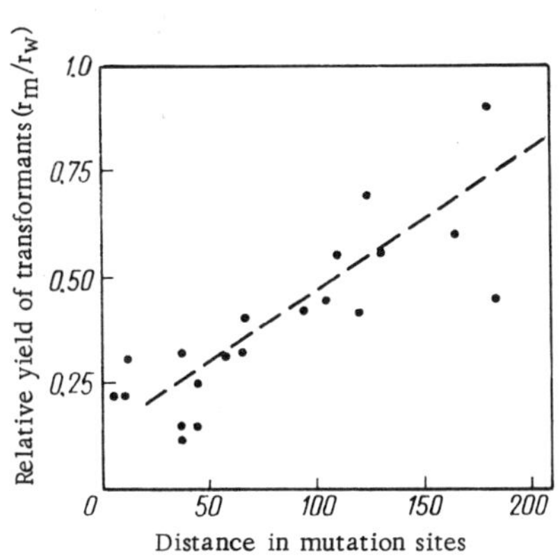

FIGURE 3. Relative yield of transformants as a function of the distance between the deletions in the donor and recipient:

r_m — yield for DNA with deletions; r_w — yield for wild type DNA.

Two series of experiments were performed in order to prove this assumption. In one of the series we determined the yield of transformants as a function of the recombination distance, which was limited by deletions in the denatured donor DNA on one side of the rII^+ marker. As shown in Figure 3, the yield of transformants rises steadily as the distance between the deletions in the donor and recipient increases to 150—180 mutation sites. However, the curves in Figures 2 and 3 do not coincide. This means that the predominance of am^+ transformants cannot be attributed to the fact that the average area of integration of denatured DNA is less than 50 sites (the beginning of the plateau). The experiment confirms that the mutation N6 falls into the pairing region but is later subjected to correction processes.

The second series of experiments comprised a clonal analysis of transformants from single bacterial spheroplasts, i. e. analysis of the particles descended from the original recombinant molecule. The experimental procedure has been described previously (Vinetskii and Aleksandrova, 1968).

Analysis of transformant genotypes in clones from single spheroplasts

Number of particles per clone	Pure clones, %		Mixed clones, %	Pure clones, %		Mixed clones, %
	am^+	am		am^+	am	
	Native DNA			Denatured DNA		
2	13	32	55	72	6	22
3	7	48	45	78	11	11
4	7	27	67	84	—	16
5	—	14	86	100	—	—
6	—	25	75	61	—	39
7	—	14	86	54	—	38
8	—	—	—	66	—	34
9	—	43	57	60	—	40
10 or more	—	13	83	65	—	35

Note. The small cross in Figure 2 indicates the distance between mutation N6 and the nearer margin of the deletion used in this experiment.

The results of the clonal analysis (see table) show that transformation with native DNA yields mostly mixed clones while in experiments involving denatured DNA mostly pure clones are obtained. In the latter case transformants with the am^+ genotype predominated. It follows that correction takes place before the first replication of the recombinant DNA begins and is directional, i. e. toward the recipient genotype.

The factors initiating the degradation of donor DNA remain unknown. Among the possible factors are local unpairedness at the mutation site and the presence of unpaired areas of single-stranded transforming DNA outside the overlapping region.

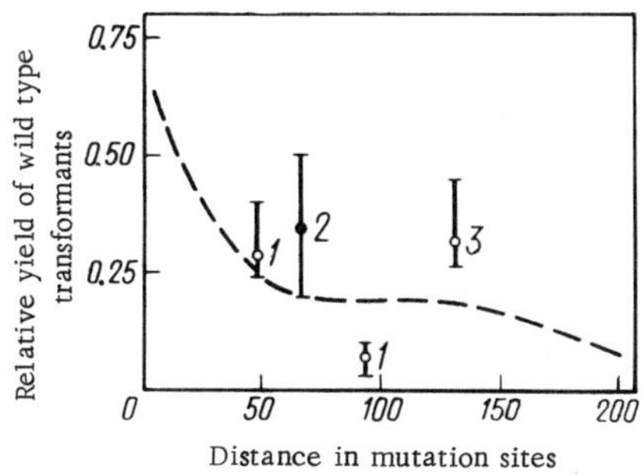

FIGURE 4. Proportion of wild type transformants as a function of the distance between the mutation and the deletion in the genome of the recipient phage.

Mutations in the recipient phages:
1) ts113 (B9-B10); 2) ts820 (A2h3); 3) amN6 (AZad).

To answer this question we performed experiments using denatured donor DNA from wild type phage and recipient bearing various deletions and the point mutation N6. In contrast to the preceding experiments, therefore, the mutation in the heterozygous area in this case was located in the recipient DNA molecule rather than in the donor DNA. If the mutational lesion alone is excised, the progeny should include wild type transformants; on the other hand, if the donor chain undergoes degradation regardless of the localization of the mutation in it, the recipient (mutant) genotype should predominate. The results obtained are given in Figure 4. Transformants with the mutant genotype obviously predominate in the progeny, which means that bases in the donor strand are corrected to the mutant recipient genotype.

The differences in the degree of correction between the sites of the above mutations are not large. At any rate they are smaller than the difference in the efficiency of transformation in bacteria between areas of enhanced and depressed integration, the nature of which is associated with correction processes (Gray and Ephrussi-Taylor, 1967). Analysis of transformation at a greater number of sites in phage T4 may reveal regions characterized by a more pronounced correction of heterozygotes.

BIBLIOGRAPHY

Bresler, S.E., R.A. Kreneva, and V.V. Kushev. — Molec. and Gen. Genet., Vol.102 (1968), 257.
Champe, S.P. and S. Benzer. — Molec. Biol., Vol. 4 (1962), 288.
Gray, T.S. and H. Ephrussi-Taylor. — Genetics, Vol. 57 (1967), 125.
Guerrini, F. and M.S. Fox. — Proc. Natn. Acad. Sci. U.S.A., Vol. 59 (1968), 429.
Vinetskii, Yu. P. and N. M. Aleksandrova. — Doklady AN SSSR, Vol. 183 (1968), 948.

UDC 575:576.8

THE COMPETENCE OF TRANSFORMATION SYSTEMS

K.I. MARKOV, M.S. LOBUTOVA and V.P. IVANOV
Higher Medical Institute, Sofia

A culture of cells in the competent stage was divided into two fractions by centrifugation in a urografin gradient. The light fraction is 100 times richer in competent cells than the heavy fraction. The nature of competence is discussed on the basis of certain physico-chemical and physiological differences between the microorganisms in the two fractions.

One of the important aspects of genetic transformation in microorganisms is the nature of competence — the peculiar brief physiological state of recipient cells which enables them to interact with the donor DNA molecule.

In bacterial populations grown under conditions promoting maximal competence, only a small percentage can be transformed for the donor marker (Young, 1967). The nature of competence cannot be studied without a method for separating competent from incompetent cells.

One of the methods for isolating competent cells is zone centrifugation in a density gradient (Singh and Pitalle, 1968; Cahn and Fox, 1968; Hadden and Nester, 1968). This divides the cells into two fractions with most of the competent cells being found in the light fraction.

The purpose of our work was to fractionate the competent cells of a bacterial population and determine their physico-chemical and physiological properties.

MATERIALS AND METHODS

The bacterial strains Bac. subtilis 168 and Staphylococcus aureus were used. To obtain competent cells, the recipient strain Bac. subtilis 168 str-s was grown as described by Anagnostopoulos and Spizizen (1961). Before or after fractionation, the competent culture was brought into contact for one hour at 37° with donor DNA bearing the str-r marker.

The staphylococci were grown in meat peptone broth with constant aeration for 12 hours and were transformed for the str-r marker.

DNA from a donor strain of Bac. subtilis was obtained according to Saito and Miura (1963), and DNA from the streptomycin-resistant staphylococcus by a method developed in our laboratory.

RESULTS

Initially we fractionated the cells in linear sucrose gradients with concentrations ranging from 4.5 to 18%.

This procedure did enable fractionation of the cells but the layers were very diffuse and the results somewhat inconsistent. We therefore changed to a stepped gradient of urografin 76, composed of three layers corresponding to concentrations of 50, 32, and 12.5%.

Centrifugation for 90 minutes at 6,000 rpm divided the cells into two clearly defined fractions situated at the borders between the layers (Figure 1). In the case of *Bac. subtilis*, the light fraction contained about 1/100 of the cells.

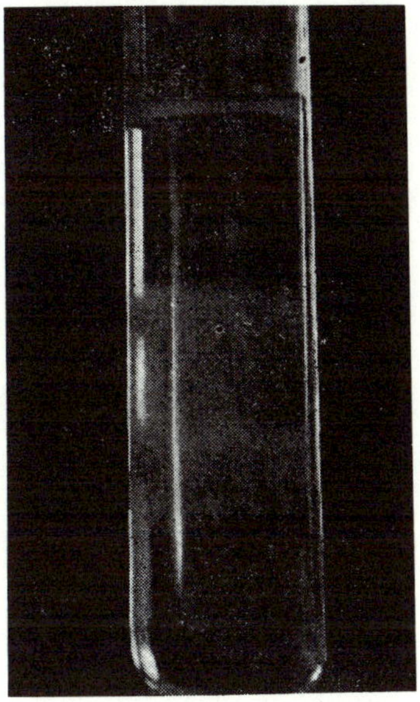

FIGURE 1. Fractionation in a urografin gradient.

Transformation experiments involving the transfer of the str-r marker showed that the upper, lighter fraction is 100 times richer in competent cells (transformation frequency $1.3 \cdot 10^{-3}$) than the lower, heavier fraction (frequency $1.1 \cdot 10^{-5}$).

Transformation of *S. aureus* yielded similar results.

Subsequently it was established that the resistance of the cells of the light fraction of the competent culture to streptomycin, penicillin and gentian violet is respectively 2, 5, and more than 10 times greater than that of the cells of the heavy fraction of the same culture (Figure 2).

It is known that one of the main indexes of the state of the surface of the bacterial cell is its electric potential. We therefore used three methods permitting the most comprehensive characterization of the physico-chemical properties of the surface of the cells in the light and heavy fractions.

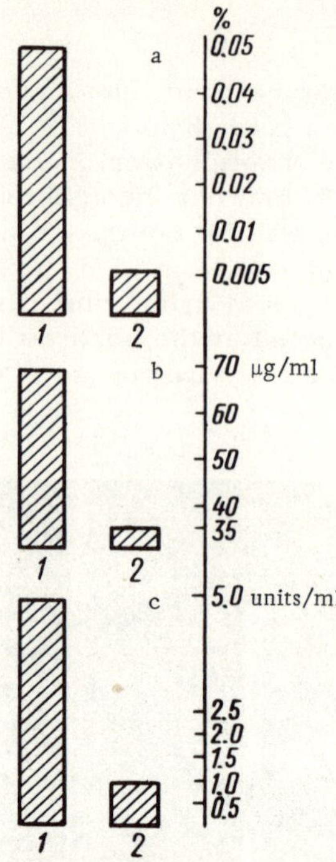

FIGURE 2. Resistance to antibiotics and dyes:

1) light fraction; 2) heavy fraction; a) gentian violet; b) streptomycin; c) penicillin.

Methylene blue adsorption provides direct data on the magnitude of the electric charge of the cell (Jenson and Haas, 1963). It proved that the bacteria in the light fraction of the competent culture of Bac. subtilis absorb the dye 22% less than the cells in the corresponding heavy fraction.

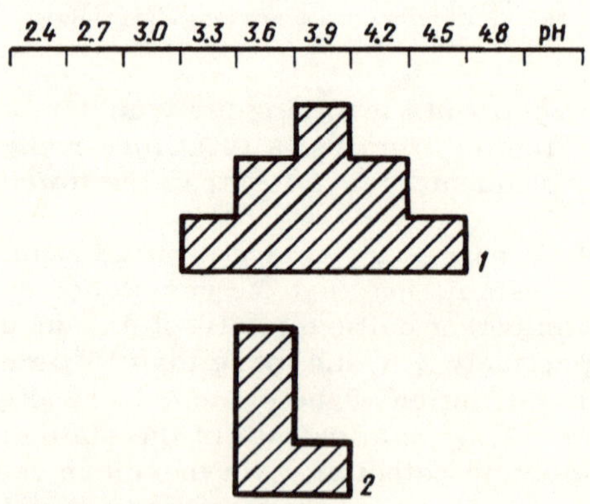

FIGURE 3. Acid agglutination:

1) light fraction; 2) heavy fraction.

Bacterial agglutination in buffers of different pH shows not only the isoelectric point of the cells but also the degree of their surface hydration.

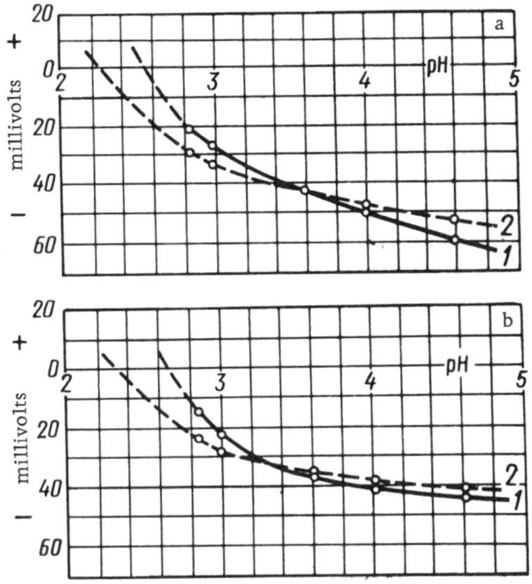

FIGURE 4. Microelectrophoretic mobility:

1) light fraction; 2) heavy fraction. a) *Bac. subtilis*; b) *S. aureus*.

We found that the cells of the heavy fraction agglutinate in a relatively narrow pH range with a peak at 3.6 while the bacteria of the light fraction agglutinate in a relatively wide pH range with a peak at 3.9 (Figure 3).

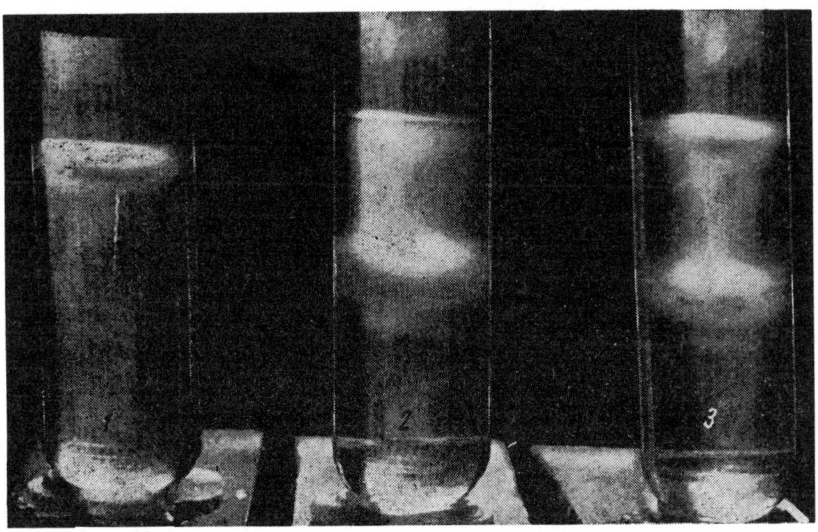

FIGURE 5. Fractionation of dividing and resting cells. Explanation in the text.

Marked differences were also revealed by comparative microelectrophoresis of cells of the two fractions of cultures of Bac. subtilis and S. aureus.

As shown in Figure 4, the electrophoretic curve of the bacteria in the light fraction falls more steeply and the isoelectric point is displaced towards alkalinity.

All these findings indicate that the cells in the light fraction of Bac. subtilis and S. aureus have a lower electric charge.

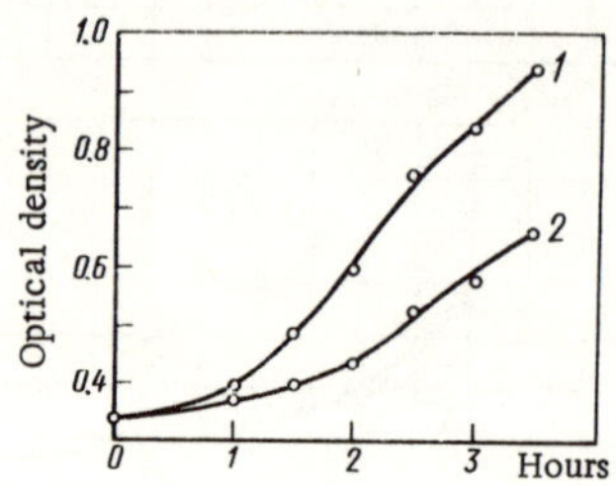

FIGURE 6. Growth curves of cells of the two fractions: 1) light fraction; 2) heavy fraction.

We examined the fractionation of S. aureus cultures at different growth phases in a urografin gradient. Cultures aged 2, 4, 6, 12 and 24 hours were used. The cells invariably split into two fractions, regardless of their age, and, consequently, of their competence (only the 12-hour cultures contained cells competent for transformation). Old cultures of Bac. subtilis and S. aureus (more than 48 hours old) yield only one heavy fraction on centrifugation in a urografin gradient.

Those observations prompted us to perform the following experiment. Three types of cultures were subjected to gradient centrifugation:

1) actively dividing, ^{32}P-labeled cells;
2) resting unlabeled cells;
3) a mixture of the two kinds of cells (Figure 5).

After centrifugation for 90 minutes the dividing cells became distributed in the light fraction and the resting cells in the heavy fraction (Figure 6). Determination of the radioactivity of the two fractions of gradient 3 confirmed this result. Transformation showed in turn that only the dividing cells are competent.

We also determined the growth curves of the cells of the two fractions obtained from a competent culture of Bac. subtilis. As shown in Figure 6, the cells of the light fraction begin to reproduce after a shorter lag phase and their growth curve rises more steeply than that of the cells corresponding to the heavy fraction. This constitutes further proof that the fraction of competent cells consists of actively dividing microorganisms.

Discussion

Our results show that the method of fractionating competent cells in a urografin density gradient enables the isolation of competent cells from the original culture, but only if the culture is grown under conditions promoting the development of competence.

A study of the properties of the purified competent bacteria showed that they possess a higher resistance to antibiotics and dyes and carry a smaller electric charge than the noncompetent cells. Although it is widely believed that bacteria capable of taking up DNA are in the resting state, our results as regards the growth kinetics of the cells from the two fractions of the urografin gradient show that the fraction containing the competent cells consists of actively dividing microorganisms. Thus the concept of the resting state of competent cells should be subjected to further verification.

BIBLIOGRAPHY

Anagnostopoulus, C. and J. Spizizen. — J. Bact., Vol. 81 (1961), 741.
Cahn, F.H. and M.S. Fox. — J. Bact., Vol. 95 (1968), 867.
Hadden, C. and E.W. Nester. — J. Bact., Vol. 95 (1968), 876.
Jenson, K. A. and F. E. Haas. — J. Bact., Vol. 89 (1963), 73.
Saito, H. and K. Miura. — Biochim. biophys. Acta, Vol. 72 (1963), 619.
Singh, R.H. and M. P. Pitalle. — J. Bact., Vol. 95 (1968), 864.
Young, F.E. — Nature, Vol. 213 (1967), 773.

UDC 575:576.858.9

THE BIOLOGICAL ACTIVITY OF PHAGE T4B DNA DURING PHAGE TRANSFORMATION

L.M. FONSHTEIN and G.G. MNATSAKANYAN
Institute of General Genetics
Academy of Sciences of the USSR, Moscow

Complementation tests in a phage transformation system containing lysozyme spheroplasts of E. coli demonstrated the potentially functional state of genes r and 32 of phage T4 donor DNA not integrated with the recipient genome. Both native and denatured DNA showed transforming activity in the system examined, although the functional activity was much more pronounced in the case of native DNA. The transformation system of phage T4 is proposed as a convenient model for the study of the function of genes in isolated DNA of this phage.

Many studies of the mechanisms of genetic recombination and the activity of isolated nucleic acids have been carried out in transformation and transfection systems. These systems have attracted attention because they use isolated nucleic acids as donors or infective agents. For a long time this approach was inapplicable to phage T4 — one of the most thoroughly studied objects from a genetic point of view — because no activity of DNA isolated from this phage was known. This state of affairs changed with the publication of Van de Pol and his colleagues (1961), describing transformation in phage T4.

The biological activity of phage T4 DNA can be assessed in a phage transformation system. The DNA of this phage is not infectious, i. e. it cannot bring about the production of infectious phage particles. Does this mean that the genes of the donor transforming DNA can only function after the integration of the DNA with the genome of the recipient phage? The present paper discusses the possibility of the expression of genes of phage T4 donor DNA not integrated with the recipient genome.

Proof of the functional state of genes of isolated DNA of phage T4 can be found in the complementation effect obtained on transformation. In this system, the existence of a complementation effect can only be determined by the increase in the yield of transformants and not by the harvest of recipient phage. In phage transformation the recipient phage infects about 10% of the lysozyme spheroplasts, whereas the DNA penetrates into approximately 1% of the spheroplasts (Vinetskii et al., 1969). Consequently, only about 0.1% of the spheroplasts are infected simultaneously with recipient phage and with donor DNA. Transformants naturally arise only within this particular pool of spheroplasts, which is smaller by about two orders of magnitude than the number of spheroplasts infected with recipient phage alone and capable of giving some harvest of this phage.

It is obvious on these grounds that the resolving capacity of the analysis can be improved considerably by using the yield of transformants, rather than the harvest of recipient phage, as the criterion of complementation.

The experiment in which we demonstrated the existence of a complementation effect in phage transformation can be outlined as follows. Donor DNA was isolated from phage T4 rIIA1368, while factors of the π-phages T4 rIIA386 and T4 rIIB638 were used as the recipients. The deletions in the rII region of the recipient genome were roughly equidistant from the deletion in the donor DNA genome (Figure 1). Hence a roughly equal yield of transformants might be expected from the following transformation variants: rIIA1368 DNA – rIIA386 recipient (variation 1), and rIIA1368 DNA – rIIB638 recipient (variation 2). With the use of lysozyme spheroplasts of Escherichia coli K12 (S), which permit the reproduction of the recipient phage, the yields of transformants in the above variants were very close, the ratio between them being 1.2:1 for both native and denatured DNA (Table 1). However, the transforming activity of denatured DNA was greater in both variants than that of native DNA. A different picture was obtained when the transformation was carried out in E. coli K12 (λ) spheroplasts, which are nonpermissive with respect to reproduction of the recipient phage. The yield of transformants was much greater when the deletions in the genomes of the donor DNA and the recipient phage were located in different cistrons. The increase in the yield of transformants in variation 2 over that in variation 1 was much greater in the case of native DNA, although it was also evident when denatured DNA was used.

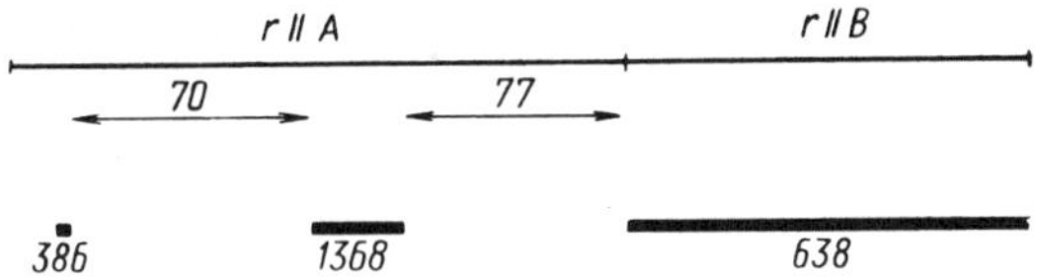

FIGURE 1. Location of the deletions in the genomes of the donor DNA and the recipients. Distances expressed in mutation sites.

It is evident from the results obtained that the yield of r^+ transformants in the transformation of rII mutants of phage T4 in K12 (λ) spheroplasts depends on the possibility of complementational interactions leading to the formation of the product of gene rII. We postulated on the basis of this conclusion that the effect of agents inducing functional damage in DNA on the yield of transformants should be more pronounced under nonpermissive conditions. The part of the phage T4 genome which is necessary for the expression of gene rII is much larger than the rII region itself (Krieg, 1959). Moreover, since fragments comparable in size to the rII gene are incorporated into the recipient genome during transformation of phage T4 (Vinetskii et al., 1969), it might be expected that UV irradiation of the phage donor DNA, resulting in damage to the rII function, should be more harmful in the case of transformation under nonpermissive conditions, where this function is essential.

TABLE 1. Transformation of phages T4 rIIA386 and T4 rIIB638 with phage T4 rIIA1368 DNA in lysozyme spheroplasts of E. coli K12 (S) and K12 (λ)*

Spheroplasts	Concentration of DNA, μg/ml	Ratio of transformant yields, variation 2/ variation 1, for DNA**		Activity of denatured DNA in comparison with native DNA, %**	
		native	denatured	variation 1	variation 2
K 12 (S)	1.0	1.2	1.2	199.6	182.9
K 12 (λ)	1.0	135.3	26.7	82.0	20.6
	0.3	420.7	8.7	292.5	9.0

* The procedure of phage T4 transformation was described earlier (Gol'dfarb et al., 1966; Fonshtein, 1970).
** Average data of 4–8 experiments.

As shown in Figure 2, the curve of UV suppression of the transforming activity of DNA in the transformation of phage rIIA386 in K12 (λ) spheroplasts differs from the curve obtained in the case of transformation of the same recipient in K12 (S) spheroplasts. The same dose of UV radiation results in different levels of residual activity in the irradiated DNA, depending on whether the complementation process was necessary to yield transformants. In K12 (λ) spheroplasts a much sharper decrease in transforming activity is observed in relation to the dose of radiation applied to the donor DNA. It appears that the curve of UV-induced suppression of the transforming activity of DNA in the transformation of rII mutants of phage T4 in K12 (λ) spheroplasts reflects not only processes associated with damage to the rII function of the donor DNA but also events involved in the incorporation of fragments of the irradiated DNA into the genome of the recipient phage.

The data obtained on the UV-induced suppression of the transforming activity of DNA agree with those concerning the effect of degradation on the transforming activity of DNA. Table 2 shows average values (from 5 experiments) of the residual activity of degraded DNA preparations in the transformation of recipients with genomes containing deletions of roughly equal length. Under conditions permissive to the recipient phage, the transforming activity of degraded preparations of donor DNA is roughly comparable to that of the original preparation, whereas in the case of transformation under nonpermissive conditions the fragmentation of DNA is accompanied by a sharp decrease in its transforming activity.

Summing up the results obtained, it may be concluded that the yield of r^+ transformants in K12 (λ) spheroplasts is derived largely from complementation and only to a minor extent from other processes. Two possibilities must be considered here: the expression of gene rII of the donor DNA prior to its integration with the genome of the recipient phage, and of transcription from an intermediate recombinant structure. Our own data together with those obtained by Bautz (1966) support the first possibility.

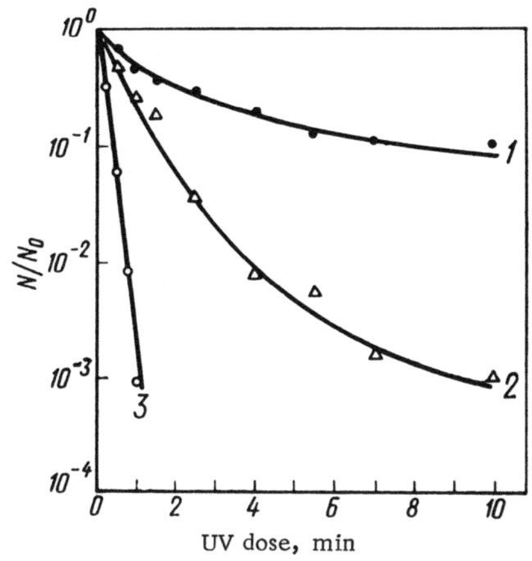

FIGURE 2. Activity of UV-irradiated DNA of phage T4 r^+ in transformation of phage T4 rII386:

N_0—transforming activity of nonirradiated DNA; N — transforming activity of UV-irradiated DNA; 1) K 12 (S) spheroplasts; 2) K 12 (λ) spheroplasts; 3) UV inactivation of free phage T4rII386.

From the data presented in Table 1 and those published in the literature (Goldberg, 1966; Gol'dfarb et al., 1966; Gol'dfarb and Sergeeva, 1968; Fonshtein et al., 1968) it follows that denatured DNA displays a high recombinational activity in this system and is therefore readily integrated with the recipient genome. At the same time the complementation effect (see Table 1) was much more pronounced in the case of native DNA. In order to obtain further proof of the likelihood of the expression of genes of the donor DNA prior to its integration with the recipient genome, we performed experiments in which the recipient was phage T4 amA453 with an amber mutation in gene 32, the product of which is essential for recombination (Tomizawa et al., 1966; Kozinski and Felgenhauer, 1967; Albertz and Frey, 1970).

TABLE 2. Transformation of phages T4 rIIA386 and T4 rIIB1299 with degraded DNA of phage T4 r^+ in E.coli K12 (S) and K12 (λ) spheroplasts (% of the original value)

Molecular weight of the DNA preparation	K12 (S) spheroplasts		K 12 (λ) spheroplasts	
	rIIA386	rIIB1299	rIIA386	rIIB1299
$2.0 \cdot 10^7$	110	120	85	81
$6.7 \cdot 10^6$	100	130	28	64
$4.2 \cdot 10^6$	110	93	6	17

The experiment was set up as follows. Our double mutant with the deletion rIIB638 and the amber mutation amA453 (recipient No. 81) was transformed in E. coli B spheroplasts with DNA of the phages T4r$^+$ amA453$^+$ (Figure 3a) and T4r$^+$ amA453 (Figure 3b). In this way a complementation test for the product of gene 32 was set up in the phage transformation system. r$^+$ transformants were selected on a lawn of E. coli CR63 (λ). The am$^+$ transformants could not be determined since the level of reversions of the amA453 mutations was relatively high for a transformation system. Recombinants could only appear as a result of the transcription of gene 32 from the donor DNA. Given that both the donor DNA and the recipient phage had a mutation at gene 32 (see Figure 3b), r$^+$ transformants could not arise. The formation of such recombinants, however, was possible if the donor DNA was wild type with respect to gene 32 (see Figure 3a). The above mentioned types of DNA also transformed the recipient rIIB638, amA453$^+$, which is wild type with respect to gene 32. In this case, transformants must obviously arise in all the transformation variants. Preliminary experiments showed that in the case of transformation of the recipients No. 81 and rIIB638 in E. coli CR63 spheroplasts, which suppress the amber effect in the recipient genome, both the DNA preparations used in the experiment (T4r$^+$ amA453$^+$ and T4r$^+$ amA453) possessed transforming activity.

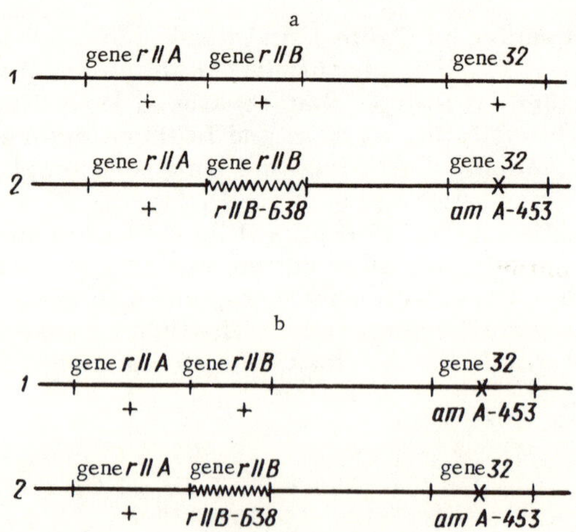

FIGURE 3. Scheme of the complementation test for the product of gene 32 in lysozyme spheroplasts of E. coli B.

Complementation effect with respect to the product of gene 32: a) possible, b) impossible; 1) donor, 2) recipient.

The results obtained (Table 3) were as expected. The data presented in Table 3 can be interpreted as follows. The use of DNA wild type for gene 32 leads to complementation with respect to the product of this gene

TABLE 3. Transformation of phages T4 rIIB638 and T4 rIIB638 amA453 with the DNA of phages T4 r^+amA453$^+$ and T4 r^+amA453

Components of transformation system		State of secondary structure of donor DNA	Number of r^+ transformants per ml in the experiments*							Activity of denatured DNA as percentage of that of native DNA
recipient phage	donor DNA from phages		1	2	3	4	5	6	7	
T4 rIIB638	T4 r^+ amA453$^+$	Native	209	152	1,410	1,225	56	198	665	215.3
		Denatured	297	262	1,360	1,825	337	328	1,196	
	T4 r^+ amA453	Native	—	—	—	—	37	86	345	292.5
		Denatured	—	—	—	—	151	192	849	
T4 rIIB638 amA453 (No. 81)	T4 r^+ amA453	Native	31	58	561	183	8	17	7	1.3
		Denatured	0.7	0.3	4	2	0	0	0.3	
	T4 r^+ amA453	Native	—	—	—	—	0	0	0	—
		Denatured	—	—	—	—	0	0	0	

* Average of 3 replicates.

and consequently to the formation of r^+ recombinants. Now since the amA453 mutation prevents the integration process and the formation of recombinants, it is evident that gene 32 of the donor DNA was expressed prior to the integration of the latter with the genome of the recipient phage.

Of particular interest is the ratio between the transforming activities of native and denatured DNA in the transformation of the recipients used in the experiment. The results of the transformation of phage rIIB638 demonstrate once again the high recombinational activity of denatured DNA in the transformation system studied. In the case of recipient No. 81, when the formation of recombinants required the functioning of gene 32 of the donor DNA, the transforming activity of denatured DNA was much lower than that of the native preparation. This means that the native DNA is functionally active in the system examined. The residual functional activity of denatured DNA cannot be explained at present.

The data presented in this communication indicate the possibility of the expression of genes rII and 32 of donor DNA which is probably not integrated with the recipient genome. The transformation system of phage T4 may be considered a convenient model for further research into the possibility and nature of the autonomous transcription of genes of isolated donor DNA.

BIBLIOGRAPHY

Alberts, B.M. and L. Frey. — Nature, Vol. 39 (1970), 1313.

Bautz, F.A. — Genetics 3, No. 5 (1966), 913.

Fonshtein, L.M. — Genetika 6, No.5 (1970), 94.

Fonshtein, L.M., S.A. Sergeeva, and D.M. Gol'dfarb. — Genetika 4, No. 10 (1968), 112.

Goldberg, E. B. — Proc. Natn. Acad. Sci. U.S.A. 56, No.5 (1966), 1457.

Gol'dfarb, D.M., A.V. Avdeeva, S.V. Blinova, S.A. Sergeeva, and G.A. Levina. — Genetika, Vol. 7 (1966), 148.

Gol'dfarb, D.M. and S.A. Sergeeva. — Genetika, 4, No. 1 (1968), 100.

Kozinski, A.W. and Z.Z. Felgenhauer. — J. Virol. 1, No. 6 (1967), 1193.

Krieg, D. — Virology, Vol. 8 (1959), 80.

Tomizawa, J., N. Anraku, and Y. Iwana. — J. Molec. Biol., Vol. 21 (1966), 247.

Van de Pol, J., G. Veldhuisen, and J. Cohen. — Biochim. biophys. Acta 48, No. 2 (1961), 417.

Vinetskii, Yu. P., N.M. Aleksandrova, and I.D. Avdienko. — Genetika 5, No. 5 (1969), 148.

UDC 575.24:576.8

ULTRAVIOLET-INDUCED MUTATIONS AND RECOMBINATIONS IN BACTERIA

EVELYN M. WITKIN
New York University

The relationships between UV-induced mutagenesis and recombinations in bacteria are discussed. According to the hypothesis proposed, UV-induced mutations are mostly associated with primary DNA lesions which persist in the course of replication. It is shown that the recA system does not mutate on UV irradiation. The appearance of mutations requires not only the product of gene $recA^{+}$ but also that of gene $exrA^{+}$. It is concluded that UV-induced mutations arise as a result of a recombination event which also involves the product of gene exr. Therefore, mutations at this locus affect both the UV-induced mutagenesis and the frequency of recombination.

As a rule, the appearance of a chemical lesion in DNA under the influence of a mutagen is merely the initial stage of a complex chain of processes which may bring about a stable mutational change. It is known that the biological effect of ultraviolet irradiation is associated with the dimerization of adjacent pryimidine bases situated in the same DNA strand. The presence of a pyrimidine dimer in replicating DNA disrupts the continuity of replication but does not prevent the resumption of replication at some point on the DNA beyond the dimer. As a result, the daughter DNA strands are not continuous (Rupp and Howard-Flanders, 1968); they contain gaps opposite each pyrimidine dimer in the template strand. These gaps (about 10^3 nucleotides long) represent secondary UV-induced lesions arising from the replication of irradiated DNA containing unrepaired primary lesions.

Cells of wild-type Escherichia coli strains can survive UV doses inducing thousands of pyrimidine dimers per chromosome. This high resistance to UV is due to the existence of two distinct systems of dark repair. The first system, excision repair, acts upon primary UV-induced lesions prior to replication. The second system, recombinational repair, operates at the level of secondary UV-induced lesions (i. e. gaps in the daughter strands) after replication of the DNA. At low UV doses, all the pyrimidine dimers formed are eliminated from the DNA by the excision system and secondary lesions do not appear. Such doses of UV cause neither lethal nor mutagenic effects. At higher doses, although 95—99% of the pyrimidine dimers are eliminated from DNA before replication, the number of primary lesions remaining at the start of replication may be quite large, and each of these lesions causes a gap in a daughter DNA strand. It was found that all the gaps must be repaired if the cell is to survive. Consequently, the resistance of any strain to unexcised pyrimidine dimers depends on the efficiency of its system of recombinational repair.

Thus, rec^+ strains, which possess a normal capacity for recombination, can repair 100—200 gaps and therefore can survive if at the start of replication their DNA contains 100—200 unexcised dimers or other primary lesions blocking replication.

The mechanism of excision repair is by now widely known and will therefore be outlined here only briefly. It comprises the following stages: 1) cleavage of the DNA strand on one side of the dimer by a UV-specific endonuclease; 2) excision of a short single-stranded fragment by a second endonuclease, cleaving the strand on the other side of the dimer; 3) widening of the gap formed as a result of the degradation of the DNA by exonuclease; 4) repair synthesis by DNA polymerase, using as the template the sector of the opposite DNA strand corresponding to the incomplete segment to be repaired; 5) restoration of the continuity of the DNA by ligase, which closes the single-strand break. Stages 2, 3 and 4 may be accomplished by a single enzyme (Kelli et al., 1969).

The mechanism of recombinational repair remains largely unknown (see, for example, Howard-Flanders, 1968). It seems to involve the formation of a hetero-duplex structure between two daughter molecules wherever one of them has a gap. Thus the intact section of one sister molecule provides the information which is missing in the other molecule, and repair is probably brought about by a two-strand exchange. In contrast to the gaps which appear in the course of excision repair, the gaps in the daughter strands opposite unexcised dimers cannot be repaired by the system of repair synthesis of DNA since the DNA lesion which originally prevented normal replication apparently also blocks the action of DNA polymerase.

The excision repair system of E. coli can be impaired by mutations at the loci uvrA, uvrB and uvrC; damage to any of these genes leads to an inability to bring about a primary break at the DNA lesion. In rec^+ uvr mutants the system of recombinational repair operates normally and the tolerance of the cell towards dimers not excised by the beginning of replication therefore remains unchanged. In this case, however, much lower doses of UV are sufficient to create the maximum acceptable number of dimers since not one dimer can be eliminated from the DNA by the excision system. Thus, uvr^+ and uvr^- strains differ only in their capacity for eliminating primary UV-induced lesions from DNA.

Mutations at the recA locus completely destroy the capacity for genetic recombination, including recombinational repair, but do not affect the excision repair system. Double mutants with the uvr^- recA genotype can repair neither primary nor secondary lesions and UV doses inducing a single pyrimidine dimer in each DNA strand are sufficient to inactivate such strains (Howard-Flanders, 1968). These data indicate that E. coli cells do not possess any system for the repair of pyrimidine dimers and the resulting secondary lesions other than the two mechanisms mentioned above.

For a number of years, our laboratory has been engaged in a study of the role of repair systems in UV-induced mutagenesis. The comparison of mutagenesis in bacterial strains isogenic in every respect except their sensitivity to UV proved especially fruitful. A comparison of four such strains with the wild type bacterium is given in the following table.

These results have been published in detail (for references see Witkin, 1969b) and provide the basis for our model of UV mutagenesis. Before presenting the new data, we should like to discuss these data and some of the conclusions arising from them.

Effect of different mutations of E. coli on sensitivity to UV irradiation, capacity for repair and level of UV mutagenesis (Witkin, 1969b)

Genotype	UV sensitivity	Excision repair	Recombinational repair	UV mutagenesis
+	1.0	1.0	1.0	1.0
uvr⁻ (ABC)	10 – 20	0	1.0	10–50
recA	100	1.0	0	0
lon	10	1.0	1.0	1.0
exrA (lex)	2 – 3	1.0	0.3–0.5	0

It is evident from the table that uvr⁻ mutations, which impair the excision repair system, increase the UV sensitivity and UV-induced mutability to approximately the same degree. First to arrive at this conclusion was Hill (1965), who studied the induction of suppressor mutations for the ochre codon in one of the uvr⁻ strains; his data were later confirmed in various mutation systems in many other uvr⁻ strains (Witkin, 1966, 1967; Bridges and Munson, 1966; Kondo and Kato, 1966). These findings indicate that excision repair is an extremely accurate process; according to our own data, the excision system can repair at least 10^6 dimers without committing a single error in the process (Witkin, 1969b). Errors of the excision system, if they occur at all, clearly contribute at most a negligible fraction of UV-induced mutations.

On the other hand, unexcised dimers are highly mutagenic. According to our calculations, the probability that such mutations will be manifested is approximately 10^{-3} per unexcised dimer.

On the basis of these results, it appears that UV-induced mutations arise mainly from unrepaired primary lesions persisting in the DNA through at least one replication. The simplest mechanism, according to which the pyrimidine dimer is responsible for the replication error, must be ruled out given the convincing evidence that pyrimidine dimers block replication. This means that UV-induced mutations probably results from errors during the repair gaps in the daughter strands. For this reason we examined the recombinational repair system as a possible source of errors in DNA.

One of the most interesting mutants studied by us, exrA, is sensitive to UV and X-ray irradiation but shows a rather high level of residual resistance to UV. The UV sensitivity of this strain corresponds to one third of the wild type level. Its capacity for recombination is likewise reduced to some extent (to 30–50% of the norm), which suggests that the UV sensitivity may be due to disturbance of the recombinational repair system. The most striking characteristic of this strain is that UV-induced mutations are totally lacking (Witkin, 1967). Mutations induced by X rays and thymine starvation are similarly absent (Bridges et al., 1968), but spontaneous mutations and those induced by nitrogen base analogs remain at the normal level.

We concluded on the basis of these findings that gaps in the daughter strands can be repaired by two mechanisms: 1) an error-prone repair system which requires the product of the exr^+ gene; 2) a less efficient but faultless repair system operating in exr^- strains. It is possible that both types of post-replicational repair include recombination and that the exr^- mutants lack the stage generating errors. On the other hand, another possibility is that some gaps in exr^+ strains are repaired by an error-generating mechanism which does not require recombination, for example, by the incorporation of random bases into the gap by the action of terminal nucleotidyltransferase. To solve this problem it is necessary to study the UV mutability of recA strains, which are totally incapable of recombination. If some gaps in daughter strands can indeed be repaired by a mechanism independent of recombination, it follows that recA cells surviving irradiation must contain UV-induced mutations. In fact recA strains do not mutate at all when irradiated with UV (Witkin, 1969a). Consequently, the products of genes $recA^+$ and $exrA^+$ are equally necessary for the occurrence of UV-induced mutations.

The table also contains data on the lon mutations, which increase the sensitivity to UV but do not affect the repair of DNA lesions. This example shows that not all UV-sensitive strains become more prone to mutation under the influence of UV; lon^+ and lon^- strains have the same mutation frequency at a given UV dose although their survival rates differ considerably (Witkin, 1967). This result would be expected in the case of strains in which UV sensitivity arises from a lesion in an extrachromosomal system unrelated to the DNA repair — in lon mutants the system of septum formation.

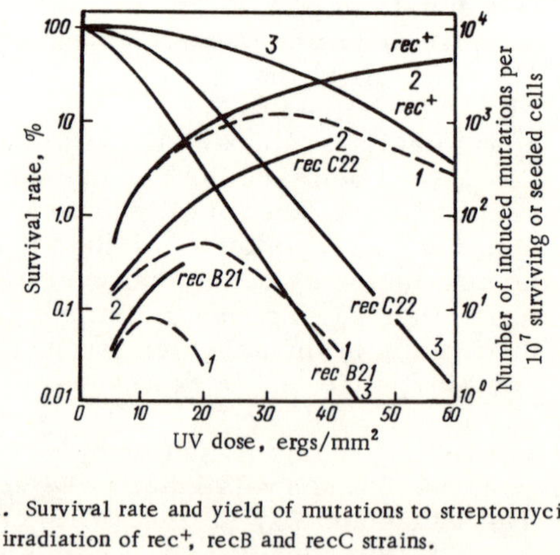

FIGURE 1. Survival rate and yield of mutations to streptomycin resistance after UV irradiation of rec^+, recB and recC strains.

Frequency of induced mutations: 1) per 10^7 seeded bacteria; 2) per 10^7 surviving cells; 3) survival rate.

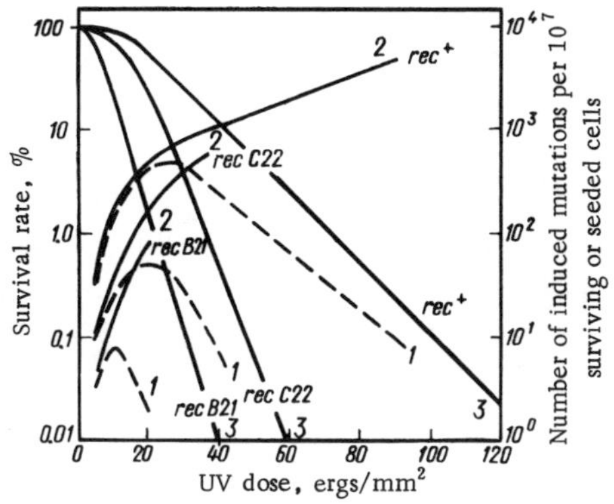

FIGURE 2. Survival rate and yield of mutations to tryptophan prototrophy (mainly by ochre suppressors) after UV irradiation of rec$^+$, recB$^-$ and recC strains.

Frequency of induced mutations: 1) per 10^7 seeded bacteria; 2) per 10^7 surviving cells; 3) survival rate.

We conjectured on the basis of these observations that UV-induced mutations usually arise as a result of errors in the recombinational repair of secondary UV-induced lesions in the DNA, i. e. gaps in the daughter strands opposite unexcised primary lesions. According to this hypothesis recB and recC mutants, characterized by a lower level of recombination, should display a lower UV-induced mutability than rec$^+$ strains. We therefore examined the UV-induced mutability of several recB and recC strains and found that these strains show a much lower frequency of induced mutations per surviving cell in comparison with the rec$^+$ strains. This agrees with earlier data (Witkin, 1969a). The results of such a study of three strains are given below. E. coli WP2S try uvrA, derived from E. coli B/r, was used as the rec$^+$ strain. The recB and recC strains were courteously placed at our disposal by Dr. Mount; these strains were obtained by transferring recB21 or recC22 alleles from K12 donors into the thymine-dependent strain WP2S, obtained from Dr. Bridges. The rec$^-$ markers were transferred by transduction with phage P1 and UV-sensitive thy$^+$ recombinants selected. Two such transductants were used in the experiments — DM203, carrying the recB21 mutation, and DM207, possessing the recC22 allele. Both strains have the same try uvrA markers as the parental rec$^+$ strain and are largely isogenic with this.

Figure 1 shows the results of introducing UV-induced mutations to streptoymcyin resistance into rec$^+$, recB and recC strains. The survival curves demonstrate the greater UV sensitivity of the recB and recC strains, resulting from their lowered capacity for recombinational repair. The mutation frequency of the recB and recC strains is obviously much lower than in the rec$^+$ strain with the same dose of radiation.

Figure 2 shows the results of comparable tests involving the induction of suppressor mutations for the ochre codon in the same strains. It may be seen that the decrease in the mutation frequency generally coincides with the increase in sensitivity to UV.

The decline of the mutation frequency per surviving cell in the recB and recC strains can be explained as follows. Mutations in genes recB and recC lower the efficiency of recombinational repair, thus decreasing the number of gaps repaired in the daughter strands of the DNA. As a result, the cell has a lower tolerance to unexcised dimers and the average number of dimers unexcised by the beginning of replication per surviving cell is lower in the recB and recC strains than in the rec^+ strain. When we compare the rec^+ cells with the recB and recC cells which have survived equivalent doses of UV, therefore, we are dealing with populations in which the average number of dimers per surviving cell differs markedly. If the populations compared differ in their tolerance to unexcised dimers, the mutation frequency per surviving cell may vary considerably even if the mutation frequency per pyrimidine dimer in each surviving cell is exactly the same. Since our recB and recC strains also carried the uvr marker and were therefore unable to excise dimers from DNA, we could calculate the number of unexcised dimers formed in a cell by any particular radiation dose, since it is known that a UV dose of 1 erg/mm^2 induces about 6 pyrimidine dimers in each *E. coli* genome. If we assume that the increased UV sensitivity of the recB and recC strains is due to a decrease in the tolerance of the cell towards unexcised pyrimidine dimers, the ratio between the mean lethal radiation dose (permitting 37.5% survival) of the recB and recC strains and the mean lethal dose for the rec^+ strain will be an index of the decrease in the tolerance of the rec^- cells toward unexcised dimers.

At the same time, the ratio between the number of mutations per surviving cell of the recB or recC strain and the number of mutations in the rec^+ strain at each given dose of UV radiation can be regarded as an index of the decrease in mutability. If these two ratios are equal, the probability of mutation per unexcised dimer should remain unchanged. It is evident from Figures 1 and 2 that these two values, though not exactly equal, are quite close to one another. For example, the lethality index for the $recC/rec^+$ pair of strains is 0.55 while the mutability index (for the mutation from try^- to try^+) is 0.39; for the recB strain, the lethality index is 0.29 and the mutability index for the induction of suppressor mutations is 0.25. Since the capacity for recombination can be changed to a varying degree along the genome, a large number of different mutation systems must be examined in order to determine the true value of the mutability index. Our limited data nevertheless imply that the recB and recC strains have a lower mutability than the rec^+ strain solely as a result of the lower tolerance of the cell toward unexcised pyrimidine dimers and not as a result of any change in the frequency of errors during recombinational repair.

These results confirm the hypothesis that the appearance of UV-induced mutations in cells surviving the irradiation requires a recombinational event. As regards the role of the exr^+ gene product in this process, two mechanisms are possible. According to one mechanism, the product of gene exr cannot itself repair the gap in the daughter strand; its action consists in creating errors in the daughter strand independently of recombination, e. g. by the random inclusion of a base in the gap. The subsequent repair of the gap necessitates recombination. In the absence of this the lesion becomes lethal and the potential mutation is not manifested. According to the second mechanism, which agrees more closely with the evidence presented, the product of gene exr participates directly in the recombination process and ensures the mutagenicity of the latter. exrA mutations only slightly reduce the efficiency of the recombination mechanism but deprive it of the ability to produce mutations. Since the main role of recombination is to ensure variability, a mutagenic recombinational mechanism can be regarded as advantageous from an evolutionary viewpoint; the mutagenicity of meiotic recombination has been clearly demonstrated in yeasts (Magni, 1963).

The genetic organization of E. coli evidently provides a choice between accuracy and efficiency in recombinational repair. In the course of natural selection efficiency has gained the upper hand. Whenever such a system ceases to be advantageous, a single mutation can shift the equilibrium towards the more accurate system.

This article is dedicated to Professor T. Dobzhansky in honor of his 70th birthday.

BIBLIOGRAPHY

Bridges, B.A., J. Law, and R.J. Munson. — Molec. and Gen. Genet., Vol. 103 (1968), 266.
Bridges, B.A. and R. J. Munson. — Biochem. Biophys. Res. Commun., Vol. 22 (1966), 268.
Hill, R.F. — Photochem. Photobiol., Vol. 4 (1965), 563.
Howard-Flanders, P. — A. Rev. Biochem., Vol. 37 (1968), 175.
Kelli, R., M. Atkinson, J. Huberman, and A. Kornberg. — Nature, Vol. 224 (1969), 495.
Kondo, S. and T. Kato. — Photochem. Photobiol., Vol. 5 (1966), 827.
Magni, G.E. — Proc. Natn. Acad. Sci. U.S.A., Vol. 50 (1963), 975.
Rupp, W.D. and P. Howard-Flanders. — J. Molec. Biol., Vol. 31 (1968), 291.
Witkin, E.M. — Science, Vol. 152 (1966), 1345.
Witkin, E.M. — Brookhaven Sympos. Biol., Vol. 20 (1967), 17.
Witkin, E.M. — Mutation Res., Vol. 8 (1969a), 9.
Witkin, E.M. — A. Rev. Genet., Vol. 3 (1969b), 525.

UDC 575:576.8

ESCHERICHIA COLI K12 MUTATIONS LEADING TO NONINDUCIBILITY OF LAMBDA PROPHAGE

R. DEVORET, M. BLANCO and A. BAILLONET
Laboratoire de radiobiologie, Centre des Faibles Radioactivités, C.N.R.S.
Laboratoire de radiobiologie, Institut National des Sciences et Téchniques Nucléaires, Gif-sur-Yvette, France

A method is proposed for the enrichment of lysogenic bacterial cultures with strains in which prophage induction is impossible. Mutants defective as regards prophage induction have been isolated. All these mutations are pleiotropic and affect either the repair or the replication of DNA or both processes.

INTRODUCTION

Wild type E. coli K12 cells contain λ prophage which is incorporated in the bacterial chromosome. If such cells are exposed to inducing agents such as UV or X rays, carcinogens, etc., or are subjected to thymine starvation, the λ prophage passes into a vegetative state. Under such circumstances the cells lyse and release numerous infective virus particles. This phenomenon, discovered by Lwoff and his colleagues (1950), was named lysogenic induction.

According to the theory of Jacob and Monod (1961), lysogenic induction results from the action of the inducer on a repressor, which in the case of lambda phage is the product of the viral gene C1. The inducer either stops the synthesis of the repressor or unites with this and so inactivates it.

Further investigations showed that thermal inactivation of the repressor leads to loss of immunity and triggers the vegetative development of the prophage (Sussman and Jacob, 1962; Lieb, 1966). The phage repressor was isolated not long ago (Ptashne, 1967). However, all attempts to identify the inducer failed.

The mechanism of prophage induction will be better understood once the inducer or at least the bacterial genes responsible for its formation in the cell can be identified.

In order to determine the role of the host genes in lysogenic induction it is necessary to isolate bacterial mutants in which different stages of this process are disrupted. We therefore developed a procedure for the selective enrichment of cultures with such mutants based on the fact that thymine starvation is lethal to differing degrees in the case of inducible and noninducible lysogenic cells. The efficiency of this method is quite high (Devoret and Blanco, 1970); the mutants isolated by us included not only the expected ones for gene recA (Brooks and Clark, 1967) but also three new types of bacterial mutants in which the induction of wild type lambda prophage is defective. One of these mutations closely resembles in its properties the exrA mutation, in which prophage induction has also been shown to be impossible (Donch et al., 1970).

We describe here the principle of the selection method and some properties of the mutants obtained.

RESULTS

Principle of the method for enriching cultures with nonindubible lysogenic cells

For the isolation of noninducible bacterial mutants we took advantage of the fact that the lethal effect of thymine starvation is much more rapid in the case of lysogenic thy$^-$ (λ) cells than in nonlysogenic cells or cells bearing the noninducible prophage λ ind$^-$ (Sicard and Devoret, 1962). In other words, noninducible bacteria are more resistant to thymine deficiency than cells in which prophage induction is possible. These observations prompted the idea that thymine starvation could be used as a selective method for enriching populations of inducible thy$^-$ bacteria with noninducible mutants. The proportion of such mutants will increase as the duration of thymine starvation is extended.

Determination of the enrichment coefficient

Cells of two strains, one of them lysogenic for wild type λ phage (GY158 (λ)$^+$), the other lysogenic for its noninducible mutant (GY158 (λ ind$^-$)$^+$ were subjected to thymine starvation under identical conditions.

TABLE 1. Effect of prophage inducibility on the survival of cells subjected to thymine starvation

Duration of incubation, min	Number of colonies per ml		Enrichment factor (b/a)
	GY158 (λ) + (a)	GY 158 (λ ind$^-$) + (b)	
0	$1.2 \cdot 10^8$	$2.6 \cdot 10^8$	2.16
30	$2.3 \cdot 10^8$	$2.9 \cdot 10^8$	1.25
60	$2.6 \cdot 10^8$	$2.3 \cdot 10^8$	0.88
90	$3.4 \cdot 10^7$	$7.4 \cdot 10^7$	2.18
120	$8.0 \cdot 10^5$	$1.2 \cdot 10^7$	15.00
150	$2.0 \cdot 10^4$	$3.2 \cdot 10^6$	160.00

Note. The cells were grown in a synthetic medium containing thymine (40 μg/ml) to a density of 10^8 cells/ml and then centrifuged, washed in buffer and resuspended in thymine-free medium. After regular intervals of incubation the cells were seeded on nutrient agar in Petri dishes. The GY158 strain bears the following genetic markers: F$^-$try4 leu8 thil thy A34 dra 34 ura49 sup E lacYI ton A101.

Table 1 shows the results obtained, which illustrate the effect of prophage inducibility on the survival of cells exposed to thymine starvation. After a 150-minute starvation, the number of viable cells in the population of the GY 158 (λ ind$^-$)$^+$ strain is 100 times greater than in the GY 158 (λ)$^+$ strain.

A threefold repetition of the starvation with intermittent culturing of the surviving cells increases the proportion of mutants to be expected 10^6 times.

Increase in the proportion of cells resistant to thymine-free induction

The GY158 (λ)$^+$ culture, after treatment with 2-aminopurine, was subjected to three successive cycles of thymine starvation. Each cycle comprised the following three stages:

1) cell growth: the culture was grown in a synthetic medium to a density of 10^8 cells/ml;

2) lysogenic induction by thymine starvation: the cells were centrifuged, washed in synthetic medium, resuspended and incubated for 150 minutes in the same medium without thymine;

3) cell lysis as a result of repeated addition of thymine: thymine (40 μg/ml) was added to the culture and incubation was continued for another 120 minutes. The culture was placed overnight in the refrigerator; the next morning it was centrifuged and the sediment was resuspended in fresh medium.

TABLE 2. Changes in the bacterial count (colonies per ml) during three consecutive cycles of thymine starvation

Duration of incubation in thymine-free medium, min	First cycle	Second cycle	Third cycle
0	$9.3 \cdot 10^7$	$5.9 \cdot 10^7$	$1.3 \cdot 10^8$
60	$5.1 \cdot 10^7$	$9.3 \cdot 10^7$	$2.2 \cdot 10^8$
90	$2.0 \cdot 10^6$	$1.6 \cdot 10^7$	$1.2 \cdot 10^8$
120	$5.5 \cdot 10^5$	$1.3 \cdot 10^6$	$3.8 \cdot 10^7$
150	$3.0 \cdot 10^4$	$9.3 \cdot 10^5$	$2.1 \cdot 10^7$

Table 2 shows cell survival after the first, second and third cycles of thymine starvation. After three starvation cycles the number of viable cells is 10^3 times greater than in the culture exposed to a single starvation cycle.

Classes of mutants selected

Out of the total of 18,000 colonies examined 21 mutants were selected, all retaining immunity to superinfecting phage λ and displaying a low level of spontaneous and UV-induced phage production in comparison with wild type lysogenic cells.

In order to determine whether this phenotype results from a mutation of the bacterium or of the prophage, each of the selected mutants was subjected to heteroimmune superinfection with the hybrid phage λb2imm434 (Signer and Beckwith, 1966) to eliminate the prophage from the cells. The mutants thus freed of prophage were again lysogenized with wild type phage λ and the level of spontaneous and UV-induced phage production was once more determined. In addition, the phage obtained from each of the original mutants (whenever this could be isolated) was used to lysogenize the C600 strain. This was done because prophage could be isolated from some of our original mutants. We concluded on the basis of these experiments that two of the mutants obtained actually bear defective prophage (Table 3).

TABLE 3. Mutant phenotypes

Class of mutants	Number of mutants	Resistance to		EP	Reproduction of λ pd bio10
		UV	X rays		
I	6	+++	+++	+	+
II	3	++—	+++	—	
III	1	+——	+——	+	+
Rec	9	———		+	—
ts mutation of prophage	1	+++	+++	+	+
Totally defective prophage	1	+++	+++	+	+

Note. EP — elimination of prophage on superinfection with λb2imm434. Resistance to radiation is designated as follows: the number of " + " signs indicates the growth of colonies as a function of the total radiation dose, i.e. 300 ergs/mm^2 for UV and 15 kr for X rays. For example, the " ++— " sign in the third column indicates that not one colony is formed at doses higher than 600 ergs/mm^2; the " ——— " sign in the fourth column means that not one colony is formed even at 15 kr.

It was also necessary to identify the bacterial mutants containing lesions in the recA and exrA genes. To this end we determined the resistance of each mutant toward UV and X rays. An additional test was used to establish the recA phenotype. Cells of each mutant, previously freed of prophage, were infected with phage λ pd bio10 which is distinguished by a very low efficiency of plating on bacteria with a defective recA gene (Signer et al., 1969). As for the bacterial mutants which are sensitive to UV and X rays and cannot sustain the growth of phage λ pd bio10, we concluded that their phenotype results from a lesion of the recA gene. Moreover, some of these mutants were used as recipients in conjugation tests and proved to be indeed incapable of recombination (George, personal communication). All these data indicate that this group of mutations, though as yet unmapped, can be assigned to the gene recA.

The phenotypes of the mutants obtained are shown in Figure 3. Nineteen of these mutants bear mutations in the bacterial chromosome and 9 strains are recA mutants.

It is evident from the data presented that our selection method gives quite a high yield of recA mutants and can therefore be used to isolate mutants of the recA gene. Moreover, the high proportion of recA mutants among the strains incapable of inducing prophage λ confirms the important role of the product of gene recA in the process of lysogenic induction. However, mutations of this type are not the only ones impairing prophage induction.

Characteristic of three new types of mutants

1. Bacteria with the phenotype itdl have the following properties. In the case of cells lysogenic for phage λ or 434, neither UV irradiation nor thymine starvation causes induction of prophage. After either of these treatments the cells remain immune to superinfection with homoimmune phages. The nonlysogenic derivatives of the itdl strain show an increased resistance to thymine starvation. A decrease in the viability of these cells takes place only after 3.5 hours of thymine starvation. Such resistance to thymine starvation cannot be attributed to a lowering of the growth rate since special experiments have shown that the itdl cells grow only 25% more slowly than the parental strain.

The resistance of the itdl bacteria to UV and X rays was the same as that of the wild type strain. Moreover, the normal course of the integration of the F' factor with the chromosome proves that the recA gene of such cells functions normally.

A surprising property of the itdl bacteria is that λ phage forms transparent plaques on them. This means that the itdl mutation also disturbs the lysogenization process. Reactivation (Weigle, 1953) of UV-induced lesions in the itdl cells occurs in the same way as in the wild type. Examination of the level of DNA degradation on thymine starvation showed that in the itdl cells this process is practically absent when compared with the wild type. On the other hand, DNA degradation after UV irradiation is far more intensive in the itdl cells than in the parental strain. This result appears surprising in view of the normal survival rate of itdl cells after UV irradiation. Moreover, the level of DNA degradation in the mutant corresponds to the survival level of the cells, which is not at all the case in the wild type. It may be postulated that in the parental strain death of the cells results from two processes — DNA degradation and filament formation, whereas the death of the itdl mutant cells is a direct consequence of DNA degradation. This assumption is supported by the finding that the wild type cells become longer after UV irradiation and produce filiform structures on thymine starvation. Filamentous forms have not been found in either case in cultures of itdl bacteria.

It was found that the period preceding the resumption of DNA synthesis after UV irradiation and the latent period following thermal induction of prophage are slightly longer in the mutant cells than in the wild type cells.

These observations imply that the itdl mutation affects in some way the mechanism controlling DNA replication.

2. The lysogenic cells with the itd2 phenotype are distinguished by the fact that the prophage cannot be removed from them by superinfection with phage λb2imm434. It turned out that this phenotype does not result from some bacterial or phage mutation making delysogenization impossible, but from the fact that no lambda-like phages of the 434, 21 or 82 types can grow in itd2 cells. Nevertheless, the addition of these phages at a high multiplicity kills the mutant cells. In contrast to the lambda-like phages, phage T2 develops normally in itd2 cells.

The transfer of F-lac$^+$ episomes into itd2 lacYl recipient cells enables these to form lac$^+$ colonies.

The mechanism whereby the growth of lambda-like phages is inhibited in itd2 cells is now being studied.

3. The phenotype designated itd3 closely resembles in its properties the exr2 mutant described by Donch and his colleagues (1970). The itd3 cells are sensitive to UV and X rays (Table 3); moreover, UV reactivation and UV-induced mutations of lambda prophage have not been detected in them.

On the other hand, the extremely low level of spontaneous phage production in the itd3 cells distinguishes these from the exrA $(\lambda)^+$ and lexA $(\lambda)^+$ strains (Howard-Flanders, personal communication). In addition, attempts to map the itd3 mutation beside the malB and metA loci by means of transduction with phage P1 have failed. Phage P1 grown in itd3 cells readily yields malB$^+$ and metA$^+$ transductants. However, a simultaneous transduction of sensitivity to UV or X rays has not been obtained. More accurate mapping of the itd3 mutation is under way.

DISCUSSION AND CONCLUSION

The discovery of the indirect induction of lambda prophage by UV irradiation (Borek and Ryan, 1960) triggered a series of studies aiming to reveal the genetic and biochemical processes involved from the formation of primary lesions by the inducing agent to the formation of the inducer (Devoret and George, 1967; Devoret, et al., 1965; Monk, 1970). These studies have led to the following conclusions:

1. The lysogenic induction caused by UV irradiation cannot be attributed either to lesions of the prophage itself or to a direct inactivation of a repressor.

2. UV induction results from the formation of pyrimidine dimers in an autonomous replicating unit (replicon), for example in the sex factor or in the bacterial chromosome. Nonreplicating DNA containing UV lesions cannot induce prophage.

3. Enzymatic repair systems — excision of pyrimidine dimers from DNA, or recombinational repair — do not by themselves bring about lysogenic induction.

4. Secondary lesions of DNA, such as single-strand gaps arising from the replication of DNA containing pyrimidine dimers, exert an indirect influence on the prophage via the cytoplasm and the cell membrane.

All these data indicate the existence of a pathway starting from lesions in the DNA and ending with the formation of the inducer.

What cellular factors bring about the formation of inducer in response to lesions in replicative DNA? All the available data on this topic are genetic in nature. Three types of bacterial mutations affecting the induction of prophage have been discovered recently: T44 (Goldthwait and Jacob, 1964) recA (Brooks and Clark, 1967; Hertman and Luria, 1967), and exrA (Donch et al., 1970). Bacteria bearing the T44 mutation are induced by transfer from low to high temperature. None of the known inducing agents can produce prophage in recA and exrA cells.

These observations suggest that the products of genes recA and exrA are responsible for the formation of the inducer (Tomizawa and Ogawa, 1967). Moreover, the possibility that the recA gene is under the control of the T44 gene cannot be ruled out. Despite the increasing number of genes known to be involved in induction, however, we still have no clear conception of the interrelationships of these genes. The available genetic and biochemical data are clearly insufficient.

Nevertheless, certain conclusions can be made concerning the nature of the mutations disturbing induction. All the mutations isolated have a pleiotropic effect. Apart from influencing the induction of prophage, they play an important role in the replication of repair of DNA (or both these processes). For example, mutants which are noninducible by thymine starvation are also noninducible by UV irradiation.

The three new types of mutation described here may prove a valuable aid to deciphering the biochemical pathway leading to lysogenic induction.

We are grateful to M. Radman and J. George for running some experiments and for valuable discussions. A. Baillonet is particularly indebted to G. Simonet for the advice and hospitality given her in his laboratory.

We thank D. Foiret for technical assistance and A. Dedieu for the maintenance of the collection of bacterial and phage strains.

BIBLIOGRAPHY

Borek, E. and A. Ryan. — Biochim. biophys. Acta, Vol. 41 (1960), 67.

Brooks, K. and A.J. Clark. — J. Virol., Vol. 1 (1967), 283.

Devoret, R. and M. Blanco. — Molec. and Gen. Genet., Vol. 107 (1970), 272.

Devoret, R. and J. George. — Mutation Res., Vol. 4 (1967), 713.

Devoret, R., M. Monk, and J. George. — Z. Bakteriol., Vol. 196 (1965), 193.

Donch, J., J. Greenberg, and M.H.L. Green. — Genet. Res., Vol. 15 (1970), 87.

Goldthwait, D. and R. Jacob. — C.r. hebd. Séanc. Acad. Sci., Paris, Vol. 259 (1964), 661.

Hertman, I. and S.E. Luria. — J. Molec. Biol., Vol. 23 (1967), 117.

Jacob, F. and J. Monod. — J. Molec. Biol., Vol. 3 (1961), 318.

Lieb, M. — J. Molec. Biol., Vol. 16 (1966), 149.

Lwoff, A., L. Siminovitch, and N. Kjeldgaard. — Annls. Inst. Pasteur, Paris, Vol. 79 (1950), 815.

Monk, M. — Molec. and Gen. Genet., Vol. 106 (1970), 14.
Ptashne, M. — Proc. Natn. Acad. Sci. U.S.A., Vol. 57 (1967), 306.
Sicard, N. and R. Devoret. — C.r. hebd. Séanc. Acad. Sci., Paris, Vol. 255 (1962), 1417.
Signer, E.R. and J.R. Beckwith. — J. Molec. Biol., Vol. 22 (1966), 33.
Signer, E.R., K.F. Manli, and M.-A. Brunstetter. — Virology, Vol. 39 (1969), 137.
Sussman, R. and F. Jacob. — C.r. hebd. Séanc. Acad. Sci., Paris, Vol. 254 (1962), 1517.
Tomizawa, J. and T. Ogawa. — J. Molec. Biol., Vol. 23 (1967), 247.
Weigle, J.J. — Proc. Natn. Acad. Sci. U.S.A., Vol. 39 (1953), 628.

UDC 575:576.8

THE EFFECT OF CHANGES IN THE CELL SURFACE ON CONJUGATION IN E S C H E R I C H I A C O L I K12

K.I.MARKOV and V.T. KANTARDZHIEV
Higher Medical Institute, Sofia

Donor and recipient cells of E. coli were subjected to treatment causing specific changes in the bacterial surface without lowering the number of viable cells. It is shown that compounds which alter the electric charge of the donors and recipients lower the recombination frequency, whereas treatment with trypsin produces the opposite effect. The acquisition of streptomycin resistance lowers the capacity for conjugation.

In previous publications we have shown that the acquisition of resistance to antibiotics in microbes is accompanied by certain physico-chemical changes in the bacterial wall (Markov and Kantardzhiev, 1964, 1969). It is known that the establishment of contact between conjugating cells requires not only specific structures determined by the presence of the sex factor (Brinton, 1965) but also chemical structures on the surface of the bacterial cell. We therefore conducted a comparative study of the conjugation process in cells which had become streptomycin resistant without being in contact with the antibiotic and in str-r cells grown in the presence of streptomycin. For this purpose both donor and recipient cells were treated with chemical agents of which the mechanism of action is known. The frequency of the transfer of prototrophy and alteration of the chemical potential were used as the criteria of changes in the surface of the parental cells.

MATERIALS AND METHODS

The bacterial strains E. coli K12: HfrH and F^- (Markov and Kantardzhiev, 1964) were used. Streptomycin-resistant strains were obtained according to Lederberg (1952) or by transfer of the original strain to streptomycin-containing media. The former strains were termed "indirect mutants."

Treatment of the bacterial cells. Cells at a concentration of 10^9/ml were treated for one hour at 37° with one of the following agents: 0.25% trypsin (Difco), pH 8.0; 40% petroleum ether; a 0.01% solution of sodium dodecylsulfate; $2.5 \cdot 10^{-4}$ M periodate solution. At these concentrations the above agents do not lower the cell count. The lipid content of the cells was increased by repeated passages in media containing 20% glycerol.

The electrokinetic potential of the cells was measured in a specially designed flat horizontal chamber (Markov and Dimitrov, 1969).

RESULTS

As shown in the table, treatment of donor and recipient cells with the above indicated agents, except trypsin, decreases the number of recombinants.

Treatment with trypsin enhances the capacity of the cells for conjugation.

Figures a and b show the electrophoretic patterns of the treated cells. All the agents used shift the isoelectric point of the donor cells towards acidity. The change observed in the isoelectric point of the recipient cells is similar in direction but less distinct. Acid agglutination of the cells gave comparable results. The behavior of bacteria sensitive and resistant to streptomycin on conjugation is illustrated by the following data.

Conjugation of *E. coli* cells treated with various agents

Treatment of the cells	Number of recombinants · 10^2 on treatment of cells		
	recipients	donors	recipients and donors
0.25% trypsin, pH 8.0	151 ± 4	128 ± 5	159 ± 4
40% petroleum ether	24 ± 9	31 ± 10	20 ± 5
0.01% dodecylsulfate	39 ± 9	47 ± 8	22 ± 10
$2.5 \cdot 10^{-4}$ M periodate	11 ± 6	8 ± 4	6 ± 7
20% glycerol	48 ± 8	58 ± 7	37 ± 8
No treatment	—	—	108 ± 4

It is evident from these results that both bacterial cells in which streptomycin resistance has been increased by direct contact with the antibiotic and "indirect" mutants cultured repeatedly in a streptomycin-containing medium lose their capacity for conjugation. Treatment of these strains with trypsin partially restores this capacity. The capacity for conjugation of the "indirect" mutant HfrH resembles that of the parental type and is not lowered by treatment with trypsin.

Conjugation mixture	Number of recombinants · 10^2	Conjugation mixture	Number of recombinants · 10^2
HfrH × F^-	100	HfrH (IM) (str) × F^-	0
HfrH (T) × F^-	129	HfrH (IM) (str) × F^-	23
HfrH × F^- (str)	0	HfrH (IM) (str) × F^- (str)	0
HfrH × F^- (str) (T)	69	HfrH (IM) (str) (T) × F^- (str) (T)	2
HfrH (IM) × F^-	81	HfrH (A) × F^-	0
HfrH (IM) (T) × F^-	86	HrfH (A) (T) × F^-	21
HfrH (IM) × F^-	0	HfrH (A) × F^- (str)	0
HfrH (IM) × F^- (str) (T)	11	HfrH (A) (T) × F^- (str) (T)	2

Note. IM — "indirect mutant"; A) cells adapted to streptomycin; T) cells treated with trypsin; str) cells repeatedly cultured in a streptomycin-containing medium.

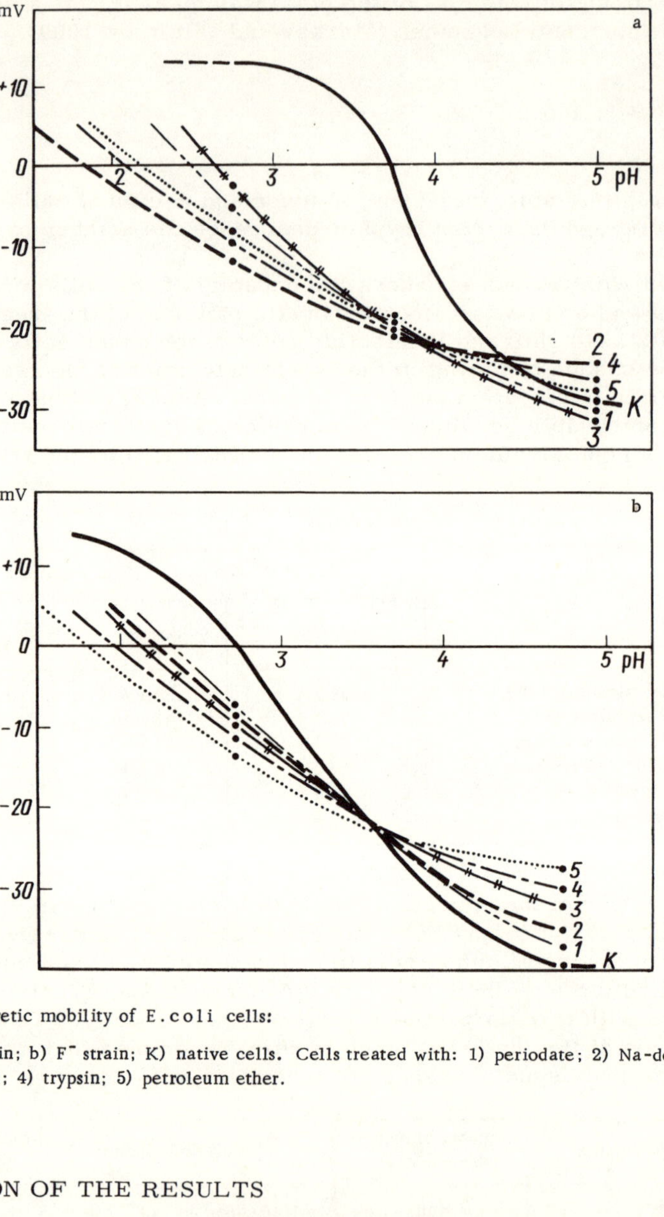

Electrophoretic mobility of E. coli cells:

a) Hfr strain; b) F^- strain; K) native cells. Cells treated with: 1) periodate; 2) Na-dodecylsulfate; 3) glycerol; 4) trypsin; 5) petroleum ether.

DISCUSSION OF THE RESULTS

Treatment of male and female cells of E. coli with trypsin, sodium dodecylsulfate, periodate or petroleum ether, as well as their enrichment with lipids by culturing in 20% glycerol, causes marked changes in the electrokinetic potential of both donors and recipients. It is known that trypsin and dodecylsulfate act on protein components of the bacterial surface. However, the resulting changes in the bacterial surface are evidently different, since dodecylsulfate treatment sharply lowers the conjugational capacity of the cells whereas trypsin enhances it. This suggests that conjugation requires particular protein components which are broken down by dodecylsulfate but not by trypsin.

It is assumed that conjugation takes place with the aid of pili which have a proteinaceous structure. On the other hand, treatment of donor and recipient cells with periodate solution, which acts on polysaccharides, markedly lowers the recombination frequency. This means that the pili are not the only structures on the bacterial surface involved in conjugational contact.

Treatment with petroleum ether, which destroys the surface lipids of the cells, reduces the capacity for conjugation. Its action probably disrupts the general composition of the cell surface and damages the cell physiologically. Extraction of a specific lipid may also be involved since the enrichment of the bacterial cell surface with lipids, by culturing the cells in media containing glycerol, lowers the recombination frequency. This effect probably results from the formation of a new superficial layer which covers the cell receptors.

The "indirect" str-r mutants have the same capacity for conjugation as the wild type strain. The capacity for conjugation is lost in the adaptive str-r strains and in the "indirect" streptomycin-resistant mutants grown repeatedly in streptomycin-containing media. It can be assumed on these grounds that the changes observed in the "indirect" mutants grown in contact with streptomycin-resemble those displayed by the adapted cells. The partial restoration of the conjugational capacity in streptomycin-resistant parental cells after treatment with trypsin indicates that this agent specifically stimulates the conjugational capacity of sexually differentiated cells, probably by inducing the release of specific chemical components onto the surface.

CONCLUSIONS

1. The conjugation frequency depends on the chemical structure of the surface layer of the parental cell.

2. Treatment with trypsin increases the recombination frequency, probably by exposing cell receptors located deep inside the bacterial surface.

3. E. coli cells which have become streptomycin-resistant by direct contact with the antibiotic are altered in such a way during this process that they cease to conjugate.

The surface changes can be partly corrected by treatment of the cells with trypsin.

BIBLIOGRAPHY

Brinton, C.C. — Trans. N.Y. Acad. Sci., Vol. 27 (1965), 1003.

Lederberg, I. and E. Lederberg. — J. Bact., Vol. 63 (1952), 339.

Markov, K. and D. Dimitrov. Recent Advances in Medical Technique and Diagnostics. Sofia, 1969. (Bulgarian).

Markov, K. and V. Kantardzhiev. — Gigiena, No.4, p.27, Bulgaria, 1964.

Markov, K. and V. Kantardzhiev. — In: "Yubileini trudove VMI." Sofia, 1969.

UDC 575.24:576.8

RECOMBINATION-DEFICIENT MUTANTS OF ESCHERICHIA COLI K12

A.P.PEKHOV and S.N. RUDNEVA
Research Laboratory of Experimental Immunobiology
Academy of Medical Sciences of the USSR, Moscow

The sensitivity of certain rec⁻ mutants to UV, methylmethanesulfonate and mitomycin was examined. The results obtained confirm the idea of a relationship between genetic recombination and the repair of radiation-induced lesions in DNA. It is further suggested that the rec genes determine functions involved in the repair of DNA lesions induced by chemical mutagens.

In 1965 Clark and Margulies (1965) reported the isolation, from E. coli JC411F^- cultures, of two recombination-deficient mutants which when crossed by the standard procedure with Hfr donors yielded genetic recombinants at a frequency 1,000 times lower than recipient cells of the original culture.

During the five intervening years, recombination-deficient mutants have been isolated from several strains of E. coli (Fuerst and Siminovitch, 1965; Howard-Flanders and Theriot, 1966; Van de Putte et al., 1966; Ogawa et al., 1968; Rudneva and Pekhov, 1969), Salmonella typhimurium (Wing et al., 1968; Bohme, 1967) and Bac. subtilis (Okubo and Romig, 1966; Prozorov, 1966).

Recombination-deficient mutants of bacteria have become favorite models for the study of various current topics of molecular genetics, particularly the role of genetic recombination in the repair of DNA lesions and in mutagenesis (spontaneous and induced).

According to the conception of Howard-Flanders and Boyce (1966), the recombination-deficient bacterial mutants must be incapable not only of forming genetic recombinants but also of repairing DNA lesions induced by UV radiation.

Studies of recombination-deficient mutants of E. coli (Clark and Margulies, 1965) have shown that they are more sensitive to radiation than the original cultures. Some recombination-deficient mutants isolated from other strains of E. coli similarly display a great sensitivity to UV radiation (Howard-Flanders and Theriot, 1966; Van de Putte et al., 1966; Ogawa et al., 1968).

E. coli mutants bearing simultaneously a mutation in the uvr locus, which determines the capacity for excising thymine dimers, and a mutation in the recA locus, which determines the capacity for genetic recombination, are far more sensitive to radiation than mutants bearing only one of these mutations (Howard-Flanders, 1968).

The discovery of an increased UV sensitivity in some recombination-deficient mutants of E. coli K12 has led to the conclusion that both these properties result from a single mutation and that survival depends on the capacity of the cells for genetic recombination. In other words, the UV resistance of the cells is based on a recombination system which operates independently of the excision of pyrimidine dimers from damaged DNA; cells resistant to UV radiation must be normal not only with respect to the excision of dimers from DNA prior to its replication but also as regards the capacity for recombination, by which the post-replicative repair of the gaps formed by unexcised dimers is accomplished (Howard-Flanders, 1968; Witkin, 1969).

However, the above observations can also be interpreted otherwise—namely, by assuming that the recombinational deficiency and UV sensitivity of the bacteria result from two mutations. This might explain why some recombination-deficient mutants can excise pyrimidine dimers from DNA (Clark et al., 1966), while certain excision-deficient strains are not necessarily recombination-deficient (Howard-Flanders and Theriot, 1966). It may be postulated that lack of capacity for recombination and inability to repair DNA lesions stem from two mutations. If this assumption is correct it may be possible to identify the rec genes which are not engaged in the repair of UV-induced lesions of DNA.

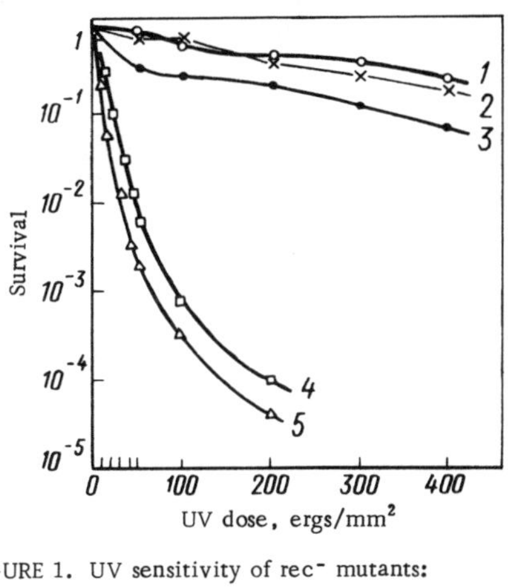

FIGURE 1. UV sensitivity of rec⁻ mutants:

1) rec+; 2) rec⁻75; 3) rec⁻15; 4) rec⁻10; 5) rec⁻48.

According to a recent widely quoted hypothesis, UV-induced mutations arise largely from errors introduced into DNA during the recombinational repair of single-strand gaps (Witkin, 1969). The main principle of this hypothesis is that if the product of gene exr+ introduces errors in DNA by a repair mechanism unrelated to recombination, the rec⁻ strains should have a normal UV mutability.

Kondo (1968) found that spontaneous mutagenesis is enhanced in one rec⁻ strain of E. coli and reduced in another rec⁻ strain. Witkin (1969) similarly observed a reduced spontaneous mutagenesis in the strain E. coli recC. According to her data, UV irradiation of the recA strain does not induce mutations, whereas cells of the recB and recC strains do mutate after such treatment, but at a lower frequency than those of the rec^+ strain. It follows from these observations that a normal UV-induced mutability requires the presence of the products of all three rec gene loci and is therefore dependent on the recombination system. It is not known, however, whether the rec genes affect the mutation spectrum.

In order to obtain data on the role of the rec genes in the repair of DNA lesions, we examined the UV sensitivity of four recombination-deficient mutants of E. coli P678F⁻ isolated in this laboratory. The mutants rec 10, rec 15, rec 48 and rec 75, with recombinational deficiency indexes equal to $1.5 \cdot 10^3$, $8.7 \cdot 10^2$, $1.0 \cdot 10^3$, and $6.1 \cdot 10^2$ respectively were used. As shown by the dose-survival curves of the rec^- mutants after UV treatment (Figure 1), the cells of the rec 75 mutant have the same sensitivity as those of the original strain (the LD_{37} was 360 ergs/mm^2 in both cases). The mutant rec 15 proved slightly more sensitive, with an LD_{37} of 50 ergs/mm^2. The rec 10 and rec 48 mutants are far more sensitive than the original rec^+ strain with LD_{37} values of 7 and 5 ergs/mm^2, respectively.

The recombination-deficient mutants of E. coli P678F⁻ examined behaved similarly with respect to methylmethanesulfonate, nitrous acid and mitomycin C. The sensitivity of the rec 75 mutant to all these mutagens and that of the rec 15 mutant to methylmethanesulfonate were the same as in the original rec^+ strain.

As in the case of UV irradiation, the rec 10 and rec 48 mutants proved more sensitive to the above mutagens than the original rec^+ strain.

Thus, not all of the examined recombination-deficient mutants are sensitive to UV and other mutagens.

In order to determine the role of genetic recombination in bacterial mutagenesis and in particular the effect of the rec genes on the mutation spectrum we first of all examined the frequency of back mutations in the rec^- mutants from auxotrophy to prototrophy with respect to threonine and leucine. For this purpose we selected colonies of the rec 48, rec 15 and rec 75 mutants grown after UV irradiation at a dose which allows only about 1% of the cells to form colonies, and determined their requirements for threonine and leucine. The results obtained appear in the following table.

Frequency of back mutations in cultures of E. coli P678F⁻ rec⁻

Strain	Number of colonies examined	Frequency of back mutants, %	
		T +L -	L +T -
rec 48	2,100	16.8	23.7
rec 15	2,100	15.9	24.7
rec 75	2,100	7	23.1
rec^+	2,100	11.6	29.2

The change in the recombinational capacity of the cells is obviously accompanied by a change in the mutation frequency.

We also examined the spontaneous and induced segregation of lactose-negative variants of recombination-deficient F-merodiploids. This study was based on the data of Willetts (1967), who found that the mechanism of the induced haploidization of merodiploids involves mutations of the genes controlling the replication of the sex factor.

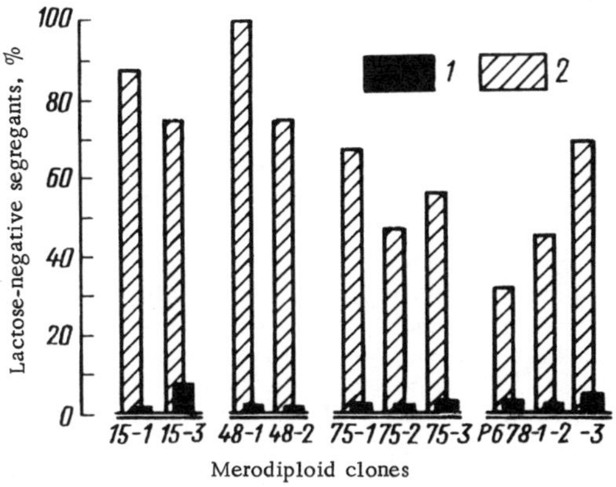

FIGURE 2. Frequency of lac$^-$ segregants in cultures of F-merodiploids of rec$^+$ and rec$^-$.

Appearance of segregants: 1) spontaneous; 2) after treatment with acridine.

As shown in Figure 2, all the merodiploids examined bearing the rec$^-$ and lac$^+$ alleles in the chromosomes and the lac$^+$ allele in the chromosomal segment included in the F' sex factor segregate lactose-negative variants both spontaneously and after acridine treatment. Here the frequency of haploidization of the rec$^-$ and rec$^+$ merodiploids is not markedly different. The results obtained show that the spontaneous and induced segregation of lac$^-$ cells from the merodiploids does not depend on the activity of the recombination system of the latter.

Our data on the increased UV sensitivity of some of the rec$^-$ mutants examined confirm the existence of a relationship between genetic recombination and the repair of UV-induced DNA lesions. They also show that the recombination mechanism can repair DNA lesions induced by methylmethanesulfonate, nitrous acid, and other mutagens. The rec 75 mutant, however, is aberrant in this respect since its sensitivity to the mutagens is the same as in the original rec$^+$ strain. These findings suggest that the recombination system and that engaged in the excision of pyrimidine dimers can operate independently of one another. If this assumption is correct, the independence of the two systems would become evident in the case of damage to a stage in recombination unrelated to DNA repair.

The interpretation of our data concerning the role of genetic recombination in mutagenesis, notably in the haploidization of F-merodiploids bearing the rec^- allele. One of the possible explanations of the observed independence of merodiploid haploidization from the recombination system is that the influence of the latter system on mutagenesis is restricted to certain mutations only. This hypothesis, however, requires further experimental verification.

Thus, despite the data available on the mutagenic role of the products of the rec genes, the nature of the relationship between mutations and genetic recombination remains obscure.

BIBLIOGRAPHY

Bohme, H. — Biochem. Biophys. Res. Commun. 28, No.2 (1967), 191.

Clark, A. J., M. Chamberlin, R.P. Royce, and P. Howard-Flanders. — J. Molec. Biol., Vol.19 (1966), 442.

Clark, A.J. and A.D. Margulies. — Proc. Natn. Acad. Sci. U.S.A., Vol. 53 (1965), 2.

Fuerst, C.R. and L. Siminovitch. — Virology 27, No. 3 (1965), 449.

Howard-Flanders, P. — A. Rev. Biochem., Vol. 37 (1968), 175.

Howard-Flanders, P. and R.P. Boyce. — Radiat. Res., Suppl., Vol. 6 (1966), 156.

Howard-Flanders, P. and L. Theriot. — Genetics, 53, No.6 (1966), 1137.

Kondo, S. — Proc. 12th Intern. Congr. Genet., Vol. 2 (1968), 126.

Ogawa, H., K. Shimada, and J. Tomizawa. — Molec. and Gen. Genet., Vol. 101 (1968), 227.

Okubo, S. and W.R. Romig. — J. Molec. Biol., Vol.15 (1966), 440.

Prozorov, A.A. — Genetika, Vol. 8 (1966), 98.

Rudneva, S.N. and A.P. Pekhov. — Byull. Eksp. Biol. Med., Vol. 7 (1969), 100.

Van de Putte, P., H. Lwenk, and A. Rorsch. — Mutation Res., Vol. 3 (1966), 381.

Willetts, N.S. — Biochem. Biophys. Res. Commun. 27, No. 1 (1967), 112.

Wing, J.P., M. Levine, and H.O. Smith. — J. Bact. 95, No.5 (1968), 1828.

Witkin, E. — Mutation Res. 8, No. 1 (1969), 9.

UDC 575:576.8

THE BIOLOGICAL ACTIVITY OF ISOLATED DNA OF *ESCHERICHIA COLI* IN RECIPIENT SPHEROPLASTS

O.N. RUDCHENKO and B.N. IL'YASHENKO
N.F. Gamaleya Institute of Epidemiology and Microbiology
Academy of Medical Sciences of the USSR, Moscow

The activity of isolated donor DNA was assessed according to the restoration of the capacity for protein synthesis in spheroplasts of auxotrophic strains. Native and denatured DNA were found to be equally active in the spheroplasts, and no competition between them was detected. The manifestation of the genetic information of DNA requires a brief synthesis of DNA, probably for the completion of a double-stranded structure since directly after penetration into the spheroplasts the donor DNA is in a single-stranded state. The function of the donor DNA in the spheroplasts is equally efficient in rec^+ and rec^- strains.

A widely used model for determining the biological activity of isolated nucleic acids is the genetic transformation of bacteria, characterized by the hereditary acquisition by the recipient of certain characters of the donor. We propose a procedure for revealing the biological activity of bacterial nucleic acids in the first generation of the recipient. Since the behavior of the new character in the recipient progeny is not traced by this method, we will adopt the term phenotypic transformation or phenotypic alteration of the recipient, induced by the donor DNA. The main difference between the new system and genetic transformation lies in the manner in which the activity of the donor DNA is revealed. Two methods are proposed for determining the manifestation of traits of the donor DNA in spheroplasts, used here as the recipients. The first method consists in the detection of "transformed" spheroplasts by means of phage 1 φ 7 infectious DNA. In the second method, protein synthesis in "transformed" spheroplasts is determined by the incorporation of a ^{14}C-amino acid into the acid-insoluble sediment of spheroplasts in a minimal medium lacking one of the essential amino acids.

For both methods the scheme of experiments to determine the function of donor DNA in spheroplasts can be outlined as follows.

Spheroplasts of an auxotrophic strain were placed in minimal medium lacking an essential growth factor. Donor DNA from a prototrophic strain and infectious DNA of phage 1φ7 or a ^{14}C-amino acid (nonessential) were added. Finally, the harvest of phage 1φ7 in the spheroplasts or the level of incorporation of the ^{14}C-amino acid in the acid-insoluble sediment were determined.

Tables 1 and 2 show the results of these experiments.

TABLE 1. Induction by donor DNA of the replication of phage 1 φ7 in spheroplasts of E. coli 15 TUH, auxotrophic for histidine

DNA*	Medium	Number of infected spheroplasts per ml**	Titer of phage 1φ7 †	Harvest of phage 1 φ7 per spheroplast
1 φ 7	Complete	$5.2 \cdot 10^5$	$4 \cdot 10^5$	80
1 φ 7	Without histidine	$5 \cdot 10^3$	$1 \cdot 10^2$	$2 \cdot 10^{-2}$
1 φ 7 + Hfr	Without histidine	$5.4 \cdot 10^3$	$1.2 \cdot 10^4$	2.7
1 φ 7 + (Hfr + + DNase)	Without histidine	$5 \cdot 10^3$	$1.2 \cdot 10^2$	$2.4 \cdot 10^{-2}$
1 φ 7 +15 T⁻	Without histidine	$5.1 \cdot 10^3$	$3 \cdot 10^4$	6
1 φ 7 +15 TUH	Without histidine	$5.1 \cdot 10^3$	$0.8 \cdot 10^2$	$1.6 \cdot 10^{-2}$

* Concentration of phage 1 φ7 DNA — 0.1 μg/ml, bacterial DNA — 10 μg/ml.
** Determined after incubation for 10 minutes at 37°.
† Determined after incubation for 90 minutes.

DNA from the prototrophic strain clearly restored the prototrophy of the spheroplasts with respect to the growth factor missing from the medium. As a result, the spheroplasts acquired the ability to synthesize protein in a minimal medium and reproduce phage 1 φ 7. As shown by the following data, this effect cannot be attributed to any contamination of the DNA with the missing growth factor or to any other nonspecific action of the DNA on the spheroplasts:

1. Treatment with DNase destroyed the activity of the DNA.
2. DNA from a homologous (auxotrophic) strain proved inactive.
3. DNA was active only in spheroplasts, never in cells, even if the permeability of these was altered by EDTA or by rapid freezing and thawing (Table 3).

TABLE 2. Induction by donor DNA of the incorporation of ^{14}C-leucine into E. coli 15 TUH spheroplasts in a histidine-free medium

DNA	Medium	DNase, μg/ml	Incorporation of ^{14}C-leucine, pulses/min*
—	Complete	—	24,900
—	Complete	10	25,100
—	Histidine-free	—	9,280
15T⁻	Histidine-free	—	25,800
15T⁻	Histidine-free	10	7,800
15 TUH	Histidine-free	—	6,700
Hfr	Histidine-free	—	24,300

* Incubation with ^{14}C-leucine for 60 minutes at 37°. Concentration of bacterial DNA 10 μg/ml. A 0.5 μcurie dose of radioactivity was introduced into each sample. Radioactivity was determined in the acid-insoluble sediment of the spheroplasts.

If the donor DNA-induced synthesis of phage 1φ 7 and incorporation of the ^{14}C-amino acid in the minimal medium depend directly on the number of "transformed" spheroplasts, an idea of the frequency of spheroplast alteration under the influence of the donor DNA can be obtained by comparing the phage yield and the incorporation of the label in complete medium and in the presence of the donor DNA.

As shown in Tables 1, 2, and 3, the frequency of donor DNA-induced alteration of the spheroplasts exceeds by far the normal transformation frequency observed in transformable bacterial species. The experiments with ^{14}C-amino acids demonstrate, for example, that the presence of donor DNA enables all or nearly all spheroplasts to synthesize protein in a minimal medium. It appears on these grounds that in our experiments the donor DNA operates in an autonomous state, without being integrated into the recipient chromosome. This view was confirmed in experiments with spheroplasts of rec strains of E. coli as well as in experiments with acridine dyes. Comparison of the activity of donor DNAs in rec$^+$ and rec$^-$ recipients revealed no differences for any of the markers tested (Table 4). Similar results were obtained when donor DNAs from rec$^+$ and rec$^-$ strains were introduced into recipients of both kinds. Consequently, the function of donor DNA of E. coli in spheroplasts does not depend on the recipient systems involved in recombination and in this respect is reminiscent of the behavior of episomal genes. The autonomous state of the donor DNA in the spheroplasts is also evident from experiments with acridine dyes. It is known that DNA not linked with the cell chromosome (episomal, viral, transforming and donor DNA during conjugation) is highly sensitive to acridine dyes. In the last two cases sensitivity persists only until integration of the DNA into the chromosome of the recipient. In our experiments, acridine dyes (acriflavine and ethidium bromide) inhibited the activity of the donor DNAs and the sensitivity of the DNA to the dyes persisted for up to 30 minutes after contact with the spheroplasts (Table 5).

TABLE 3. Activity of E. coli Hfr DNA in cells of E. coli 15 TUH with different degrees of damage to the cell wall

Recipient	Donor DNA*	Medium	Incorporation of ^{14}C-leucine, pulses/min**
	—	Complete	3,420
Cells	—	Without histidine	540
	Hfr	Without histidine	510
Cells + EDTA $2 \cdot 10^{-3}$M	—	Complete	3,610
	—	Without histidine	720
	Hfr	Without histidine	680
Cells after freezing and thawing	—	Complete	2,650
	—	Without histidine	370
	Hfr	Without histidine	320
	—	Complete	3,270
Spheroplasts	—	Without histidine	380
	Hfr	Without histidine	2,830

* Concentration of Hfr DNA — 10 μg/ml.

** Incubation with ^{14}C-alanine for 60 minutes.

TABLE 4. Activity of $15T^-$ DNA in spheroplasts of rec^- strains of E. coli

Spheroplasts	DNA*	Medium	Incorporation of ^{14}C-leucine, pulses/min**
	–	Complete	8,832
rec A13	–	Without histidine	960
	$15T^-$	Without histidine	11,840
	–	Complete	11,520
rec A1	–	Without histidine	3,600
	$15T^-$	Without histidine	10,800

* Concentration of DNA – 10 μg/ml.
** Incubation with ^{14}C-leucine for 60 minutes.

It appears, therefore, that for at least 30 minutes the donor DNA operates autonomously in the spheroplasts.

TABLE 5. Effect of acriflavine on the activity of E.coli Hfr DNA in E.coli PA209 spheroplasts

DNA* (10 μg/ml)	Medium	Addition of acriflavine (10 μg/ml)	Incorporation of ^{14}C-leucine, pulses/min*
–	Complete	–	8,540
–	Complete	Simultaneously with DNA	8,600
–	Without histidine	–	3,200
Native Hfr DNA	Without histidine	–	10,000
	Without histidine	Simultaneously with DNA	3,500
	" "	After 10 minutes	3,670
	" "	After 30 minutes	5,050
Denatured Hfr DNA	Without histidine	–	9,600
	Without histidine	Simultaneously with DNA	3,000
	" "	After 10 minutes	3,000
	" "	After 30 minutes	3,500

* ^{14}C-alanine was added at the same time as acriflavine. Incubation with ^{14}C-alanine for 60 minutes.

The following experiments were carried out in order to determine the fate of the donor DNA in the spheroplasts. Donor ^{14}C-DNA was used in all experiments.

Analysis of the kinetics of the penetration of donor DNA into the spheroplasts showed that both native and denatured DNA enter very rapidly into the spheroplasts. About 50% of the label becomes inaccessible to DNase within 2 minutes, and incorporation reaches a peak after five minutes. However, these data may simply reflect the transition of the DNA into a state which is resistant to DNase. It was found that native donor DNA (or at least part of it) becomes single-stranded soon after (or during) penetration into the spheroplasts. This condition may persist if the spheroplasts are infected with DNA under conditions which do not allow DNA synthesis (incubation in a thymine-free medium). The experiment was carried out as follows. After contact with the DNA, the spheroplasts were washed free of extracellular DNA and then destroyed; DNA was extracted by the phenol method. The extracted DNA was passed through a column of methylated bovine serum albumin or kieselguhr (MAK), and the distribution of the radioactive label was determined. Under conditions which prevent DNA synthesis (absence of thymine) there are two distinct radioactivity peaks corresponding to native and denatured DNA, whereas pure donor DNA yields a single peak (see graph). This means that at least part of the native donor DNA in the spheroplasts is single-stranded. The single-stranded state appears to be transitory, since under conditions which allow normal DNA synthesis in the spheroplasts the radioactivity peak produced by denatured DNA accounts for less than 10% of the total label in the spheroplasts.

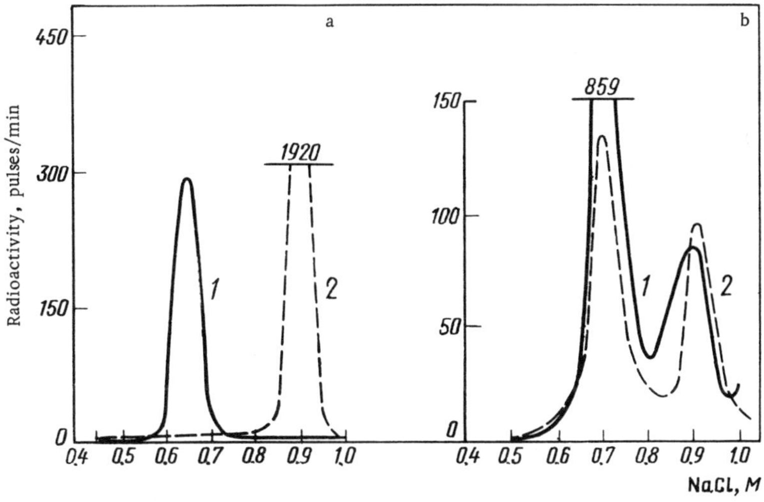

Distribution of radioactivity on MAK column fractionation of the DNA isolated from the membrane fraction of spheroplasts incubated with donor ^{14}C-DNA:

a) donor ^{14}C-DNA before incubation with the spheroplasts: 1) native DNA; 2) denatured DNA;
b) ^{14}C-DNA isolated from the membrane fraction of the spheroplasts: 1) ^{14}C-DNA incubated with the spheroplasts in a thymine-free medium; 2) ^{14}C-DNA incubated with the spheroplasts in a medium containing thymine.

We were interested in the fate of the donor DNA in the spheroplasts. If most of the DNA does not become integrated into the recipient chromosome, the question arises whether this DNA remains in a free state, or breaks down, or unites with certain cellular structures. First we tried to find out whether the donor DNA is bound to the cellular membranes of the recipient in view of recent evidence that the membranes participate in major metabolic processes in the cell.

We applied the procedure used by many workers for detecting a link between the growing point of the chromosome and the cell membranes or the "rapidly sedimenting fraction" of cell lysates. The spheroplasts were destroyed by osmotic shock under relatively mild conditions and were then centrifuged in a sucrose density gradient. When the centrifugation was performed so that free DNA still remained in the top layer, DNA label could be detected in the fraction collected in the 35% sucrose layer. Electron microscopy of this fraction by A. F. Bykovskii showed that it consists of a mixture of inner and outer membranes.

The bond between the DNA and the "rapidly sedimenting fraction" of the cell lysate was severed by deoxycholate and dodecylsulfate.

Thus, the data obtained suggest that at least part of the donor DNA in the spheroplasts is bound to rapidly sedimenting structures. We assume that these structures are the cell membranes. The functional role of this complex is being studied.

Further research into this problem may explain some peculiarities of the functioning of isolated bacterial DNAs in an autonomous state, particularly the functioning of specific genes of donor DNA.

UDC 575:576.8

INTERRUPTION OF THE INJECTION OF THE CHROMOSOME OF PHAGE T4B INTO THE BACTERIAL CELL

M.I. MEKSHENKOV and R. D. GUSEINOV
Institute of General Genetics
Academy of Sciences of the USSR, Moscow

It is demonstrated, firstly, that different particles of phage T4r^{+} introduce DNA fragments containing different sets of genes into cells of E. coli B, which means that the chromosome population of this phage contains cyclically permuted genes; secondly, that the artificially fragmented genome is infectious.

The interruption of the chromosomal transfer from donor to recipient is a widely used procedure in bacterial conjugation. This approach has illuminated both the genetic topography of the bacterial chromosome and the recombination process.

Such a procedure is inapplicable to phage infection, where the transfer of the phage DNA molecule into the cell normally proceeds very rapidly (within one minute).

We found that low temperature slows down the injection of the DNA molecule of adsorbed phage into the cell, so that the process can be interrupted by treatment of the infected cells in a mixer.

This method of interrupting the injection of phage DNA into the cell was used to analyze the organization of the phage T4 chromosome.

MATERIALS AND METHODS

The work was carried out with the bacterium Escherichia coli — phage T4r^{+} system, a tryptophan-requiring strain. Concentrated phage was obtained according to Swanstrom and Adams (1951). The determination of the phage titer and the number of infective centers, as well as the preparation of rabbit antiserum against phage T4 were as described by Adams (1959). ^{32}P-labeled phage was obtained by culturing the phage in cells grown in a medium containing radioactive phosphorus. The concentrated phage had a specific radioactivity of 1,100 pulses/sec. The background of free radioactive phosphorus did not exceed 0.01%.

Hershey's medium (Hershey and Chase, 1952) was used, with an addition of 1% caffeine and with an increased concentration of the following ions: Ca^{++} — 10^{-3} M and Mg^{++} — 10^{-2} M (medium 1).

Adsorbed phage was separated from the cells by a 40-second treatment in an Omnimixer (14,000 rpm), produced by Ivan Sorvall Inc.

RESULTS AND DISCUSSION

Slowing down and interruption of the injection of DNA

Cells grown in meat peptone broth to the middle of the logarithmic phase were centrifuged and the sediment resuspended in medium 1 (to a density of $2-5 \cdot 10^8$ cells/ml) previously cooled to the necessary temperature. The cell suspension was incubated for 5 minutes at this temperature, after which it was infected with phage previously diluted in the same medium to give a multiplicity of 0.1 (the multiplicity, designated m, indicates the number of particles added per cell). Adsorption lasted 2 minutes. Afterwards antiserum (diluted 1:10) was added to the mixture, causing the inactivation of unadsorbed phage within 60 seconds of incubation. Starting from this moment, which was arbitrarily taken as zero, samples were taken at the desired intervals and diluted 100-fold in medium 1 (without tryptophan) previously cooled to the required temperature. The samples were then treated in the mixer. The titer of infected cells and the amount of ^{32}P-DNA precipitated with the cells were determined in both treated and untreated samples. To determine the amount of precipitated radioactive phosphorus the samples were subjected to threefold centrifugation, followed by resuspension first in medium 1, then in meat peptone broth and finally again in medium 1. The cell suspension obtained was passed through a No. 2 Millipore filter. The quantity of ^{32}P-DNA in pulses/sec/μg DNA, i.e. the ratio between the radioactivity of the sample and the amount of DNA transferred throughout the injection period, expressed in μg, was calculated from the amount of radioactivity on the filter.

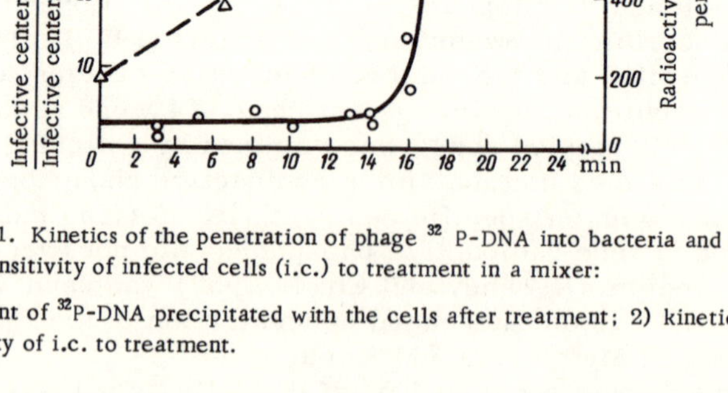

FIGURE 1. Kinetics of the penetration of phage ^{32} P-DNA into bacteria and kinetics of the sensitivity of infected cells (i.c.) to treatment in a mixer:

1) amount of ^{32}P-DNA precipitated with the cells after treatment; 2) kinetics of the sensitivity of i.c. to treatment.

The experiments showed that at low temperature most of the DNA of the adsorbed phage in the mixer-treated samples is detached from the cell together with the phage envelope, while the infected cells lose the capacity to form plaques. This effect of temperature is reversible: the sensitivity of the infected cells to such treatment immediately disappears if the sample is placed for a short time at 37°.

Figure 1 shows the results obtained at a temperature of 7° and a multiplicity of infection equal to unity. Although the transfer of ^{32}P-DNA from the adsorbed phage into the bacteria is linear with time (curve 1), the number of infective centers resistant to treatment does not increase (curve 2) until the ^{32}P-transfer curve has reached a plateau, i.e. until the injection of all the phage DNA into the cell is completed. This means that the fraction of ^{32}P-DNA found in the bacteria at any time consists of DNA fragments originating from a single phage particle and not entire molecules from different particles, since in the latter case the number of infective centers would increase proportionally to the amount of ^{32}P-DNA transferred. Thus, the period during which the infected cells remain sensitive to treatment in the mixer corresponds to the time required for the transfer of an individual DNA molecule into the cell. Figure 2 shows the experimentally determined duration of this period of temperature sensitivity. The activation energy, calculated from this curve on the basis of the Arrhenius equation, is equal to 16,000 calories/mole of nucleotide pairs. Hence the period of DNA injection at 37° equals 1.1 minute, which corresponds to a transfer at the rate of 3,000 nucleotide pairs per second. The data obtained indicate that at 7° the time required for DNA injection is 18 minutes and the transfer rate is 195 nucleotide pairs per second.

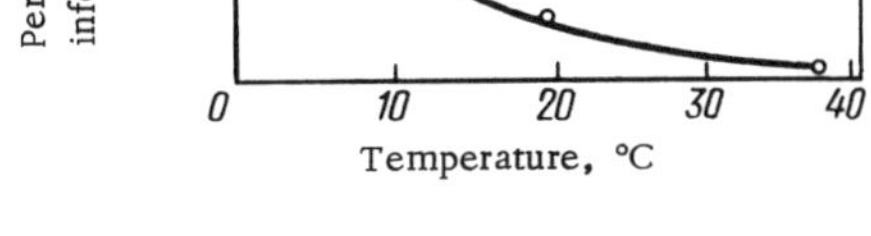

FIGURE 2. Effect of the temperature of the incubation mixture on the period of sensitivity of infected cells to treatment in a mixer; m = 0.1.

It follows from these results that at low temperatures the injection of phage DNA into the bacterium is slowed down and the transfer of the DNA molecule into the cell can be interrupted by treatment in a mixer, so that only a fraction of the phage genome enters the cell.

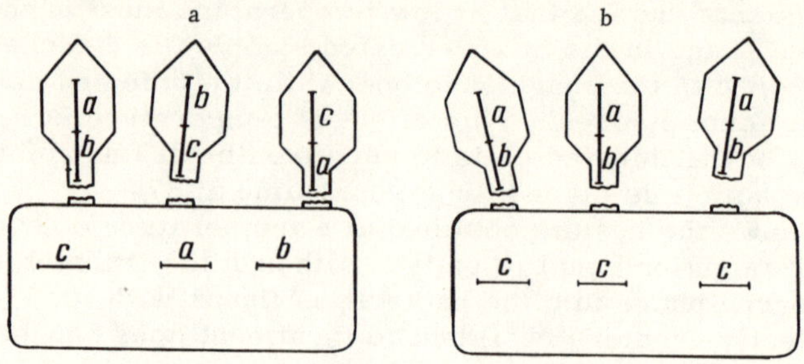

FIGURE 3. Diagram showing the results of interrupting the injection of DNA in phages characterized by chromosomal populations bearing genes in different sequences:

a) cyclic permutations of genes: a b c : b c a; b) the chromosomes of all phage particles have identical gene sequences: a b c.

The sequence of genes in the chromosomes of phage particles

Experiments interrupting the injection of phage DNA into the cell provide a novel approach to the cyclic permutations of genes in the chromosomes of phage T4 populations. In this way we determined whether the DNA fragments obtained by the cell from different phage particles in the case of multiple infection contained identical or different sets of genes. As shown in diagrams a and b (Figure 3), the fragments which have entered the cell before the injection is interrupted may or may not contain collectively a complete set of phage genes, depending on whether the DNA molecules of the phage particles have different or identical gene sequences.

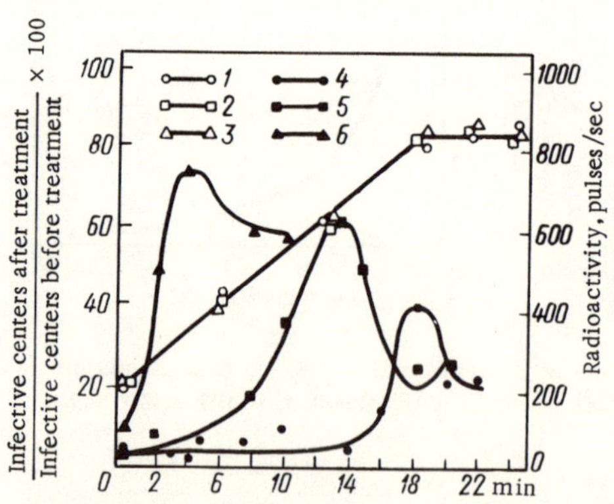

FIGURE 4. Sensitivity of infected cells (i.c.) to treatment in a mixer and kinetics of the penetration of ^{32}P-DNA into bacteria at different multiplicities of infection (at 7°C).

Kinetics of the penetration of ^{32}P-DNA at: m = 0.1 (1), m = 2.2 (2); m = 22 (3); sensitivity of i.c. to treatment at: m = 0.1 (4); m = 2.2 (5); m = 22 (6).

Accordingly, the period of time necessary for the transfer of a complete set of phage genes into the cell will either decrease or remain constant with an increase in the multiplicity of infection. If the fragmented genome can cause an infection, the period during which the infected cells are sensitive to treatment in a mixer will accordingly decrease or remain constant.

The results obtained are shown in Figure 4. The period during which the infected cells are sensitive to treatment in a mixer is seen to decrease — from 18 minutes at m = 0.1 to 13 minutes at m = 2.2 and 4 minutes at m = 22. In accordance with diagram a (Figure 3) the shortening of the period of sensitivity to treatment in a mixer can indeed be attributed to the fact that the time necessary for the penetration of DNA fragments with different gene sets into the cell is shorter than that required for the transfer of an entire DNA molecule. An increase in the multiplicity of infection does not accelerate the injection of DNA, as demonstrated by the identical slope of the curves of the transfer of ^{32}P-DNA and the coincidence of the moments at which transfer is computed at all multiplicities of infection.

The following conclusions can be made on the basis of the results obtained.

The population of phage T4 is characterized by the presence of cyclical gene permutations in the chromosomes of the phage particles.

The artificially fragmented genome of phage T4 is infectious, i.e. it can induce the development of phage progeny.

BIBLIOGRAPHY

Adams, M. Bacteriophages. — (Interscience) Wiley, 1959.

Hershey, A.D. and M. Chase. — J Gen. Physiol., Vol. 36 (1952), 39.

Swanstrom, M. and M.H. Adams. — Proc. Soc. Exp. Biol. Med., Vol.78 (1951), 372.

REPAIR OF DNA LESIONS

UDC 575:676.8

MOLECULAR MECHANISMS OF DNA REPAIR

V.P PARIBOK and E.A. VAL'DSHTEIN
Institute of Cytology
Academy of Sciences of the USSR, Leningrad

Recent studies have revealed the molecular nature of certain processes of DNA repair: photoreactivation, multistage repair and post-replicative repair. Some of these processes are highly specific; others have a broad spectrum of action. The cell repairs both lethal and mutagenic lesions. It is not known whether the repair mechanism is the same in both these cases. A major problem is to evaluate the biological significance of repair processes. The nature of irreparable lesions of DNA is of particular interest. Repair processes have been found to be widespread in living organisms.

There is no doubt at present that special enzymatic systems of the cell repair a considerable part of the structural DNA lesions induced by radiation and other agents (see reviews: Val'dshtein, 1970; Howard-Flanders, 1968). The mechanism of action of some of these systems is already known; for example the photoreactivation enzymes and polynucleotide ligase. These repair systems consist of a single stage and are therefore highly specific and accurate. The specificity and accuracy of multistage repair mechanisms, on the other hand, is a complex and still poorly understood topic. This applies not only to the largely obscure process of post-replicative repair but also to the excision-substitution mechanism, which has been under intensive study for a number of years. In the latter case the difficulties are associated with the initial stage of repair — the elimination of the structural lesion from the DNA.

It was unknown until a short time ago whether the excision of dimers involves one or several enzymes. An answer to this question was obtained in experiments with *Escherichia coli* mutants incapable of excising dimers in which the dimers were eliminated by an additional exposure to ionizing radiation (Paribok et al., 1968, 1969). We concluded on these grounds that the excision of dimers in normal strains consists of two consecutive events: incision and excision, each performed by a specific nuclease cleaving the phospho-ester bonds on either side of the dimer. Such enzymes UV endonuclease and UV exonuclease — have now been described and isolated from *Micrococcus lysodeikticus* cells (Takagi et al., 1968; Grossman et al., 1968).

The role of the additional exposure to ionizing radiation in our experiments can be explained as follows. Ionizing radiation causes breaks in the DNA strands and induces a wave of enzymatic degradation which reaches the dimer. The exposure of the dimer imitates the action of one of the nucleases participating in excision. The dimer becomes accessible to the second enzyme which breaks the bond on the other side of the dimer and thus releases this from the DNA. The wave of degradation can spread further and thus eliminate subsequent dimers with the participation of the second enzyme. If this model is correct and the wave of degradation spreads in a polar sense, additional exposure to ionizing radiation may be used as a method to separate mutants differing from one another in the nucleolytic activities involved in the excision of dimers. The additional irradiation leads to elimination of dimers only in those mutants in which the enzyme acting on the side of the spreading wave of degradation is inactive.

Excision of dimers and DNA degradation induced by ionizing radiations in UV-irradiated* uvr⁻ mutants

E. coli strain	Genotype	Radioactivity (%) 60 minutes after exposure to ionizing radiation, krad					
		0	5	10	0	5	10
		dimer fraction			acid-insoluble fraction		
K12 AB1886	uvrA	100	82	65	100	67	60
K12 AB1884	uvrC	100	99	97	100	86	68
$B_{s-1}\gamma$ R**	uvrB	100	—	67	100	—	65

* UV dose — 860 ergs/mm^2.

** The $B_{s-1}\gamma$ R strain was exposed to α rays.

As shown in the table, UV irradiation alone does not bring about DNA breakdown and an elimination of dimers. When UV is combined with ionizing irradiation, DNA breakdown is observed in all strains but dimers are eliminated in only two of the strains examined — E. coli K12 AB1886 uvrA and $B_{s-1}\gamma$ R uvrB. Dimers are not eliminated in the K12 AB1886 uvrC mutant. These findings can only be regarded as preliminary. The identification of the genes associated with the incision and excision enzymes requires the analysis of numerous uvr⁻ mutants originating from a single strain.

It is noteworthy that additional exposure to X rays at a dose rate of 1.6 krad/min decreases the lethal effect caused by UV irradiation in the E. coli strains $B_{s-1}\gamma$ R and K12 AB1886 but not in the K12 AB1884 mutant (Figure 1). The reactivation effect is small and requires the combination of low doses of UV and X rays. The original strain K12 AB1157 uvr⁺ shows no reactivation effect. Moreover, UV-irradiated cells of the latter strain are more sensitive to further exposure to X rays than cells not treated with UV. The sensitizing effect of a preliminary UV irradiation was described earlier by Haynes (1964) in experiments with E. coli B/r.

The reactivation effect was observed in strains in which additional irradiation brought about the elimination of dimers. This coincidence of results at the molecular and cellular levels suggests that reactivation is due to the removal from the DNA of a certain number of dimers lethal to the cell. The view that the reactivating effect of ionizing radiation results from a stimulation of post-replicative repair appears less acceptable since reactivation occurred not only in the AB1886 strain but also in the $B_{s-1}\gamma$ R strain, which is defective not only in the uvr gene but also in the exr gene.

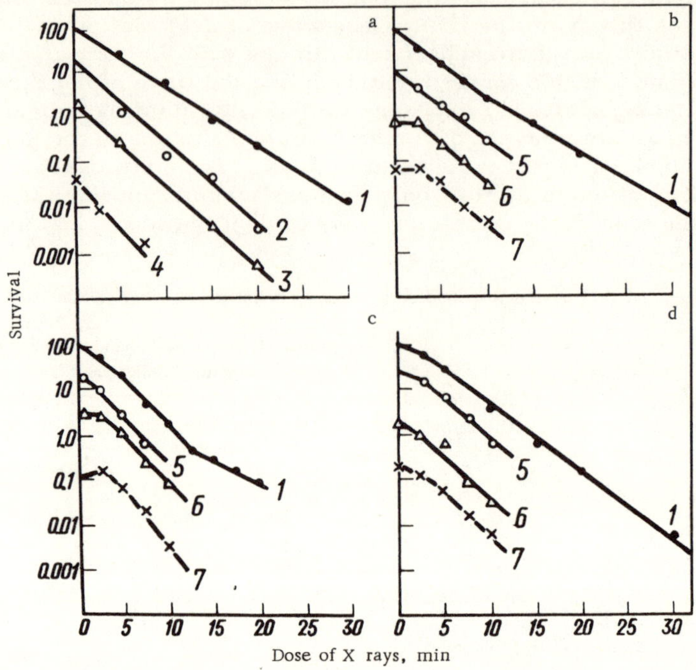

FIGURE 1. Survival of different strains of E. coli after consecutive exposure to UV and X rays: a) K12 AB1157; b) K12 AB1886 uvrA; c) B_{s-1} γ R; d) K12 AB1884 uvrC; UV doses (ergs/mm^2): 1) no irradiation; 2) 335; 3) 500; 4) 670; 5) 14; 6) 28; 7) 42.

Such double mutants are more sensitive to UV than the uvr mutants. This increased sensitivity can probably be attributed to a reduced capacity for post-replicative repair (Witkin, 1969).

Incision thus emerges as a primary event which triggers a repair involving excision and substitution. The endonuclease performing the incision is therefore responsible for recognition and for the accuracy of the operation during this stage of the repair process. The incision enzyme and the whole repair mechanism are widely believed to be of low specificity. This mechanism is often regarded as an error-correcting one on the assumption that it recognizes some secondary defect in the DNA double helix rather than a particular base modification. If this is actually so, the incision enzyme may erroneously break the strand opposite the lesion and thus lead to mutations (Setlow, 1964) or chromosomal aberrations (Dubinin and Soyfer, 1969).

The accuracy of incision endonuclease has been studied in our laboratory by N. V. Tomilin in experiments with artificial DNA hybrids containing dimers in one strand and a radioactive label in the other strand (Paribok and Tomilin, 1970). The hybrids were incubated with purified endonuclease isolated from M. lysodeikticus. As shown in Figure 2, the molecular weight of the DNA strand lacking dimers does not differ from that of the corresponding strand in a hybrid formed by nonirradiated partners. This means that the enzyme attacks only the dimer-containing strand and does not act on the opposite strand. The result is independent of whether the dimers lie in the light or heavy DNA strand. Significantly, the number of breaks in the dimer-containing strand was close to the number of dimers in this.

Similar data have been obtained by Carrier and Setlow (1970) in a study of the activity of endonuclease toward UV-irradiated and nonirradiated strands of hybrid DNA. It is furthermore known that the M. lysodeikticus enzyme which breaks down methylmethanesulfonate-treated DNA does not cause breaks in strands opposite a methylated base (Strauss and Robbins, 1968).

Data obtained at the molecular level provide no indication of any errors in the repair process at the excision stage. Genetic experiments lead to the same conclusion. According to Witkin (1969), not more than one mutation arises per 10^6 dimers eliminated.

Subsequently, the activity of the enzyme was tested with respect to single-stranded DNA containing dimers. As shown in Figure 3, the endonuclease efficiently breaks down single-stranded UV-irradiated DNA also although it does not act on denatured DNA. This finding disagrees with the hypothesis that locally denatured areas of double helical DNA are recognized.

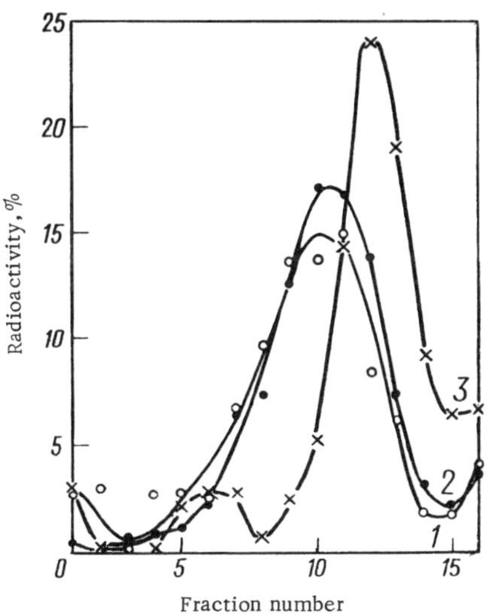

FIGURE 2. Sedimentation pattern showing the effect of UV-endonuclease on hybrid DNA composed of UV-irradiated and nonirradiated strands.

The incubation mixture consisted of 0.2 ml hybrid DNA (2 μg), 0.1 ml 0.2 M Tris-HCl, 0.16 M EDTA (pH 7.4), and 0.1 ml of the enzyme (70 μg protein). After incubation for two hours at 37°, 0.2 ml of the mixture was placed on a 5—18% gradient of alkaline sucrose centrifuged for 3 hours at 48,000 rpm in a Spinco L265K, rotor SW50L.

1) ^{14}CL-^{12}CH$\widehat{TT}$; 2) ^{14}CL-^{12}CH; 3) ^{14}CL$\widehat{TT}$-^{12}CHTT; L — light strand; H — heavy strand; TT — dimer.

In order to determine whether the enzyme recognizes particular base changes or is less specific and detects secondary-structure disturbances induced by various base modifications, we tested the activity of the endonuclease with respect to DNAs treated with different mutagens.

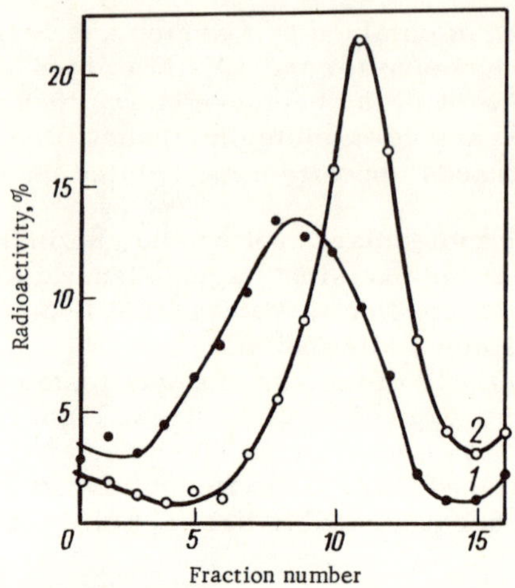

FIGURE 3. Effect of UV-endonuclease on UV-irradiated single-stranded DNA (sedimentation pattern).

Test conditions as in Figure 2; 1) ^{14}CHT̂T + H_2O; 2) ^{14}CHT̂T + + enzyme.

The endonuclease recognizing the dimers proved inactive toward hydroxylamine-treated DNA (see Figure 4). The enzyme was also inactive with respect to control DNA kept at the same temperature and ionic strength but without hydroxylamine.

The slight shift in the sedimentation profile toward lesser molecular weight in the sample containing the enzyme did not disappear in the presence of excess DNA irradiated with a large dose of UV. Consequently, this shift cannot be attributed to a UV-specific endonuclease. Our enzyme preparation is apparently contaminated with an endonuclease activity attacking hydroxylamine-treated DNA.

Treatment with hydroxylamine (HA) causes fragmentation of DNA. This fragmentation possibly results from the oxidative action of hydroxylamine, which is usually evident at low concentrations of the agent (Rhaese and Freese, 1968).

In previous experiments with E. coli B and K12 cells we were able to show that hydroxylamine induces lesions which are lethal for the B_{s-1} and K12 AB2463 rec$^-$ strains but are nonlethal for the X-ray-resistant strains B/r, B_{s-1} γ RK12 AB1157 and K12 AB 1886 (Isabaeva et al., 1967; Isabaeva, 1967). We concluded on these grounds that hydroxylamine induces single breaks in DNA strands which can be corrected given a fully functional repair system. The hcr system does not participate in the repair of hydroxylamine-induced DNA lesions, as demonstrated by model experiments according to which modified cytosine molecules are not recognized by dimer-specific endonuclease. In phage T4, the frequency of lethal and mutational events induced by treatment with hydroxylamine does not depend on the condition of gene v, which controls the excision of dimers (Freese and Freese, 1966). The survival of phages ϰ and λ in the presence of hydroxylamine is the same regardless of whether the phages replicate in hcr^+ or hcr^- host cells (Winkler, 1965; Soyfer, 1969).

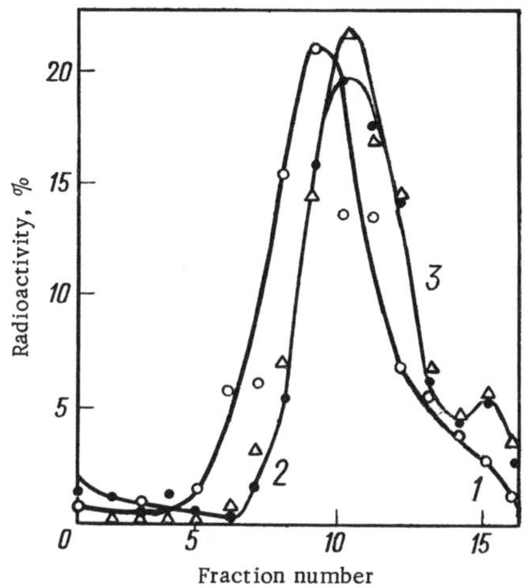

FIGURE 4. Sedimentation pattern showing the effect of UV-endonuclease on hydroxylamine (HA)-treated DNA.

Incubation mixture composed of 0.1 ml ^{14}C-DNA previously treated with HA, 0.1 ml 0.2 M Tris-HCl, 0.16 M EDTA (pH 7.4), 0.1 ml of the enzyme (50 μg protein) or H_2O and 0.1 ml UV-irradiated ^{12}C-DNA or H_2O. After incubation for one hour at 37° the mixture was centrifuged as described in Figure 2.
1) HA-DNA + H_2O; 2) HA-DNA + enzyme; 3) HA-DNA + enzyme + ^{12}C-UV-DNA. ^{12}C-UV-DNA was added to (1) and (2) after incubation. Conditions of HA treatment: 0.1 ml ^{14}C-DNA, 0.5 ml 2M $NH_2OH \cdot HCl$ and 0.4 ml 2M NaCl, 0.01 M EDTA (pH 5.2); incubation for 60 minutes at 70°. HA was eliminated by passage through a Sephadex G50 column.

Experiments with phage T1 in our laboratory gave similar results. The phage survived equally well in E. coli strains capable and incapable of repairing DNA lesions induced by UV and X rays. Moreover, mutation of the gene hcr in recipient cells of Bac. subtilis does not affect the transforming activity of hydroxylamine-treated DNA (Bresler et al., 1970).

In experiments with DNA treated with nitrogen mustard we revealed a distinct degradation as a result of exposure to a crude extract from M. lysodeikticus. This finding agrees with the data of Reiter and Strauss (1965), who obtained a marked decrease in the transforming activity under comparable experimental conditions. The crude extract also caused extensive degradation of DNA treated with methylnitronitrosoguanidine and methylmethanesulfonate. The latter finding corresponds to the results of Strauss, who worked with an enzyme preparation extracted from the same bacterial species (Strauss and Robbins, 1968). Thus, the extract shows a nuclease activity towards alkylated DNA. It is difficult to say whether this enzyme is identical with the endonuclease which recognizes dimers or a totally different enzyme. It is only known that E. coli contains an endonuclease which is active toward methylated DNA but not toward UV-irradiated DNA (Friedberg and Goldthwait, 1968).

BIBLIOGRAPHY

Bresler, S.E., V.L. Kalinin, and D.A. Perumov. — Mutation Res. 9, No. 1 (1970), 1.

Carrier, W.L. and R.B. Setlow. — J. Bact. 102, No. 1 (1970), 178.

Dubinin, N.P. and V.N. Soyfer. — Mutation Res. 8, No. 2 (1969), 353.

Freese, E.B. and E. Freese. — Genetics 54, No. 5 (1966), 1055.

Friedberg, E.C. and D.A. Goldthwait. — Cold Spring Harbor Sympos. Quant. Biol., Vol. 33 (1968), 271.

Grossman, L., J.C. Kaplan, S.R. Kushner, and I. Mahler. — Cold Spring Harbor Sympos. Quant. Biol., Vol. 33 (1968), 229.

Haynes, R.H. — In: "Physical Processes in Radiation Biology," p. 51. New York, 1964.

Howard-Flanders, P. — A. Rev. Biochem., Vol. 37 (1968), 175.

Isabaeva, M.K. — Tsitologiya 9, No. 12 (1967), 1490.

Isabaeva, M.K., E.A. Val'dshtein, and V.D. Zhestyanikov. — Tsitologiya 9, No. 12 (1967), 1481.

Paribok, V.P. and N. V. Tomilin. — Doklady AN SSSR 195, No. 2 (1970), 489.

Paribok, V.P., E.A. Val'dshtein, E.E. Novikov, and N.V. Tomilin. — Synopses of Reports 9th All-Union Conference on Biological Effect of Ultraviolet Radiation, p.38. Sochi, 1—5 October, 1968. (Russian)

Paribok, V.P., E.A. Val'dshtein, and N.V. Tomilin. — Tsitologiya 10, No. 1 (1969), 93.

Reiter, H. and B. Strauss. — J. Molec. Biol. 14, No. 1 (1965), 179.

Rhaese, H.-J. and E. Freese. — Biochim. biophys. Acta 155, No. 2 (1968), 476.

Setlow, R.B. — J. Cell. Comp. Physiol., 64, Suppl. 1,2 (1964), 51.

Soyfer, V.N. — Microbiol. Genet. Bull., Vol. 31 (1969), 18.

Strauss, B.S. and M. Robbins. — Biochim. biophys. Acta 161, No. 1 (1968), 68.

Takagi, Y., M. Sekiguchi, S. Okubu, H. Nakajama, K. Shimada, S. Yasuda, T. Nishimoto, and H. Yoshihara. — Cold Spring Harbor Sympos. Quant. Biol., Vol. 33 (1968), 219.

Val'dshtein, E.A. — In: Protsessy reparatsii pri luchevom porazhenii. Moscow, Atomizdat, 1970.

Winkler, U. — Z. Naturf. 20b, No. 9 (1965), 864.

Witkin, E.M. — A. Rev. Microbiol., Vol. 23 (1969), 487.

UDC 575:576.8

REPAIR SYSTEMS IN BLUE-GREEN ALGAE: A STUDY OF MUTANTS SENSITIVE TO INACTIVATING FACTORS

S.V.SHESTAKOV, V.D. ZHEVNER and T.N. MITRONOVA
Moscow State University

A group of mutants distinguished by their sensitivity to UV and other inactivating factors was isolated from the blue-green unicellular algae *Synechocystis aquatilis* Sanw. and *Anacystis nidulans*. The uvs mutants of *S. aquatilis* include a mutant which is highly resistant to alkylating agents and defective with respect to photoreactivation, showing a high level of spontaneous and UV-induced mutagenesis, a mutant with a low level of mutability and mutants showing a cross sensitivity to ionizing radiation, methylmethanesulfonate, nitrosoguanidine and mitomycin C. Recipient cells of uvs mutants of *A.nidulans* repair UV-induced lesions in transforming DNA less efficiently than the wild type cells. Study of the properties of the various uvs mutants suggests that blue-green algae possess repair systems analogous to the mechanisms of excision and post-replicative repair.

Blue-green algae are highly resistant to radiation and show a low level of mutagenesis, both spontaneous and induced by radiation or certain chemical agents (Shestakov, 1970; Shestakov and Mitronova, 1971). It may be postulated that the genetic stability of the Cyanophyta is based on the repair of lethal and mutagenic lesions. Blue-green algae possess an efficient system of enzymatic photoreactivation (Wu et al., 1967; Werbin and Rupert, 1968) and there are indications that they are equipped with mechanisms comparable to dark repair (Singh, 1968; Stevens and Van Baalen, 1969).

One of the main approaches to the evaluation of repair processes in microorganisms consists in the analysis of mutants sensitive to various inactivating agents. We isolated a group of UV-sensitive mutants from the unicellular, obligate photosynthetic algae *Synechocystis aquatilis* Sanw. (Gromov and Avilov, 1969) and *Anacystis nidulans*. The conditions of cultivation, irradiation and treatment of the algae with chemical agents as well as the mutagenesis induced by nitrosoguanidine (NG) and nitrosomethylurea (NMU) have been described earlier (Stoletov et al., 1965; Bozhukova and Zhevner, 1970; Shestakov et al., 1971). Seventeen UV-sensitive (uvs) mutants of *S. aquatilis*, induced by NMU and NG, were isolated by the replica method after threefold UV selection on solid media. Four of these mutants (m12, m6, m5 and n21) were studied in greater detail. Of the uvs mutants of *A. nidulans* examined in this work, the Fil 11 strain was isolated from NG-induced morphological mutants with impaired cytokinesis (Shestakov and Zhevner, 1969), the Br8 strain was found among spontaneous mutants resistant to the antibiotic bruneomycin and the An60 mutant was isolated after selection for UV sensitivity.

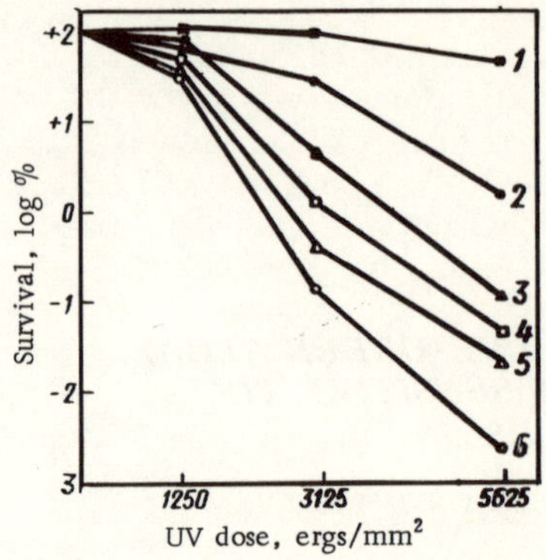

FIGURE 1. Sensitivity of wild type and mutant cells of S. aquatilis Sanw. to UV radiation:

1) wild type; 2) n21; 3) m18; 4) m5; 5) m6; 6) m12.

Figures 1 and 2 show dose-effect curves for wild type cells and some mutants of S. aquatilis and A. nidulans exposed to UV irradiation.

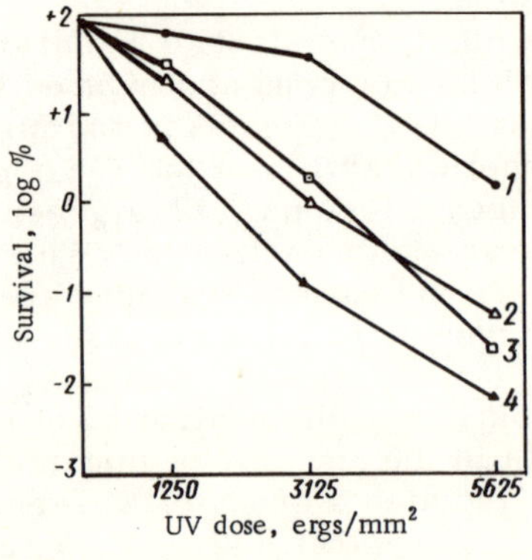

FIGURE 2. Sensitivity of wild type and mutant cells of A. nidulans to UV radiation:

1) wild type; 2) Fil 11; 3) An60; 4) Br8.

Comparison of the curves for the survival rate and the lethal effect of UV at equal doses (3125 or 5625 ergs/mm^2) shows that the mutants examined differ in UV sensitivity. As in other microorganisms, the UV resistance of blue-green algae is apparently controlled by a complex polygenic system. This is also suggested by the different sensitivity of the uvs mutants to ionizing radiation and alkylating agents (Table 1). The discovery in A. nidulans and S. aquatilis of uvs mutants with an increased sensitivity to X rays and simultaneously to methylmethanesulfonate (MMS), mitomycin C (MitC) and nitrosoguanidine suggests that blue-green algae possess repair systems comparable to the mechanisms of excision and post-replicative repair (Rörsch et al., 1966; Howard-Flanders, 1968). Because of the different course of the survival curves it is impossible to carry out a strict quantitative comparison of the lethal effects of the above inactivating agents on the mutants examined, or of the sensitivity of each mutant to the different agents. It is evident from the data of Table 1, however, that the mutants with a cross sensitivity to the inactivating agents are not identical in their characteristics (compare the lethal effects at similar survival levels in the wild type cells — 65% for radiation and 7% for chemical agents). Thus, the m5 mutant is highly sensitive to MMS but is more resistant to NG and especially to MitC, whereas the n21 mutant is less sensitive to UV but extremely sensitive to X rays, NG and MitC. The A. nidulans mutants show less pronounced differences in sensitivity to UV and X rays, apparently because of the lower initial level of the radio-resistance of this species in comparison with S. aquatilis (Bozhukova and Zhevner, 1970). The differences in the cross resistance of the uvs mutants possibly result from disturbances of different stages of a given repair mechanism, or disturbances of different specific systems controlling resistance not only to UV but also to other inactivating factors.

TABLE 1. Sensitivity of uvs mutants of S. aquatilis and A. nidulans to inactivating factors

Strain	Survival of wild type/survival of mutant, %				
	UV	X rays	MMS	NG	MitC
		S. aquatilis			
m12	325	1.1	0.7	0.12	0.11
m6	180	1.2	0.8	0.4	0.23
m5	54	110	230	9	1.8
n21	17	660	70	35	230
		N. nidulans			
Fil 11	3.5	4.8	2.8	1.2	6
Br8	26	7	—	—	0.08
An60	3.1	1.9	—	—	90
Survival of wild type, %	65	65	7	7	7

The m12 and m6 mutants, which are the most sensitive to UV, differ markedly from the S. aquatilis strains sensitive to X rays. Cells of the mutant m12, and to a lesser extent those of m6, are more resistant to alkylating agents than wild type cells. This property may be due to a change in the level of photoreactivation. As shown in Table 2, the m12 mutant lacks any capacity for photoreactivation in vivo, whereas in the m6 mutant the efficiency of photoreactivation is lowered. Assuming that a functional relationship exists between the mechanisms of enzymatic photoreactivation and dark repair (Hanawalt, 1968; Fedorova, 1969), it appears that the increased resistance of the m12 and m6 mutants to alkylating agents is due to an activation of the excision system, compensating for the absence or weakening of photoreactivation; the enhancement of the photoreactivating sector in the pleiotropic mutant n21 can be attributed to a disturbance of the dark repair mechanisms. However, the inverse correlation between resistance to alkylating agents and capacity for photoreactivation is not a general rule: in the m5 mutant (as well as in the Fil 11 mutant of A. nidulans) the pleiotropic increase in sensitivity to various factors is accompanied by a simultaneous lowering of the efficiency of photoreactivation.

TABLE 2. Efficiency of photoreactivation and mutability of uvs mutants of S. aquatilis

Property	Wild type	m12	m6	m5	n21
Efficiency of photoreactivation*	0.68	0	0.46	0.32	0.76
Spontaneous mutability**	0.8	6.2	0.1	1.4	1.2
UV-induced mutability †	2.6	1,370	1.8	3.6	9.1

* Photoreactivated sector $1 - A/B$, where A indicates incubation of the cells for 6 hours in a phosphate buffer pH 7.0 in the light, B—in the dark; $A(B) = \sqrt{N_0/N} - 1$, where N_0 is the titer of viable cells in the control and N the titer after irradiation with a UV dose of 1,250 ergs/mm^2. Calculated according to Setlow et al. (1968).

** Total number of eryr mutants per 10^8 viable cells (for procedure see Shestakov and Mitronova, 1971).

† At a UV dose of 2,400 ergs/mm^2.

The m6 mutant, though phenotypically similar to the m12 mutant with respect to sensitivity to inactivating factors, differs fundamentally from it not only in its capacity for photoreactivation but also in its rates of spontaneous and UV-induced mutagenesis, determined according to the frequency of mutants resistant to erythromycin (see Table 2). The m12 mutant shows very high levels of spontaneous and especially UV-induced mutagenesis, whereas the level of spontaneous mutability in the m6 mutant is much lower than in the wild type. As in the m5 mutant, the decrease in the UV-induced mutagenesis in the m6 mutant is more conspicuous at equilethal irradiation doses. Possibly a mutator gene is active in the cells of the m12 mutant. On the other hand, the higher levels of UV-induced and spontaneous mutagenesis in the m12 mutant may result from an absence of photoreactivation. Photoproducts in the DNA, leading to lethal and mutational lesions, evidently accumulate in the genome of blue-green algae not only with UV irradiation but also on exposure to visible "white" light, necessary for the cultivation of the algae. These premutational lesions, normally repaired by the photoreactivation mechanism, become manifest in the cells of a mutant devoid of photoreactivation.

The changes in the rates of UV-induced mutagenesis in various uvs strains can be regarded as indirect evidence that blue-green algae possess a recombinational (post-replicative) repair mechanism which is involved in the emergence of UV-induced mutations (Witkin, 1969).

In A. nidulans, the recombinational system of genetic transformation (Shestakov and Nguen Than Khyen, 1970) can be used directly for the assessment of repair mechanisms. Disturbances in the repair systems of bacteria are identified by the ability of recipient cells to restore the transforming capacity of UV-irradiated DNA (Setlow et al., 1968). As shown in Figure 3, the "survival" of transforming DNA of A. nidulans after UV irradiation is very high if wild type cells participate in the transformation and much lower if the Fil 11 or An60 strains are used as the recipients. The lowering of the transforming activity of UV-irradiated DNA in the cells of the Fil 11 mutant can be attributed in part to a decrease in the efficiency of photoreactivation. Not so in the An60 mutant, which has a normal level of photoreactivation; here we may be dealing with a disturbance in a different intracellular mechanism repairing UV-induced lesions in transforming DNA. It is noteworthy that the Fil 11 and An60 mutants show a pleiotropic sensitivity to various inactivating factors (see Table 1) and a lowered capacity for transformation. These mutants (like the exr^- or rec^- mutants of E. coli) possibly contain defects in certain recombination mechanisms, the disturbance of which lowers the capacity of the cell for repair (Witkin, 1969).

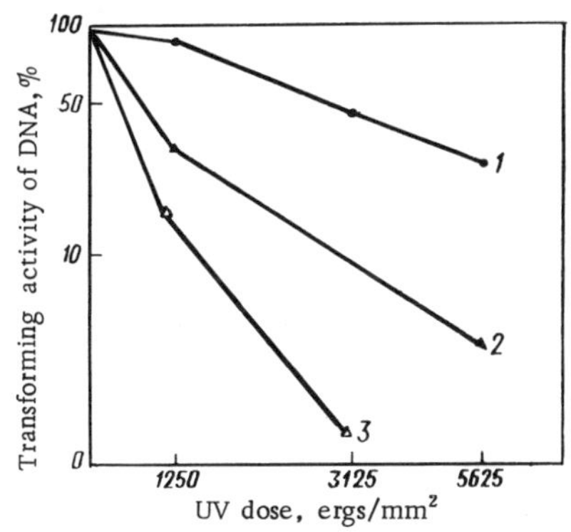

FIGURE 3. UV-induced inactivation of transforming DNA of A. nidulans:

1) wild type; 2) An60; 3) Fil 11. Conditions of the experiment: 10 μg of donor DNA from an A. nidulans mutant resistant to 100 μg erythromycin was incubated in 1.2 ml of phosphate buffer pH 7.2 with recipient cells (density of suspension $3 \cdot 10^7$ cells/ml) taken from 3-day-old cultures of the wild type strain or the Fil 11 or An60 mutants for 40 minutes at 37°, illumination 2,000 lux; 0.1 ml of the suspension was inoculated in dishes containing a solid medium; erythromycin at final concentration of 5 μg/ml of medium was introduced under the agar after 2 days and after 7 days of culturing, the number of ery clones was determined. DNA was irradiated in a citrate-salts solution in quartz cuvettes. The transforming activity of nonirradiated DNA with respect to each recipient was taken as 100%.

The following conclusions can be drawn concerning the properties of the various uvs mutants of blue-green algae. 1. Blue-green algae possess a potent photoreactivation system which confers effective protection against lethal and mutational lesions induced by solar radiation. 2. Apart from photoreactivation, blue-green algae are also equipped with other repair systems which control the sensitivity of the cells to UV radiation, ionizing radiation and alkylating agents. 3. The rates of the mutation process in blue-green algae are genetically determined and depend on the operation of the photoreactivation system and other intracellular processes, including repair and recombination.

As a result of the possibility of isolating and analyzing various uvs mutants, blue-green algae emerge as interesting and promising models for research into the specificity, interaction, evolution and biological significance of the different repair systems and the mechanisms of the mutation process in photosynthesizing organisms.

BIBLIOGRAPHY

Bozhukova, E.E. and V. D. Zhevner. — Vest. MGU (Moscow State University), biol. ser., Vol.5 (1970), 97.

Fedorova, I.V. — Genetika, Vol. 12 (1969), 120.

Gromov, B.V. and I.A. Avilov. — Fiziologiya Rastenii 16, No. 6 (1969), 1088.

Hanawalt, P.C. — Photophysiology, Vol. 4, p.203. New York, Academic Press, 1968.

Howard-Flanders, P. — A. Rev. Biochem., Vol. 37 (1968), 175.

Rörsch, A., P. Van de Putte, I. Mattern, H. Zwenk, and M. Van Sluis. — In: "Genetical Aspects of Radiosensitivity," p.105. Vienna, 1966.

Setlow, J.K., D.C Brown, M. Boling, A. Mattingly, and M.P. Gordon. — J. Bact. 95, No. 2 (1958), 546.

Shestakov, S.V. — In: "Taxonomy and Biology of Blue-green Algae," p. 47, 1970.

Shestakov, S.V., E. E. Bozhukova, and V.D. Zhevner. — In: "Teoriya khimicheskogo mutageneza," p.75. Moscow, "Nauka," 1971.

Shestakov, S.V. and T. N. Mitronova. — Biologicheskie Nauki, Vol.2, p.98. MGU (Moscow State University), 1971.

Shestakov, S.V. and Nguen Than Khyen. — Molec. and Gen. Genet., Vol. 107 (1970), 372.

Shestakov, S.V and V.D. Zhevner. — In: "Biologiya sinezelenykh vodoroslei," No. 2, p.110. MGU (Moscow State University), 1969.

Singh, H.N. — Radiat. Bot., Vol. 8 (1968), 355.

Stevens, S.E. and C. Van Baalen. — J. Phycol., Vol. 5 (1969), 136.

Stoletov, V.N., V.D. Zhevner, D.V. Garibyan, and S.V Shestakov. — Genetika, Vol. 6 (1965), 61.

Werbin, H. and C.S. Rupert. — Photochem. Photobiol., Vol. 7 (1968), 225.

Witkin, E.M. — A. Rev. Genet., Vol. 3 (1969), 525.

Wu, J.H., R.A. Lewin, and H. Werbin. — Virology, Vol. 31 (1967), 657.

UDC 575:576.8

THE PARTICIPATION OF REPAIR ENZYMES IN MUTAGENESIS AND REPLICATION

V.N. SOIFER
Institute of General Genetics
Academy of Sciences of the USSR, Moscow

Hydroxylamine-induced lesions of phage in the E. coli strains hcr^+ and hcr^- were found to be irreparable, which implies that changes in the secondary structure of DNA may trigger dark repair reactions. The relationship between repair and replication in bacteria is demonstrated. Evidence is given showing the similarity between dark repair in microorganisms and mammalian cells.

By the early 1960s radiation genetics possessed a vast amount of experimental evidence proving that the classic equations of the target theory become invalid under extreme conditions. The mutation frequency peak obtained at high doses of radiation (Soifer, 1966a), the changes in the mutation frequency on fractionation and prolongation of the dose of chemical and radiation mutagens (Dubinin and Tarasov, 1969), the total absence of induced mutagenesis in some bacterial lines (e.g. the exr^- lines of Escherichia coli B), the deviation from first order kinetics on exposure to small doses (Glembotskii and Yarmonenko, 1969) — all these findings stressed the need for an integral explanation of the genetic effects of radiation. Hopes for the establishment of a common platform for the explanation of the observed "anomalies" were awakened with the discovery of the dark repair systems. A similar approach could be used to interpret the mysterious "synergic" action of combinations of mutagens. The synergism effect, discovered by Auerbach, has been the object of numerous studies, some of them in the phage-bacterium system. In an attempt to explain this effect, D. M. Gol'dfarb proposed in 1966 that one of the members of such a pair of mutagens (for example, acridine in the case of the hydroxylamine-acridine pair) inhibits the dark repair reaction and thus enhances the mutagenic action of the second mutagen (hydroxylamine).

Research into the role of repair systems in mutagenesis was thus expected to solve many difficulties encountered by the mutation theory. Some of these expectations have been realized, but others have to be discarded. The conception of the dark repair of hydroxylamine-induced lesions falls into the latter category.

On the advice of D. M. Gol'dfarb, we began in 1966 a series of experiments on the combined application of hydroxylamine (a 1 M solution of hydroxylamine hydrochloride in a medium of high osmotic pressure) and acridine to phage T2 passaged in B and B/2 cells. The phage was treated with hydroxylamine extracellularly, after which it was used to infect E. coli B and E. coli B/2 cells. Acridine was applied during the intracellular cycle of phage replication. The mutations examined were those altering the spectrum of the lytic effect of the phage — namely the h mutations (for the procedure used to detect these see Ardashnikov et al., 1963; Soifer, 1966b). The frequency of h mutations in the case of combined treatment was on the average 20 times as high as that obtained with hydroxylamine alone and 80 times as high as with acridine alone (see below).

Treatment	Number of tests	Mutation frequency
No treatment (spontaneous background)	12	$(4.9 \pm 1.2) \cdot 10^{-6}$
Hydroxylamine (in vitro) 2 hours	10	$(1.4 \pm 0.6) \cdot 10^{-4}$
Acridine (in vivo) 1.5 hours	4	$(2.6 \pm 1.0) \cdot 10^{-5}$
Hydroxylamine (in vitro) 2 hours + acridine (in vivo) 1.5 hours	4	$(2.3 \pm 1.8) \cdot 10^{-3}$

Preliminary tests showed that neither dialysis of the incubation mixture of phage with hydroxylamine nor a stopping mixture (Freese et al., 1961) exerts any influence on the mutation frequency. Thus, our experiments revealed a slightly greater synergic effect (Soyfer, 1970) than those of D. M. Gol'dfarb and D. L. Belyaev (1966).

In order to verify the main postulate of Gol'dfarb's hypothesis, namely, that the synergism can be attributed to the action of one member of the mutagenic pair, we examined the reparability of hydroxylamine-induced lesions by dark repair enzymes. This work was done with phage λ and cells of E. coli K12 uvr^{+} and E. coli K12 uvrA6. These objects were chosen on the basis that enzymes of the bacterial host perform the dark repair of phage λ. The choice of bacteria isogenic for all genes, with the sole exception of gene uvrA6, and the use of phage λ make it possible to test the triggering of the dark repair reaction in a phage-bacterium system containing hydroxylamine-treated phage.

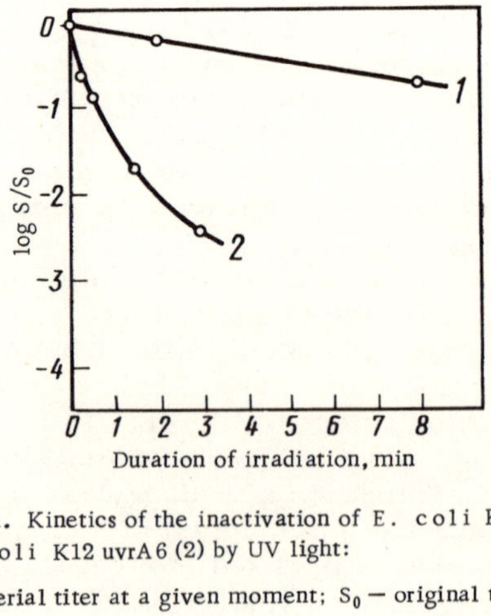

FIGURE 1. Kinetics of the inactivation of E. coli K12 uvr^{+} (1) and E. coli K12 uvrA6 (2) by UV light:

S — bacterial titer at a given moment; S_0 — original titer.

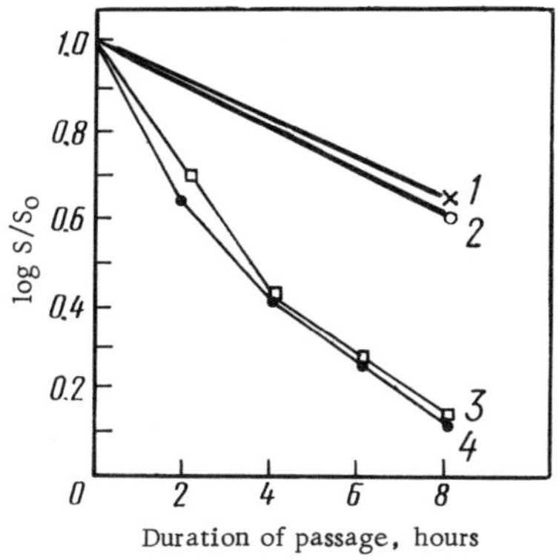

FIGURE 2. Survival of phage λ treated with a 1M solution of hydroxylamine (HA) and afterwards passaged in hcr^+ and hcr^- cells:

1) phage treated with NaCl, passage in hcr^+cells; 2) phage treated with NaCl, passage in hcr^- cells; 3) phage treated with HA, passage in hcr^+ cells; 4) phage treated with HA, passage in hcr^- cells.

The bacteria E. coli K12uvr$^+$ and E. coli K12 uvrA6 (obtained from Professor P. Howard-Flanders, U. S. A.) differed markedly from one another as regards UV sensitivity (Figure 1). Infection of these cells with phage λ previously treated with a 1M hydroxylamine solution in 1M NaCl revealed no difference in the survival of phage passaged in uvr$^+$ and uvr$^-$ cells (Figure 2), which in our view (Soyfer, 1969a; Soifer, 1970a) rules out the hypothesis that hydroxylamine triggers the dark repair reaction. The irreparability of hydroxylamine-induced lesions in E. coli uvr$^+$ and E. coli uvr$^-$ was also stressed by M. K. Isabaeva (1965).

The data obtained provided a clue to another problem — namely, the nature of the DNA lesions recognized by the dark repair system. Two contradictory mechanisms have been proposed to explain the manner in which the repair systems recognize DNA lesions: 1) each type of lesion has its own specific enzymes which perform the excision and repair of the damaged area; 2) what is recognized is not a specific type of lesion but a change in the structure of the DNA (most probably the secondary structure); the endonuclease begins the repair at the site on the damaged molecule where the structure of this is impaired. In view of the chemical aspects of the interaction between hydroxylamine and DNA (see Soifer, 1969, pp. 305—314) we suggested (Soyfer, 1969a, b) that hydroxylamine-induced lesions do not alter the secondary structure of DNA. With the confirmation of our assumption by Smyrt (1970) it became clear that our data on the irreparability of hydroxylamine-induced lesions also have another implication — namely that lesions are not repaired in the absence of disturbances in the secondary configuation of the DNA.

A variety of events can occur during recognition of the lesions and at the beginning of their repair; in effect, repair (notably excision) can take place at the lesion or in the opposite DNA sector. The possibility of the excision of a sector of the DNA strand opposite the lesion was stressed by Holliday, Witkin, Nasim, and Auerbach (Holliday, 1970). The consequences of excision errors have been discussed in detail by Soifer and Dubinin (V. N. Soifer, 1969, 1970a, b; N. P. Dubinin and V. N. Soifer, 1969, 1970; Dubinin and Soyfer, 1969). Our model is based on the assumption that the erroneous excision of the sector opposite the lesion may be followed either by a patching of the resulting gap (during which a wrong base may be inserted opposite the lesion, causing a complete mutation), or by a second passage of the repair enzymes, this time excising the lesion proper (thus causing the breakage of both DNA strands). This mechanism might explain the mysterious appearance of complete mutations and double-stranded breaks in DNA. Our calculations (Soifer, 1970b) show that one excision error per $10^5 - 10^6$ correct excisions agrees with the frequency of complete mutations and chromosomal breaks. Since the available biochemical and physical procedures cannot reveal excision errors at such a frequency, we tried to resolve this problem by means of biological tests.* We determined the frequency of complete and mosaic mutations in bacteria with normal and defective repair systems. This work was carried out in collaboration with Yu. L. Dorokhov on the bacterial strains E. coli K12 uvr^+ and E. coli K12 uvrA6. We determined the frequency of complete and mosaic mutants incapable of assimilating glucose. These mutants were detected by the addition of triphenyltetrazolium chloride to the medium (final concentration 0.05%, pH 6.6). With this treatment all mutant colonies acquired a bright raspberry color. The procedures for the detection of these mutants have been described in the article by V. N. Soifer and Yu. L. Dorokhov (see p. 49 in this collection). Table 1 shows the results of inducing mutations by UV irradiation in uvr^+ and uvrA6 cells. With an increase in the radiation dose the frequency of complete mutations in the uvr^+ strain rises sharply in comparison with that of mosaics; this does not occur in the uvr^- strain. In view of the fact that pyrimidine dimers are excised in the uvr^+ strain we may conclude that the increase in the frequency of complete mutations results from the more frequent excision of DNA areas opposite the lesions and that the error is subsequently fixed during the reparative replication.

The greater frequency of mosaic mutations in the uvr^- strain may also be due to the existence of a greater number of nucleoids. To counter this objection we measured microscopically the length of the uvr^+ and uvr^- cells, and determined the number of nucleoids in the two strains. No significant differences between the two kinds of cell were detected in this respect.

According to a current view, the role of repair systems in the cell is not confined to the healing of lesions. It is known that recombination enzymes behave in the same manner as those engaged in repair. We believe that there are sufficient grounds for assuming that dark repair enzymes participate in mutagenesis.

* The possibility of the excision of both DNA strands in an area in which pairing has been disrupted was first noted by Hogness (1966) and later proved experimentally by Hogness and his colleagues (1966).

TABLE 1. UV induction of complete and mosaic mutations for glucose assimilation in E. coli K12 uvr$^+$ and E. coli K12 uvrA6

Duration of irradiation	Number of colonies counted	Number of mutant colonies		Mutation frequency		Ratio between frequencies of complete and mosaic mutations
		complete	mosaic	complete	mosaic	
E. coli K12 uvr$^+$ (AB1157)						
Control	8,523	2	0	$2.3 \cdot 10^{-4}$	$< 10^{-5}$	> 23
1 min	113,810	133	6	$1.2 \cdot 10^{-3}$	$5.3 \cdot 10^{-5}$	> 24
8 "	19,952	211	23	$1.1 \cdot 10^{-2}$	$1.2 \cdot 10^{-3}$	9.2
10 "	62,936	33	0	$5.3 \cdot 10^{-3}$	$< 10^{-5}$	> 530
15 "	31,692	39	0	$1.2 \cdot 10^{-3}$	$< 10^{-5}$	> 120
E. coli K12 uvrA6 (AB1886)						
Control	67,662	4	0	$5.9 \cdot 10^{-5}$	$< 10^{-5}$	> 6
30 sec	27,210	19	3	$7.0 \cdot 10^{-4}$	$1.1 \cdot 10^{-4}$	6.4
75 "	58,360	50	5	$8.6 \cdot 10^{-4}$	$8.6 \cdot 10^{-5}$	10
135 "	26,043	11	5	$4.2 \cdot 10^{-4}$	$1.9 \cdot 10^{-4}$	3.2

Repair systems apparently take part also in the replication of DNA. Okazaki and his colleagues (1968) have shown that polynucleotide ligase participates in the joining of DNA fragments during normal replication. This phenomenon has since been demonstrated in a large number of higher organisms.

The role of repair systems in replication is also evident from our studies of the reproduction of phages T4Dv$^+$ and T4Dv$^-$. The v gene is responsible for the excision of thymine dimers from the phage DNA (Yasuda and Sekiguchi, 1970); this process is impaired in the v$^-$ mutants.

Our experiments on the reproduction of phages T4Dv$^+$ and T4Dv$^-$ in wild type cells of E. coli B and in cells of E. coli B_{s-1}, mutant for genes uvr and exr yielded a somewhat unexpected result (Soyfer and Dorohov, 1969). It turned out that phage T4Dv$^-$ reproduced 2—4 times more rapidly in E. coli B_{s-1} cells. This could be due either to a greater phage harvest per cell or to the presence of a greater number of cells capable of adsorbing and producing the phage in the E. coli B_{s-1} population (in comparison with the E. coli B population). To solve this problem we examined the adsorption of phages T4Dv$^+$ and T4Dv$^-$ on E. coli B and B_{s-1} cells and the phage yield per cell (Soifer and Dorokhov, 1970a).

Determination of the rate of phage adsorption on the cells showed that the adsorption of phage v$^-$ on E. coli B_{s-1} cells is even poorer than on E. coli B cells. Yet the average yield of mutant phage per cell is twice as high in E. coli B_{s-1} as in E. coli B (Table 2).

TABLE 2. Yield of T4Dv$^+$ and T4Dv$^-$ phages from bacteria

E. coli strain	Titer of bacteria $\cdot 10^7$	Titer of phage $\cdot 10^9$	Number of dishes without plaques	Number of infected bacteria*	Average yield per cell
			T4Dv$^+$		
	5	7.5	34	18	98
B	—	1.5	27	26	53
	4	3.6	32	22.5	84.7
	3.2	7.5	29	27	51
B_{s-1}	2	1.5	31	21	117
	3.2	3.6	28	29	48.6
			T4Dv$^-$		
	2.8	1.1	36	18	110
B	2.5	2.3	38	13.5	219
	8	2.3	9	85.5	119
	2.6	3.7	32	22.5	75
	1.1	1.1	36	16.5	262
B_{s-1}	1.7	2.3	49	1	485
	—	2.3	28	27.4	204
	1	3.7	25	35	160

* Calculated according to the Poisson formula.

We found, moreover (Soyfer and Dorohov, 1970), that this effect depends on the multiplicity of infection (Table 3). Only at a multiplicity greater than 2 was the yield of v$^-$ better in the E. coli B_{s-1} cells. The presence of several phage genomes per cell clearly enhances phage replication.

To explain the effect of the multiplicity of infection we postulated that several copies of phage chromosomes are needed if replication is accompanied by a partial "spoilage" or inactivation of phage DNA molecules owing to the apperance of double-stranded breaks in them. If the E. coli B_{s-1} cells provide suitable conditions for the formation of a greater number of single-stranded breaks of DNA in comparison with the E. coli B cells, the probability of a coincidence of two single breaks at a given point on the DNA would be greater in the E. coli B_{s-1} strain. In this case the inactivation of the replicating phage in the presence of a single genome would be greater in these cells than in the cells of E. coli B, and the opposite effect would occur when the multiplicity of infection is increased.

The possibility that breaks were formed in DNA during its replication, suggested by Delbruck as long ago as 1957, was ruled out by most workers until recently. In our view this process cannot be explained satisfactorily without accepting that breaks in DNA arise during replication. The main difficulty lies in determining the manner in which DNA uncoils before the beginning of replication. According to the calculation of Hanawalt and his colleagues (1968) the ends of DNA must rotate at a rate of 15,000 rpm during replication.

TABLE 3. Effect of the multiplicity of infection on the yield of T4v⁻ from E. coli strains

B		B_s-1		Ratio between the phage yield from E. coli B_s-1 cells and phage yield from E. coli B cells
multiplicity of infection	average yield of phage per cell	multiplicity of infection	average yield of phage per cell	
0.1	227	0.1	207	0.9
0.06	77	0.08	30	0.4
0.002	116	0.003	42	0.4
0.06	91	0.003	63	0.7
0.67	40	0.65	44	1.1
4.6	15	4.5	45	3.0
20	36	8.4	35	1.0
14	74	18	278	3.8
33	80	70	236	3.0
160	143	67	232	1.6

This difficulty can be overcome by assuming that single-stranded breaks appear in the DNA prior to its uncoiling; then the DNA uncoils around these points, replicates and becomes joined by ligase. If the DNA polymerase operates at a constant rate, the rate of DNA synthesis will depend on the ratio between the rates of the cutting and sealing of the single-stranded breaks.

If this concept is correct (see also Soyfer and Dorohov, 1969) one must assume that the E. coli B_{s-1} cells are richer in endonucleases than those of E. coli B. On these grounds one would also expect that E. coli B_{s-1} bacteria should themselves reproduce more rapidly than E. coli B. Although the growth curve of viable cells of E. coli B_{s-1} was almost identical to that of E. coli B (Figure 3), closer examination revealed that the E. coli B_{s-1} population includes numerous dead cells which do not produce colonies but cause a marked increase in the optical density of the bacterial suspension in comparison with E. coli B (Figure 4).

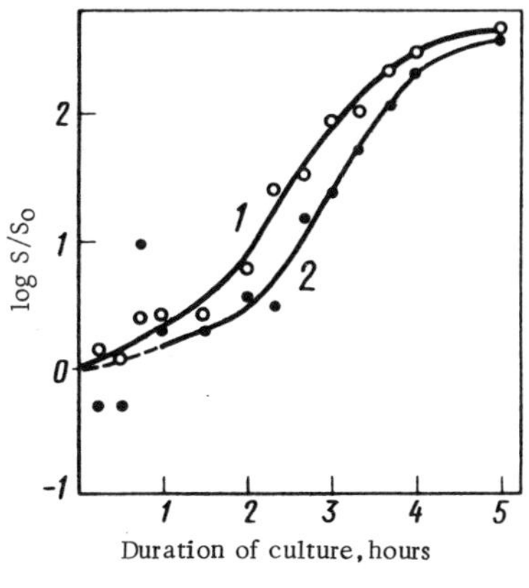

FIGURE 3. Growth curve of the titer of E. coli B (1) and E. coli B_s-1 (2) in liquid media.

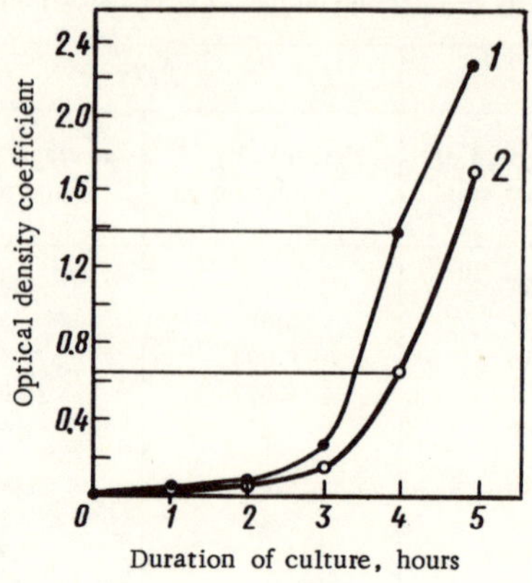

FIGURE 4. Kinetics of the increase in the optical density of E. coli B (1) in E.coli B_{s-1} (2) in a liquid medium.

In our view, these findings imply a relationship between the multiplicity of infection and the presence of single- and double-stranded breaks in the DNA.

We performed the following experiment in order to locate the sites of these breaks in the DNA. Phage T4v⁻ was grown at a low and high multiplicity of infection in cells of E. coli B and E. coli B_{s-1}, and then treated in vitro with low concentrations of hydroxylamine (10^{-2}M) in an aqueous medium. Under such conditions, as shown by Rhaese and Freese (1968), hydroxylamine in the presence of oxygen forms peroxides which induce single-stranded breaks. If the DNA of the phage released from the cells already contains single-stranded breaks which were created during replication of the DNA and have not been sealed by ligase, the coincidence of such a break with another induced by the peroxides in the other DNA strand will cause a double-stranded break and consequently inactivation of the DNA. The curve in Figure 5 show that the hydroxylamine-induced inactivation of phage $T4v^-$ grown in cells of E. coli B_{s-1} and E. coli B at low and high multiplicities of infection follows different courses. Thus, the results obtained can be arranged in the following order according to the degree of inactivation: inactivation of phage $T4v^-$ from E. coli B_{s-1} cells at a low multiplicity of infection > inactivation of the same phage from B cells at low multiplicity of infection > inactivation of phage $T4v^-$ from B_{s-1} cells at high multiplicity of infection > inactivation of phage $T4v^-$ from B cells at the same multiplicity. Thus, the data obtained demonstrate the existence of single-stranded breaks in the phage DNA.

Later it became clear that the enhancement of the replication of phage $T4v^-$ in E. coli B_{s-1} cells results from a change in the rate of DNA synthesis. This experiment was carried out as follows. At the moment of introduction of the phage, ^{3}H-thymidine was added to the adsorption mixture at a concentration of 0.3 μcuries/ml; samples were then taken at intervals, precipitated, washed in 5% trichloroacetic acid, poured on to cellulose nitrate filters and placed in a Nuclear Chicago Mark I scintillation counter for determination of the radioactivity of the acid-insoluble fraction.

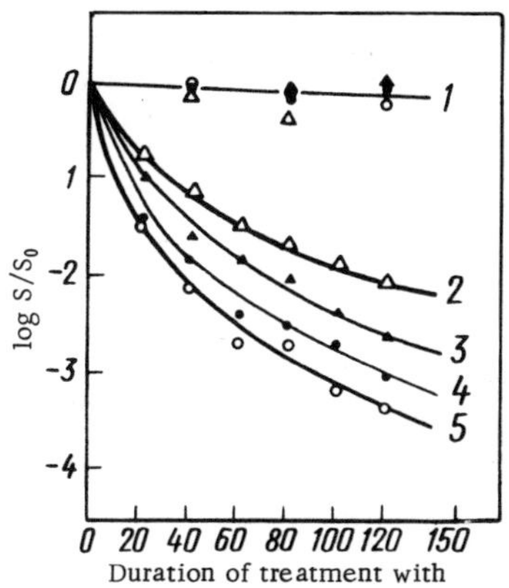

FIGURE 5. Kinetics of the inactivation of phage T4v⁻ by a dilute solution of hydroxylamine.

The phage was initially reproduced in E. coli B cells at a high (3) and low (4) multiplicity of infection and in E. coli B_{s-1} cells at a high (2) and low (5) multiplicity of infection, after which it was treated with hydroxylamine; 1) control. The high and low multiplicities are respectively 10 and 0.1. Other designations as in Figure 1.

Since the selected multiplicity of infection was 5 (and at this level lysis was delayed), the determination was continued for 120 minutes. As shown in Figure 6, the synthesis of phage $T4v^-$ DNA in E. coli B_{s-1} cells was greater than in E. coli B cells and exceeded that of phage $T4v^+$ in cells of both E. coli B and E. coli B_{s-1}.

If our scheme for the replication of DNA is correct, E. coli B_{s-1} cells must be relatively richer in endonucleases than polynucleotide ligases. This, we believe, explains why the phage reproduces 1.5—2 times better in E. coli B than in E. coli B_{s-1} at a low multiplicity of infection, when the formation of double-stranded DNA breaks inactivates the intracellular phage progeny in some cells. Kozinsky (1969) observed that the synthesis of phage-induced endonuclease in the E. coli B — phage T4 system starts 5 minutes after infection and can be inhibited at that time by chloramphenicol. By adding chloramphenicol (100 μg/ml) 5 minutes after infection, on the basis of this observation, we reduced the inactivation of phage $T4v^-$ in E. coli B_{s-1} cells and obtained a 2.5—3.5-fold increase in the phage yield in comparison with the harvest from the E. coli B cells.

Experiments with caffeine — an inhibitor of dark repair — furnished proof that the increase in the yield of phage $T4v^-$ from E. coli B_{s-1} cells results from the activity of dark repair enzymes. Thus, the addition of 0.2% caffeine solution to the E. coli B culture 5 minutes before the beginning of infection with the phage increased the yield of the latter (Soifer and Dorokhov,

1970b) to a level comparable to the phage yield from E. coli B_{s-1} cells in a medium without the inhibitor. Although caffeine decreased slightly the yield of phage T4v⁻ from E. coli B_{s-1} cells, this still exceeded the yield of wild type phage from the same cells (Table 4). Thus, our data demonstrate the importance of repair enzymes — endonuclease and polynucleotide ligase — in the control of the rate of DNA replication; they also indicate the role of the appearance and subsequent closure of single-stranded breaks in the acceleration or slowing down of replication.

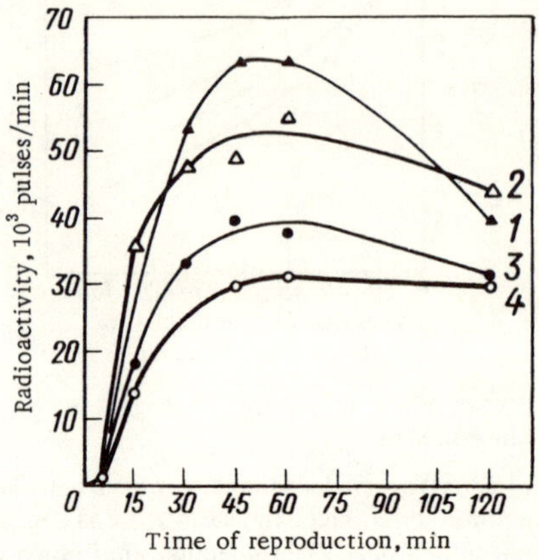

FIGURE 6. Kinetics of the incorporation of ^{3}H-thymidine in the DNA of phages T4 and $T4v^-$ grown in E. coli B and E. coli B_s-1 cells:

1) phage $T4v^-$ reproduced in E. coli B_{s-1}; 2) phage $T4v^+$ reproduced in E. coli B_{s-1}; 3) phage $T4v^-$ reproduced in E. coli B; 4) phage $t4v^+$ reproduced in E. coli B.

Repair systems comparable to the dark repair system of bacteria were unknown until recently in higher organisms. Not long ago, several workers (Regan et al., 1968; Horikawa et al., 1968; Soifer et al., 1970) demonstrated the excision of thymine dimers from the DNA of UV-irradiated tumor cells. The excision of dimers in normal human cells remained an open question.

Our experiments (carried out in collaboration with A. N. Mustafina and N. I. Yakovleva of the Institute of Poliomyelitis and Viral Encephalitides at the USSR Academy of Medical Sciences) showed that normal cells similarly possess a dimer excision system with an efficiency equal to that of its counterpart in HeLa cells (Table 5).

As in HeLa cells (Soifer et al., 1970), the excision of dimers from human embryonic lung cells was most effective during the first hours after UV irradiation; by the 10th hour the rate of dimer excision had decreased, and between the 10th and 24th hours after irradiation it remained constant.

TABLE 4. Total yield of phage per cell in different strains of E. coli in the presence of 0.2% caffeine

Inhibitor (100 μg/ml)	T4Dv⁺ from		T4Dv⁻ from	
	B	B_s-1	B	B_s-1
Not added	16	12.9	10.6	66
" "	31	39	5.4	46
Caffeine	Not determined		47.6	25.8
" "	" "		43.8	33.0

To sum up, we have tried to demonstrate here the important role of repair enzymes in the major vital processes of cells — both microbial and those of higher organisms (see also the article by V. N. Soifer and Yu. L. Dorokhov, p. 49 in this collection). Our findings show that mutagenesis, recombination and replication involve to a degree the action of enzymes which also participate in repair processes in the cell.

TABLE 5. Excision of thymine dimers from the DNA of diploid cells of human embryonic lung

Experiment	Duration of incubation	Radioactivity, pulses/min						$\hat{T}T/T$, %	Proportion of dimers remaining in DNA, %	Proportion of dimers excised from DNA, %
		total activity	background	net thymine activity	total activity	background	net thymine activity			
		Dimer sector			Thymine sector					
1	0	15,344	5,625	10,219	999,446	8,500	990,946	1.01	100	0
	9.5	5,271	1,950	3,321	468,929	4,005	465,429	0.71	70	30
	24	2,830	1,285	1,545	207,587	11,800	195,787	0.79	78	22
2	0	33,412	1,150	22,262	893,241	18,000	875,241	2.54	100	0
	9.5	14,774	4,720	10,054	676,709	6,700	670,009	1.57	62	38
	24	16,727	5,300	11,424	807,136	7,900	799,236	1.43	56	44
3	0	28,682	7,450	21,232	745,109	41,000	704,109	3.01	100	0
	6	13,832	2,400	11,432	562,122	10,000	552,122	2.06	68	32
	10	12,121	2,100	10,021	444,179	5,000	439,179	2.28	75	25
	24	14,026	3,210	10,816	460,033	9,450	450,583	2.38	79	21
4	0	21,246	3,750	17,496	545,663	12,000	533,663	3.28	100	0
	6	10,040	2,790	7,250	324,699	3,500	321,199	2.26	69	31
	10	5,767	1,800	4,687	255,952	5,000	250,952	1.81	55	45
	24	12,592	1,600	10,992	539,545	9,500	530,045	2.07	63	37

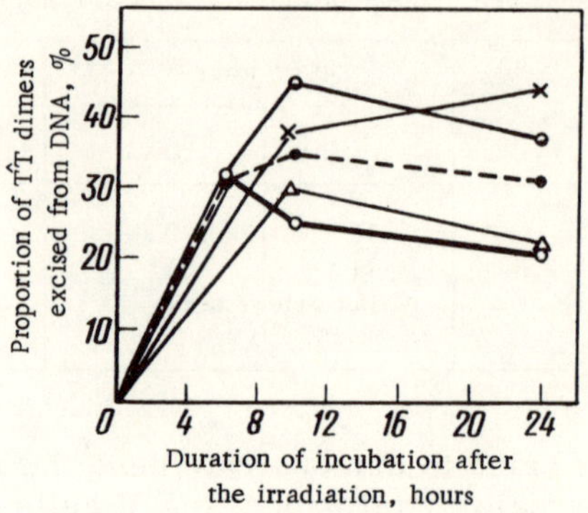

FIGURE 7. Kinetics of the excision of UV-induced thymine dimers from DNA of human embryonic lung cells in the course of dark repair. Each curve represents the results of a different experiment.

BIBLIOGRAPHY

Ardashnikov, S.N., V.N. Soifer, and D.M. Gol'dfarb. — Vest. AMN SSSR, Vol. 12 (1963), 43.

Dubinin, N.P. and V.N. Soifer. — Izvest. AN SSSR, biol. ser., Vol. 5 (1969), 637.

Dubinin, N.P. and V.N. Soifer. — Izvest. AN SSSR, biol. ser., Vol. 4 (1970), 483.

Dubinin, N.P. and V.N. Soyfer. — Mutation Res., Vol. 8 (1969), 353.

Dubinin, N.P. and V.A. Tarasov. —In: Sovremennye problemy radiatsionnoi genetiki, p.7. Moscow, Atomizdat, 1969.

Freese, E., E. Bautz, and E.B. Freese. — Proc. Natn. Acad. Sci. U.S.A., Vol. 47 (1961), 845.

Glembotskii, Ya. L. and S.P. Yarmonenko. — In: Sovremennye problemy radiatsionnoi genetiki, p.135. Moscow, Atomizdat, 1969.

Gol'dfarb, D.M. and D.L. Belyaev. — Genetika, Vol. 6 (1966), 113.

Hanawalt, P.C., D.E. Pettijohn, E.C. Pauling, C.F. Brunk, D.V. Smith, L.C. Kanner, and J.L. Kouch. — Cold Spring Harbor Sympos. Quant. Biol., Vol. 33 (1968), 187.

Hogness, D.S. — J. Gen. Physiol., Vol. 49 (1966), 29.

Hogness, D.S., W. Doerfler, J.B. Egan, and L.W. Black. — Cold Spring Harbor Sympos. Quant. Biol., Vol. 31 (1966), 129.

Holliday, R. — In: Organization and Control in Prekaryotic and Eukaryotic Cells. H.P. Charles and B.C. J. G. Knight (Eds.), p.359. Cambridge Univ. Press, 1970.

Horikawa, M., O. Nakaido, and T. Sagahara. — Nature, Vol. 218 (1968), 499.

Isabaeva, M.K. — Tsitologiya, Vol. 9 (1965), 1490.

Kozinsky, A. — Biochem. Biophys. Res. Commun., Vol. 35 (1969), 294.

Okazaki, R., T. Okazaki, K. Sakabe, K. Sugimoto, and A. Sugino. — Proc. Natn. Acad. Sci. U.S.A., Vol. 59 (1968), 598.

Regan, J.D., J. Trosco, and W. Carrier. — Biophys. J., Vol. 8 (1968), 319.

Rhaese, H.J. and E. Freese. — Biochim. biophys. Acta, Vol.155 (1968), 476.

Smyrt, K. — Synopses of Reports 7th Intern. Sympos. on Chemistry of Natural Compounds, p. 224. Riga, "Zinatne," 1970. (Russian)

Soifer, V.N. — Bull. Moscow Society of Naturalists (MOIP), biol. ser., Vol. 71 (1966a), 88.

Soifer, V.N. — Vest. AMN SSSR, Vol. 2 (1966b), 47.

Soifer, V.N. Molecular Mechanisms of Mutagenesis. — Moscow, "Nauka," 1969. (Russian)
Soifer, V.N. — Voprosy Virusologii, No.6 (1970a), 667.
Soifer, V.N. Synopses of the Development of Molecular Genetics. — Moscow, "Nauka," 1970b.
Soifer, V.N. and Yu. L. Dorokhov. —Doklady AN SSSR, Vol. 192 (1970a), 899.
Soifer, V.N. and Yu. L. Dorokhov. — Mikrobiologiya, Vol. 6 (1970b), 1068.
Soifer, V.N., L.L. Matusevich, and G.I. Goroshkina. — Radiobiologiya, Vol. 10 (1970), 275.
Soyfer, V.N. — Archs. roum. Path. exp. Microbiol., Vol. 28 (1969a), 941.
Soyfer, V.N. — Microbial Genet. Bull., Vol. 31 (1969b), 18.
Soyfer, V.N. — Microbial Genet. Bull., Vol. 32 (1970), 12.
Soyfer, V.N. and J.L. Dorohov. — Microbial Genet. Bull., Vol. 31 (1969), 18.
Soyfer, V.N. and J.L. Dorohov. — Biochem. Biophys. Res. Commun., Vol. 41 (1970), 1396.
Yasuda, S. and M. Sekiguchi. — J. Molec. Biol., Vol. 47 (1970), 243.

UDC 575:576.8

THE MECHANISM OF REPAIR OF DNA LESIONS INDUCED BY MONOFUNCTIONAL ALKYLATING AGENTS IN ESCHERICHIA COLI K12

G.B. SMIRNOV and Yu. N. FAVORSKAYA
N.F. Gamaleya Institute of Epidemiology and Microbiology, Moscow

The lethal and mutagenic effect of methylmethanesulfonate (MMS) and N-methyl-N-nitro-N-nitrosoguanidine (NG) on dar mutants of E. coli K12 was examined. A number of strains were discovered with a defect in the system repairing the DNA lesions induced by these agents. Defects in the system repairing the lesions induced by MMS and NG lead to an increase in the mutability of the strains after treatment with these compounds.

We found in previous studies that the UV-sensitive mutant E. coli K12 KMBL91 dar_2, isolated by Van de Putte and his colleagues (1965), shows an increased sensitivity to the mutagenic effect of N-methyl-N-nitro-N-nitrosoguanidine (NG) (Smirnov and Skavronskaya, 1968) and methylmethanesulfonate (MMS) (Smirnov and Favorskaya, 1970). The increased mutability of the dar_2 bacteria when exposed to NG and MMS appears to be unrelated to their UV sensitivity since, firstly, this property is lacking in other UV-sensitive mutants — of the uvr type, and secondly, we have obtained a UV-resistant revertant of the dar_2 mutant — dar_2R (Smirnov, 1969), which shows an increased mutability when exposed to NG and MMS.

We established, moreover, that treatment of the dar_2 and dar_2R strains with NG or MMS prolongs the lag period of cell division relative to that of strains possessing a normal mutability (Smirnov and Favorskaya, 1970).

We assumed on the basis of these data that the dar_2 bacteria bear a defect in the genetic determinant of a system engaged in the repair of DNA lesions induced by monofunctional alkylating agents.

We present here results which confirm this assumption.

Figure 1 shows the kinetics of the inactivation of KMBL91 dar_2 and KMBL100 dar_5 bacteria by MMS and NG.

Treatment with MMS was carried out in phosphate buffer pH 7, while NG treatment was performed in acetate buffer pH 5.0. Figure 1 shows that the dar_2 strain is more sensitive to the inactivating effect of both MMS and NG than the dar_5 strain.

The kinetics of DNA synthesis in dar_2 and dar_5 bacteria treated with equal doses of MMS and NG is shown in Figure 2. Treatment with either of the two agents temporarily blocks DNA synthesis in both strains, though for a much longer period in the dar_5 cells. The later resumption of DNA synthesis in dar_2 cells treated with the same dose of MMS or NG as the dar_5 cells indicates that the dar_2 strain has a lower capacity for the repair of NG- and MMS-induced lesions preventing DNA replication. The arrest of DNA synthesis in dar_5 cells following treatment with MMS or NG is accompanied by a partial degradation of DNA, which can be detected by the decrease in radioactivity in the acid-insoluble fraction. DNA degradation practically does not occur in the dar_2 strain (large doses of MMS provoke some slight degradation).

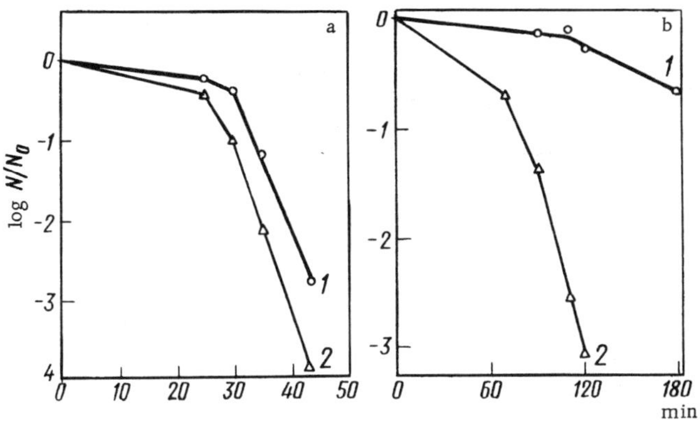

FIGURE 1. Kinetics of the inactivation of KMBL100 dar_5 (1) and KMB191 dar_2 (2) bacteria on exposure to 0.5% MMS (a) and 1 mg/ml NG (b).

The bacterial suspension treated contained 10^9 cells/ml. Survival was determined on minimal glucose-salts agar.
N_0 — original number of cells; N — number of cells after treatment with the mutagen.

We propose that the degradation of DNA in the MMS- and NG-resistant strain dar_5 represents one of the stages of a repair process which, like the system of excision repair of UV lesions, eliminates DNA lesions together with the neighboring nucleotides.

The rate of DNA synthesis after the temporary blockage is inversely proportional to the mutagen dose. Hence the conclusion that unrepaired DNA lesions induced by MMS or NG slow down the replication process. Since the rate of DNA synthesis after treatment with equal doses of the mutagens is much lower in dar_2 cells than in dar_5 cells (see Figure 2), it follows that the dar_2 cells can restore a much smaller number of the DNA lesions induced by MMS or NG than the dar_5 cells.

Our study of the kinetics of DNA synthesis and cell division in hcr^+ and hcr strains after UV irradiation has shown that the qualitative differences between these strains coincide with those observed between dar_5 and dar_2 cells treated with MMS or NG. Thus, when the hcr mutant and the hcr^+ strain are exposed to the same UV doses the hcr mutant shows a longer lag period of cell division and a longer period of arrest of DNA synthesis; moreover, the mutant does not break down DNA during the arrest period and resumes DNA synthesis at a slower rate than the hcr^+ strain.

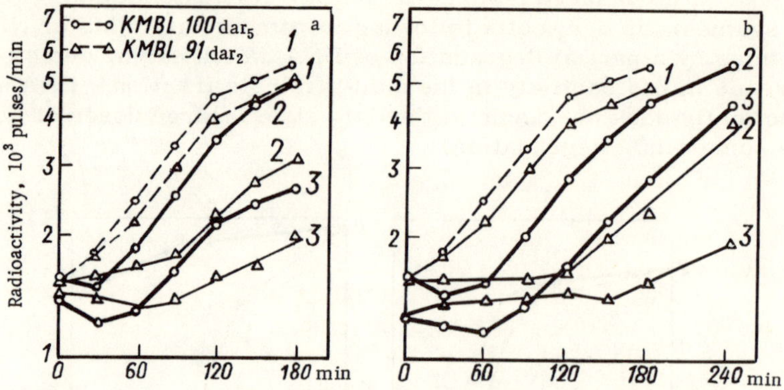

FIGURE 2. Kinetics of the incorporation of ^{14}C-thymidine in KMBL100 dar_5 and KMBL91 dar_2 bacteria treated with 0.5% MMS (a) and 1 mg/ml NG (b):

1) untreated cultures; 2) treatment for 25 min; 3) treatment for 35 min. The bacterial suspension treated contained 10^9 cells/ml.

It follows that the repair of lesions induced by monofunctional alkylating agents may be similar to the excision repair of UV-induced lesions and probably comprises the same major stages as excision repair:

1) recognition of the lesion and cutting of the damaged strand;

2) displacement of the single-stranded break as a result of degradation by exonuclease with simultaneous resynthesis on the template of the intact strand;

3) reunion of the ends of the sugar-phosphate backbone (reunion of the displaced single-stranded break).

The enzyme which performs the first stage must be specific to alkylated DNA. The most probable candidate for this role is endonuclease II, described by Friedberg (Friedberg and Goldthwait, 1968; Friedberg et al., 1969).

At present it is not known which of the repair stages is impaired in the dar_2 strain. According to the available data, however, the enzyme involved in the semi-conservative replication of DNA is defective in this strain. As shown in Figure 2, the kinetics of DNA synthesis in intact dar_2 bacteria differs somewhat from that characteristic of dar_5 cells. Moreover, we found that broth cultures of dar_2 are more sensitive to the bacteriostatic effect of 2-aminopurine (AP) than similar cultures of dar_5 cells. Finally, the dar_2 strain grown in the presence of AP mutates at a higher frequency than other strains (Smirnov and Skavronskaya, 1969).

These findings imply that the dar_2 bacteria have a defective enzyme which participates in the replication of DNA and performs the second stage (degradation-resynthesis) of the repair of lesions induced by monofunctional alkylating agents.

According to our working hypothesis, the increased sensitivity of the dar_2 bacteria to the inactivating effect of alkylating agents results from their reduced capacity for eliminating alkylated phosphate groups from DNA, while the increased mutability of this strain is due to the greater probability of errors being committed by the damaged enzyme during replication of the alkylated bases of DNA.

The available data indicate that E. coli can repair DNA lesions induced by monofunctional alkylating agents. It is highly probable that both the capacity for repairing UV-induced lesions and the capacity for restoring alkylated DNA depend on several genetic determinants. We found, for example, that the E. coli p3478 mutant, isolated by de Lucia and Cairns (1969) and noted for its almost complete lack of DNA polymerase activity, is extremely sensitive to the inactivating effect of NG and MMS.

It is noteworthy that the differences between the NG and MMS sensitivities of the p3478 mutant and those of the wild type strains are much more pronounced than the differences in UV sensitivity between hcr^- mutants and hcr^+ bacteria. Thus, the bacteria obviously possess a whole array of genetically determined mechanisms which prevent distortion of the genetic information and ensure resistance to various inactivating factors.

BIBLIOGRAPHY

Friedberg, E.S. and D.A. Goldthwait.—Cold Spring Harbor Sympos. Quant. Biol., Vol. 33 (1968), 271.
Friedberg, E.S., S.M. Hadi, and D.A. Goldthwait.—J. Biol. Chem., Vol. 244 (1969), 5879.
Lucia, P. de and J. Cairns. — Nature, Vol. 224 (1969), 1164.
Smirnov, G.B. — Zhurnal mikrobiologii, epidemiologii i immunobiologii, Vol.4 (1969), 18.
Smirnov, G.B. and Yu. N. Favorskaya. — Genetika, Vol. 6 (1970), 139.
Smirnov, G.B. and A.G. Skavronskaya. — Genetika, Vol. 4 (1968), 111.
Smirnov, G.B. and A.G. Skavronskaya. — Genetika, Vol. 5 (1969), 94.
Van de Putte, P., C.A Van Sluis, J. Dillewijn, and A. Rörsch. — Mutation Res., Vol. 2 (1965), 97.

COMPANION WORKBOOK